Air Pollution and Industry

Air Pollution and Industry

Edited by

R. D. Ross

President
Thermal Research & Engineering Corp.
Conshohocken, Pa.

Van Nostrand Reinhold Company

New York / Cincinnati / Toronto / London / Melbourne

Van Nostrand Reinhold Company Regional Offices:
New York Cincinnati Chicago Millbrae Dallas
Van Nostrand Reinhold Company International Offices:
London Toronto Melbourne

Library of Congress Catalog Card Number: 70-180160
ISBN: 0-442-27052-6

Manufactured in the United States of America
Published by Van Nostrand Reinhold Company
450 West 33rd Street, New York, N. Y. 10001
Published simultaneously in Canada by Van Nostrand Reinhold Ltd.
15 14 13 12 11 10 9 8 7 6 5 4 3 2

Library of Congress Cataloging in Publication Data
Ross, Richard D
Air pollution and industry.
(Van Nostrand Reinhold environmental engineering series)
Includes bibliographies.
1. Air—Pollution. I. Title.
TD883.R58 628.5'3 70-180160

Van Nostrand Reinhold Environmental Engineering Series

ADVANCED WASTEWATER TREATMENT, by Russell L. Culp and Gordon L. Culp

ARCHITECTURAL INTERIOR SYSTEMS—Lighting, Air Conditioning, Acoustics, John E. Flynn and Arthur W. Segil

THERMAL INSULATION, by John F. Malloy

AIR POLLUTION AND INDUSTRY, edited by Richard D. Ross

INDUSTRIAL WASTE DISPOSAL, edited by Richard D. Ross

MICROBIAL CONTAMINATION CONTROL FACILITIES, by Robert S. Runkle and G. Briggs Phillips

SOUND, NOISE, AND VIBRATION CONTROL, by Lyle F. Yerges

Van Nostrand Reinhold Environmental Engineering Series

THE VAN NOSTRAND REINHOLD ENVIRONMENTAL ENGINEERING SERIES is dedicated to the presentation of current and vital information relative to the engineering aspects of controlling man's physical environment. Systems and subsystems available to exercise control of both the indoor and outdoor environment continue to become more sophisticated and to involve a number of engineering disciplines. The aim of the series is to provide books which, though often concerned with the life cycle—design, installation, and operation and maintenance—of a specific system or subsystem, are complementary when viewed in their relationship to the total environment.

Books in the Van Nostrand Reinhold Environtal Engineering Series include ones concerned with the engineering of mechanical systems designed (1) to control the environment within structures, including those in which manufacturing processes are carried out, (2) to control the exterior environment through control of waste products expelled by inhabitants of structures and from manufacturing processes. The series will include books on heating, air conditioning and ventilation, control of air and water pollution, control of the acoustic environment, sanitary engineering and waste disposal, illumination, and piping systems for transporting media of all kinds.

Contributors

AUTHOR	CONTRIBUTION
R. D. Ross, President Thermal Research & Engineering Corp. Conshohocken, Pa.	Chapters 1 and 9
Franklin B. Flower Extension Specialist in Environmental Sciences New Jersey Cooperative Extension Service Rutgers University New Brunswick, N. J. *and* James V. Feuss, P.E. Director, Environmental Sanitation Cortland County Health Dept. Cortland, N. Y.	Chapter 2
Richard D. Grundy Professional Staff Member Committee on Public Works United States Senate Washington, D.C. Presently on assignment as Executive Secretary National Fuels & Energy Policy Study U.S. Senate Committee on Interior and Insular Affairs	Chapter 3
H. William Blakeslee *and* Louis R. Reckner Scott Research Laboratories Plumsteadville, Pa.	Chapters 4 and 5
G. J. Celenza Environmental Engineer Catalytic, Inc. Environmental Systems Division Philadelphia, Pa.	Chapter 6

E. S. Monroe
Senior Consultant
Engineering Services Div.
E. I. Du Pont De Nemours & Co.
Wilmington, Delaware

Chapters 7 and 10

J. L. Schumann
Vice-President
Belco Pollution Control Corp.
Patterson, N. J.

Chapter 8

Preface

Fifty years ago a book about air pollution would have attracted very few readers. Those that did read it would have felt that it was just about as contemporary as a book about air traffic control, abortion, or the Pill. Today when the word *pollution* is mentioned almost everyone, from the child in kindergarten to the octogenarian, appreciates that we are talking about a significant problem affecting all society.

Man's desire for comfort and wealth, along with the explosion in population, has created as a by-product a problem of wide concern. Man's own wastes have begun to menace him from the air, the water, and the soil to such a degree that he has begun to realize that unless he does something about it and soon, he will be the victim of his own excesses. This monster man he has created can only prove a difficult adversary. Victory can come only through denial of some of the very basic indulgences man assumed carried no penalty.

Like many of the other problems engendered by a rapidly increasing industrialism and a burgeoning population, air pollution can only be aggravated unless means of abatement are discovered and employed. Society, unfortunately, never really begins to cope with its problems until they seriously begin to impinge upon the way it lives. Only then are legislation and technological force brought to bear in an effort to solve the problem.

Now that pollution has become a pressing problem of the sixties and seventies, it has become a household word. As with most sociological issues, it is difficult to separate fact from emotion. Pollution thus becomes a subject for the political arena, further charging the emotional aspect and clouding the facts. The air pollution problem cannot be solved solely through technology. It is an area where engineering and scientific knowledge can help to achieve satisfactory answers but only if we first lift the veil of emotions and politics. We must deal in facts and insist that others do the same.

It is the mission of the present book to deal with industrial pollution. So why talk about the *philosophy* of pollution? Does it help us get a better grasp on the problem or finally to solve it? Certainly, knowing where the problem comes from may be the first step in solving it, but more important to the industrialist is the fact that the layman assumes that industry is the chief offender. Therefore the industrialist will have his head on the block if he cannot solve his own pollution problem.

Certainly it is true, and it can be proven statistically, that wastes from industry today contribute only about 20 percent to our total air pollution. We know that the internal combustion engine used in automobiles, buses, trucks, airplanes, and other modes of transportation contributes about 60 percent of all the air pollution in the world today. Another 10 percent or more is caused by the generation of electricity for the ultimate comfort of people. Nevertheless, the average layman is ignorant of these statistics and he generally continues to assume that industry is the culprit; and since industry has caused the pollution, it must pay the bill. Legislation which is being developed today often springs from a notion rather than technical good sense, and unless we, in industry, are willing to do more than our share of solving the problem we are going to have the entire bill for the solution handed to us. It is also reasonable to assume that since industrial air pollution is so small a part of the total problem and industry possesses a greater degree of technology that industry can solve its problems with more dispatch than other areas of society. For example, is it reasonable to suggest that John Q. Public give up his second car—or even his first—and use some form of mass transportation to get to his job in order to reduce air pollution? Is it reasonable to assume that this same man will do without air conditioning in his office or in his home, or will switch to a more expensive fuel to guarantee lower pollution levels from his heating system? Is it reasonable to assume that he will brick up all his fireplaces, burn no trash, stop smoking, and assume the high tax burden required to dispose of the trash and garbage carried from his domicile several times a week? Will he do without plastic disposables and paper items, and save all of his newspapers so they can be recycled into new newsprint? Will he throw away virtually nothing, such as aluminum cans and bottles, that can be reused again, thus drastically lessening the mountains of refuse?

The answer to this is no, because he has become spoiled by a certain standard of living and despite the black picture painted for the future he is selfish enough to wish to continue at that same level of living. In fact, the average citizen expects newer and better and more convenient ways to do things and many of these will contribute more and more to the total pollution problem. Why then should industry be expected to take the total burden on its shoulders? Why should it clean up its processes so that the only effluents which come from plants manufacturing the items the public demands will be satisfactory to put into the air and the water and the land on which we live? Perhaps because the pollution of industry is more glaring and more easily attacked by the layman. Or, perhaps because industry is considered a colossus, while we tend to think of the citizenry in terms of the individual, who could hardly contribute enough pollution to bother anyone.

What then as industrialists are we to do about the total problem? Even if we put our own house in order, how is this going to help solve everything? We must not only correct our immediate dilemmas but must spend money—our own money and federal, state, and local government funds—try to come up with new ways to alleviate problems stemming from pollution by the internal combustion engine and power generating stations which represent the comforts people want to keep. This book does not aim to give all the answers, but it is an attempt to define the problems and discuss possible means for solving them.

R. D. ROSS

Editor's Note: When the EPA came into existence in December 1970 certain agencies under the HEW disappeared. The functions of The National Air Pollution Control Administration (NAPCA), The Federal Water Quality Administration (FWQA), and the Bureau of Solid Waste Management were absorbed into the EPA. This book was written before and during this period of change so a number of references are made to these defunct agencies.

Contents

Air Pollution and Industry

1 Defining the Problem

WHAT IS AIR POLLUTION?

Air pollution has been defined in many different ways. In one sense it is the addition to our admosphere of any material which will have a deleterious effect to life upon our planet. The material might be a toxic gaseous hydrocarbon with some long-lasting effect on an organism ingesting it or perhaps a particulate irritant which could cause similar problems. It might be atomic radiation in a form that we cannot see which would be damaging to animal or plant cells. A pollutant can be anything which when put into the atmosphere either purposely or through some act of nature reduces the oxygen content or significantly changes the composition of the air.

An air pollutant does not have to be inhaled. It becomes a pollutant merely by being in the air. Smog, made up of gases and particulate matter in large quantities, and seen over many of our major cities in the last fifteen years, forms a blanket which shields some of the sun's radiation so necessary to life on earth. This blanket changes heat absorption patterns on the surface of the earth and can prohibit reradiation of the earth's heat to the sky, resulting in the "greenhouse effect" which ultimately changes weather and temperature patterns across the entire surface of the planet.

Sources of Air Pollution

Major categories of air pollution may be considered as caused by (1) transportation, (2) domestic heating, (3) electric power generation, (4) refuse burning, (5) industrial fuel burning and process emissions. It is difficult to assign accurate percentage figures to any of these classifications because of the lack of good documentation. As suggested in the Preface, it is reasonable to assume that transportation of all types, internal combustion engines and turbine engine-driven vehicles, contributes on the order of 60 percent of the total annual emissions which pollute the air. Electric power generation contributes 10 to 15 percent, domestic heating about 10 percent, industrial fuel consumption and process emissions about 20 percent, and refuse burning about 5 percent. Since this is not a static society these approximate figures are not static either. As we build and sell more automobiles the transportation portion will increase. As we have a need for more power plants the figure for this will increase also. Industrial and municipal incinerators are also in wider use today and will add to the burden unless better designs are developed.

In addition to these major emissions we have a number of minor ones, and while they are not significant, they nevertheless contribute to our overall problem. Some which should be considered are as remote as particulate matter from rubber vehicle tires. If you have ever watched a multi-ton jet airplane land at an airport you will notice smoke appear as the tires come in contact with the runway, or if you have ever seen a car traveling at high speed apply the brakes suddenly, a similar phenomenon occurs. This happens thousands of times a day in every community.

Organic compounds in perfumes, after-shave lotions, etc. tend to give off odors which are pleasing but at the same time contribute to our air pollution problem in a small way.

How often have you watched construction crews build a road or a building on a windy day? Moving vast quantities of earth in dry weather usually results in large amounts of material becoming airborne. The great dust bowl of the thirties certainly became an air pollution problem even though it had its beginning in a natural occurrence—drought. At least 50 percent of all the people in the world smoke at one time or another, whether cigarettes, pipes, or cigars, and the effluent from burning tobacco is definitely an air pollutant. Cosmic dust, hydrogen sulfide from natural sources, and even the use of the favorite aerosol can for spraying bugs in the garden or just freshening the air in one's living room contribute to the total problem.

If you have spent time at the seashore you know that the windows on your automobile become coated with a wet, sticky spray which when dried leaves salt particles on the surface. These salt particles, which become airborne near the ocean, are carried many miles inland. The decomposition of vegetation in forests, swamps, and even in the backyard compost pile add to air pollution. Such simple things as mothballs give off naphthalene at a high rate. Every time you walk down the street, rubber or leather is worn off the soles of your shoes. When you fill your car with gasoline you can usually detect a gasoline odor. This is another form of air pollution.

The list could go on and on. Many small quantities of air pollutants are released to the atmosphere every day, so there is little wonder that it is difficult to accurately establish percentages. It is sufficient to say, however, that in our own way, consciously or unconsciously, each of us is contributing to the total air pollution problem every day that we live.

Of the major sources of air pollution we have attributed about a fifth to industrial processes. However, the layman is inclined also to consider electric power generation as an industrial polluter, and in the broadest sense it is, since industry requires a large portion of the electric power output. This book, however, does not attempt to deal with pollution from electric power generation systems, either those which are fossil fuel fired or of the nuclear type. The problems in this industry are peculiar to power generation only and hence not normally under the control of industrial processing companies. We should therefore take a closer look at the major industrial sources of pollution and try to identify them.

Major industrial polluters have been classified relative to the type of industry. Generally, they can be categorized by the following list, with annual emission levels:

Petroleum Refining—8.4 billion pounds of particulates, sulfur oxides, hydrocarbons, and carbon monoxide.

Smelters—aluminum, copper, lead, zinc, etc.—8.3 billion pounds of particulates and sulfur oxides.

Iron Foundries—7.4 billion pounds of particulates and carbon monoxide.

Kraft, Pulp, & Paper Mills—6.6 billion pounds of particulates, carbon monoxide, and sulfur oxides.

Coal Cleaning & Refuse—4.7 billion pounds of particulates, sulfur oxides and carbon monoxide.

Coke (used in steel manufacturing)—4.4 billion pounds of particulates, sulfur oxides, and carbon monoxide.

Iron & Steel Mills—3.6 billion pounds of particulates and carbon monoxide.
Grain Mills & Grain Handling—2.2 billion pounds of particulates.
Cement Manufacturing—1.7 billion pounds of particulates.
Phosphate Fertilizer Plants—624 million pounds of particulates and flourides.

While there are many other sources, these constitute the largest polluters in industry and therefore present the most immediate danger and demand the most immediate interest. It is significant to note that in 1968 the average ratio of air pollution control expenditures to total capital expenditures in all industry was only 1.65 percent. It is obvious that this ratio is going to increase in the next decade and probably many industries will approach the 5 percent level.

The individual pollutants roughly can be categorized into five major areas. Three of these areas are specific and two are general, but for purposes of evaluating the total air pollution problem, we should classify pollutants as follows: carbon monoxide, sulfur oxides, nitrogen oxides, hydrocarbons, and particulates. The first three are essentially specific, while the last two could be made up of dozens of different compounds and elements. The source of these air pollutants are classified in relative terms in Table 1-1.

TABLE 1-1[a]

	POLLUTANT (IN MILLIONS OF TONS/YEAR)					
Source	*Carbon Monoxide*	*Sulfur Oxides*	*Nitrogen Oxides*	*Hydro-carbons*	*Particu-lates*	*Totals*
Transportation:						
Autos	67.3	0.3	7.0	12.7	0.7	88.0
Other	3.9	0.1	1.0	1.1	0.5	6.6
Total	71.2	0.4	8.0	13.8	1.2	94.6
Fuel Combustion:						
Power Plants	0.1	14.0	3.5	neg.	2.3	19.9
Industry	0.3	5.5	3.1	0.1	3.0	12.0
Residential	1.3	1.8	0.5	0.6	0.4	4.6
Other	0.2	0.7	0.4	neg.	0.3	1.6
Total	1.9	22.0	7.5	0.7	6.0	38.1
Processing	7.8	7.2	0.2	3.5	5.9	24.6
Solid Waste Disposal	4.5	0.1	0.7	1.4	1.2	7.9
Miscellaneous	1.2	0.6	0.2	4.2	0.4	6.6
Totals	86.6	30.3	16.6	23.6	14.6	172.8

Source: Public Health Service
[a]Table taken from *Chemical Week,* June 17, 1970.

It can be seen from this table that most of the carbon monoxide generated comes from the internal combustion engine. The majority of the sulfur oxides (sulfur dioxide or sulphur trioxide) come from fuel combustion in power plants and industry. The greatest amount of oxides of nitrogen also come from the internal combustion engine with fuel combustion running a close second. The highest percentage of particulate and hydrocarbon emissions also come from the internal combustion engine, but large emissions are noted from processing plants and other similar manufacturing operations, while particulate emission comes mainly from chemical processing applications as well as fuel combustion.

These figures indicate that industry's major job at the present time is to improve all of its combustion operations and carefully evaluate the air pollution potential of its various processes.

In addition to those pollutants which are generated as part of a manufacturing process or operation, there are even more serious consequences caused by pollutants which are not necessarily generated by industry per se but emanate from products manufactured by industry for commercial and domestic use. While industry does not have direct control over the amount of these emissions nor the ultimate use of their product, they must become more aware of the possible complications caused by use of certain materials and their overall effect on our air pollution problem, and for that matter, the total ecology of the earth.

An example of this type of pollutant is DDT, which is a chlorinated hydrocarbon produced by a number of chemical companies as an insecticide or pesticide. DDT has been in production and wide use for a number of years but only lately have ecologists and environmentalists become concerned over its indiscriminate use. While it was impossible for the original manufacturer of this chemical to determine its long-range effects on human and animal life (because there was no way to accelerate these effects in a laboratory investigation), its use is nevertheless a serious problem today, some 25 years after it was introduced in any quantity. The ecological implications are very similar to those described in Rachel Carson's book *Silent Spring.* At the time her book was published it was discredited by a number of competent people, including chemists and engineers, and considered to be an overdramatic and emotional approach to a nonexistent problem. We know now that she was right in her speculations. The federal government has recently placed a ban on the manufacturing and use of DDT. The point of these comments is that industry is being blamed for its shortsightedness as well as for continuing to produce and encourage the use of a material which has widespread ecological effects on human and animal life. Whether this could have been predicted at an earlier stage is academic. The fact is that the blame for not comprehending the problem is being charged to the chemical manufacturer. Similar situations will undoubtedly occur in the future unless we guard more carefully against long-range possibilities of this type.

Ralph Nader and others who have jumped onto the ecological bandwagon are condemning the use of tetraethyl lead as an additive to high-test gasoline on the premise that the amount of lead being distributed over the face of the earth by the internal combustion engine has reached staggering proportions and that lead is toxic and damaging to human and animal life. It is estimated that 400,000,000 pounds of lead are annually emitted by automobiles in the United States. Oddly enough the Public Health Service first anticipated this possibility back in 1925 but the adverse health effects of tetraethyl lead were then discounted because there was insufficient evidence to justify preventing the use of the additive to gasoline. Very little new scientific information was available until the 1960s. This in itself is not a condemnation of the Public Health Service but more specifically of the industries that produced the fuel and the industries that used it.

These facts only help to reinforce the argument that industrial air pollution must consider not only present emissions from plants but also must expend adequate research and development on new products which, if used widely on a commercial or domestic basis, might contribute to serious air pollution in the future.

HOW DOES AIR POLLUTION AFFECT OUR ENVIRONMENT?

Today there is overwhelming evidence that various pollutants do and will continue to affect life on this planet as we know it. Medical evidence of deleterious effects of various pollutants is mounting daily. In addition to the damage which air pollution causes in the human organism it causes similar, and in some cases more severe, damage to animals, to plant life, and even to our climate.

Health Effects

First let us consider what the major pollutants do to the human organism. Carbon monoxide is a definite contributor to heart disease. It causes fatty degeneration of blood vessels in moderate concentrations when tested on laboratory animals and therefore is presumed to have similar effects on the human body. The four main sources of carbon monoxide for human intake are cigarette smoking, automobile exhaust, fossil fuel combustion, and other types of domestic heating. It is by far the greatest air pollutant, and perhaps the most difficult to eliminate. It is estimated that 94,000,000 tons of carbon monoxide were produced in the United States in 1966; 67,000,000 tons of this total came from motor vehicles. In addition to this, forest fires and other nontechnical sources added 8.6 million tons to the figure. Carbon monoxide from natural processes also adds to the total. Carbon monoxide has been shown to have an effect on the central nervous system in concentrations as low as 10 parts per million. This level corresponds to the formation of about 2 percent carboxyhemoglobin in the bloodstream after prolonged exposure for more than 8 hours. There is evidence that oxygen transport in the bloodstream is affected when this level goes above 5 percent carboxyhemoglobin, which can be achieved from a carbon monoxide concentration of 30 parts per million in the air. Carbon monoxide levels often reach 50 parts per million and can go as high as 140 parts per million for short periods in heavy automobile traffic. Cigarette smoking can produce up to 15 percent carboxyhemoglobin in the blood. Controlled laboratory experiments have shown that animals which were fed continuously a high cholesterol diet and were exposed to high levels of carbon monoxide for long periods of time averaged two and one-half times as much cholesterol in the aorta as animals on an identical diet which were not exposed to a high carbon monoxide level.

The sulfur oxides, sulfur dioxide and sulfur trioxide, which generally come from the combustion of high-sulfur fossil fuels contribute about 24,000,000 tons to our air pollution problem each year. The normal threshold limit of exposure to sulfur dioxide over an 8-hour period is approximately 5 parts per million, which is usually compatible with most fuel-burning operations, especially when tall stacks are used for dispersion. In smelting operations this level can go as high as 30 or 40 times the threshold level for short exposure periods, and if a high concentration of sulfur dioxide is trapped in an inversion layer, as in Donora, Pa., in 1948, or in several of the famous London death smogs, there are serious results (see Table 2-1). High concentrations of sulfur dioxide are toxic to the human body but usually cause death only in cases of previous respiratory disease, such as emphysema, and generally affect older people more seriously than young people who are in good physical condition. High sulfur dioxide concentrations are easily measured, but since their source is primarily the combustion of fossil fuels, corrective measures are expensive to specify and implement.

The oxides of nitrogen contribute about 10,000,000 tons annually to air pollution in the United States and generally are found in much lower concentrations than either carbon monoxide or sulfur dioxide. While the ultimate effect on human welfare is not clearly understood, they are irritants to breathing and result in discomfort to the eyes. They are one of the major contributors to smog formation due to photochemical reactions in the atmosphere.

Since there are literally hundreds of hydrocarbon vapors which may be released into the air from both industrial and commercial applications, it is difficult to generalize their effect on the human body. Nevertheless, hydrocarbons are known to have some adverse effects on human health. Some affect the operation of certain body organs when ingested in very small quantities while others have a relatively high threshold level. Safe threshold levels can be obtained by writing the American Conference of Governmental Industrial Hygienists, 1014 Broadway, Cincinnati, Ohio 45202. Many hydrocarbons are thought to be carcinogens because certain cyclic organic compounds have produced cancer in rats and mice and are therefore considered capable of producing similar tumors in the human body. No hydrocarbon vapor should be consistently inhaled or ingested by other means over any period of time unless there is expert toxicological advice that it is permissible.

Particulate matter is damaging to the respiratory system of animals and human beings. It is possible for particulate matter to be toxic or nontoxic, depending on the source. Ingestion of toxic materials, such as mercury vapors or mercury compounds in a particulate form, can cause some direct biochemical reactions in the body. Far and away the most common type of particulate is inorganic in nature and nontoxic. Such types are ingested through the nose and the mouth and deposited in the lungs, causing a gradual buildup of the material on the lining of the lungs. Silicosis is a common disease in certain types of mining operations and in cement mills. The black lung disease, which is a form of silicosis, is prevalent among coal miners. In these cases particulate matter deposited in the lungs reduces the ability of the lungs to transfer oxygen into the bloodstream causing shortness of breath and eventual overtaxation of the heart and possibly premature death.

Ecological Changes on Animals

If we assume that virtually all of the commonly known air pollutants have an adverse effect on human well-being and life, it can be hypothesized that these will have a like effect on other animals which have a similar physiological makeup. While there does not seem to be a wealth of data substantiating the effect of carbon dioxide, sulfur dioxide, carbon monoxide, and particulate emissions on the life of specific domestic animals, it should be axiomatic that such animals, living in areas of high population density, would average shorter lives than those living far from the industrial and population centers in the world. Since man has been more interested in his own welfare, there seems to be a dearth of literature confirming that similar effects are felt by animals in the same environment, though controlled experiments have been carried out using animals rather than humans, and the effect has been essentially the same. Animals seem to be no more resistant, and in some cases, less resistant than their human counterparts.

Other ecological changes that go along with man's environment probably have a greater effect in keeping the wild animal and bird population in cities and urban areas to a minimum. There are no ready sources of natural food for the predatory animal and there is no covert to protect their intended victim. We cannot say that the bald eagle and the osprey have become victims of air pollution just because we cannot find either on Manhattan Island. Studies have been made, however, to determine the lead content in the bodies of birds and mammals indigenous to more urban areas and the findings have substantiated the fact that animals are similarly affected by lead pollution from gasoline burned in the internal combustion engine.

Perhaps more important to animal life is the indiscriminate use of deadly insecticides. Smaller animals obviously have a lower tolerance level for most of these materials than man, simply on the basis of total body weight, blood content, etc. Most of these toxic compounds are ingested through the food chain. For example; the eagle and the osprey eat fish which have been contaminated through water pollution by chlorinated hydrocarbon pesticides. The result is twofold. In some cases the fish die, in other cases they are eaten by the predatory bird and the bird, or later generations of his species, suffers the consequences. Definite correlation between the thickness of eggshells of fish-eating birds and the chlorine content in their body has been established. Evidently the chlorine or chlorinated hydrocarbon inhibits the calcium formation necessary to produce a strong egg shell for the offspring. Therefore, the egg often breaks before it hatches and the next generation is aborted before it has a chance to be born. All of this is ultimately a water pollution problem, but it often starts as air pollution because the pesticide or insecticide is spread over a particular area by spraying. Insecticides are found both in aerosol and vapor phases and obviously can be ingested through the lungs of birds and mammals, as well as through their food chain.

In most cases, wildlife is driven away from highly polluted areas because the species have no economic reasons like man to remain there and suffer the consequences, but there is little doubt that air pollution of every kind affects the pattern and longevity of the animal kingdom.

Ecological Changes on Plants

In the last few years it has become increasingly evident that air pollution is affecting vegetation, at least to the same extent that it is affecting human and animal life. The U.S. Department of Agriculture estimates that agricultural losses due to increasing air pollution levels are close to $500,000,000 per year. California alone loses $100,000,000 annually. The state of New Jersey estimates that over 57 different plant species have become damaged by increasing air pollution. Air pollution affects plants in two ways: incidence of high air pollution causes visible damage, but also, chronic sublethal levels of air pollutants contribute to the eventual destruction of the plants' physiological life process, affecting the growth, productivity, and quality of the vegetation.

Naturally the plant is affected by the level and duration of exposure, as well as meteorogical factors. For example, a tree growing in an urban area where the air pollution level is continuously high will probably die sooner or experience greater defoliation and greater depression of growth when compared to one exposed to lower pollution levels on a casual basis.

Each pollutant or combination of pollutants will produce a certain pattern of injury which leaves graphic records of the level and type of pollutant. The TR-7 Agricultural Committee of the Air Pollution Control Association has produced an excellent report with colored photographs demonstrating the effects of air pollution on various types of plants. It is entitled *Recognition of Air Pollution Injury to Vegetation and Pictorial Atlas,* edited by Jacobson and Hill, and is published by the Air Pollution Control Association, Pittsburgh, Pa. It is suggested reading for anyone who would like to know more about the effects of air pollution on plants.

Plant air pollutants can be designated as two types; first, primary pollutants which are lethal to plants as they originate from the source, and secondary pollutants which do not originate thus but which are formed through reaction of pollutants from the source. The reaction can occur at the emission point or at localities removed from the source, depending upon the rate of reactivity of the compounds involved.

The major primary pollutants are sulfur dioxide and hydrogen fluoride. The major secondary pollutants are ozone and peroxyacetylnitrate.

The injury caused by sulfur dioxide to foliage may be either acute or chronic. Acute injury occurs when the pollutant is absorbed by the plant at a rapid rate, and chronic injury is caused by long-term absorption of sulfur dioxide at sublethal levels. Vegetation can be affected when ground levels of sulfur dioxide exceed 0.5 parts per million (ppm). Vegetation is often totally destroyed in the area of smelting operations, where sulfur dioxide concentration is the greatest. In fact, in the vicinity of most smelters there is little if any plant life.

Fluorides, which come from a variety of sources, but predominantly from the manufacture of phosphate fertilizer and aluminum reduction processes, have a cumulative poisoning effect on the plant. Fluorides are usually found in the form of hydrogen fluoride or silicon tetrafluoride and these are readily absorbed by the leaves of the plant causing the leaf tissues to die at the tip. The problem with fluoride damage is that much lower concentrations, on the order of 0.5 ppm, can cause damage.

Ozone is a typical secondary pollutant. It occurs naturally in the atmosphere and is generated by electric storms. The primary source of high ozone concentrations in the lower atmosphere comes from the photochemical reaction of oxides of nitrogen and hydrocarbons which are emitted by the automobile and industry. Nitrogen dioxide reacts in the presence of sunlight to form nitric oxide plus atomic oxygen. This atomic oxygen then combines with molecular oxygen in the presence of light to form ozone. The rate of formation of ozone depends upon the type of hydrocarbon and its concentration in the atmosphere and also the concentration of the nitrogen oxide and the length of time it is exposed to sunlight. Ozone bleaches plant cells much the same as sulfur dioxide. Coniferous trees are susceptible to ozone damage which initially shows at the end of the leaf. Concentrations below 0.05 ppm will cause symptoms on sensitive species.

The other secondary pollutant, peroxyacetylnitrate, often referred to as PAN, is a member of a series of compounds which originate from the reactions of unsaturated hydrocarbons and oxides of nitrogen in the presence of light. As with ozone the amounts and exposure time to the radiating source control the amount of PAN which is developed. Natural sources of hydrocarbons such as terpenes that evolve from coniferous vegetation in the vicinity of man-made sources of hydrocarbons have the ability to react photochemically and produce PAN. This naturally complicates control efforts. Concentrations of 0.2 ppm and an exposure time of several hours is sufficient to cause injury to vegetation.

The damage to the forests, farms, and fields of our country is a frightening consequence of our inability to control air pollution. Recently it was determined that many acres of the San Bernadino Forest in California were seriously affected by smog from the Los Angeles basin. It was once expected that this area was too far away for any damage but the smog is now coming out of the basin and over the coastal range into large areas of virgin forest. This situation will happen not only in Los Angeles but in nearly every metropolitan area in the nation unless we soon develop control methods.

TABLE 1-2 MAJOR SOURCES OF PLANT-PATHOGENIC AIR POLLUTANTS

	SOURCE				
Pollutant	*Transportation*	*Industry*	*Generation of Electricity*	*Space Heating*	*Refuse Disposal*
	(million tons/year)				
Sulfur oxides (SO_2)	1	9	12	3	1
Hydrocarbons (O_3, PAN)	12	4	1	1	1
Nitrogen oxides (O_3, PAN)	6	2	3	1	1
Fluorides (HF, SiF_4)		1			
Others	2	8	4	2	2
Total	21	24	20	7	5
%	28	30	26	9	7

Source: Restated from F. A. Wood, "Sources of Plant-Pathogenic Air Pollutants," *Phyto-pathology,* 58: 8 (1968), pp. 1075–1084.

Meteorological Changes

Perhaps one of the most interesting and terrifying aspects of air pollution is its effect on the climate. To demonstrate how this can happen let us take the case of LaPorte, Ind., which is 30 miles southeast of Gary, Ind., which is predominantly a steel mill town. LaPorte is also southeast of Chicago. The amount of rainfall in LaPorte is vastly different than neighboring Gary or Chicago. Between 1946 and 1967 LaPorte averaged 47.1 in. of rainfall per year which is 19 in. or 47 percent more than the precipitation at localities upwind of Chicago. LaPorte's proximity to Lake Michigan has been discounted for this high rainfall difference. It has been noticed that LaPorte's rainfall pattern closely follows that of Chicago's haze and smog pattern and that its precipitation increases and decreases with the area's fuel use (see Fig. 2-4). Another example, Belleville, Ill., which is 10 miles southeast and downwind of St. Louis, presently gets about 7 percent more rain annually than areas upwind of St. Louis where there is less pollution. To substantiate this fact the extra rainfall is mainly during the week when the pollution from St. Louis is the heaviest.

These are just two examples of how pollution has changed weather patterns in several isolated areas. Scientists believe that dust particles in man-made pollutants are very small but they are large enough to have a strong attraction for water vapor which condenses and freezes on them, forming ice crystals which in turn form clouds. If sufficient moisture is present the cloud droplets grow in size and eventually fall as rain. Therefore as pollution levels rise in industrial or metropolitan areas rain levels and possibly snow levels will increase in these areas or in areas which are directly downwind.

While pollution can and does cause more rainfall it can also have the opposite effect when clouds become so overseeded that no rain falls. This happens when pollution creates so many dust particles that they cannot attract enough water vapor to grow to raindrop size. This has happened in many places around the world and has been documented in the sugar-producing area of Queensland, Australia. Here, precipitation has been reduced by about 25 percent during the sugarcane harvesting season. At this time the leaf is burned off during cutting, causing thick columns of black smoke to rise into the air, overseeding the atmosphere.

For the last five or six years the rain and snow pattern over the United States, especially in the northeast, has changed quite radically. It is very possible, and also probable, that this change is due to pollution patterns changing in the same area.

Pollution is also the cause in many cases of dense fog. This often happens where a plant is located in a valley and the effluent from the plant in an inversion layer forms the nuclei for the fog droplets.

There is also evidence that pollution is associated with violent weather. Thunderstorms are found more frequently in heavily polluted areas than in nonpolluted ones. Hailstorms are also directly associated with pollution. LaPorte, Ind., has had 130 days of hail during the last 14 years, or about four times as much as other weather stations in the area.

These are several of the dramatic weather changes; however, there are many much less dramatic ones which probably have more far-reaching consequences. Some scientists accept the theory that pollution, by acting as a blanket, tends to reradiate heat back to metropolitan areas. This explains why cities and suburban areas have become warmer over the years. Average annual temperature minimums in the center of Washington, D.C. run around 5 degrees higher than those in outlying areas of the city and this variation is the greatest on weekdays when pollution is at its highest level. Chicago's frost-free period is 197 days; the surrounding countryside, however, is only 160 days. To find a similar area with frost-free periods as long as that of Chicago you would have to go 400 miles to the south.

Pollution also shuts out sunlight from cities and suburban areas. There is 15 to 20 percent less sunshine in the city than there is in surrounding areas. During four winter months London has about 96 hours of sunshine compared with as much as 268 hours in the open countryside.

Today there is increasing concern that the jet aircraft is affecting our climatic conditions. The exhaust from jets in the upper atmosphere is responsible for increased cloud formations in certain areas. It has been noted that the cloud cover along the heavily traveled New York-to-Chicago jetway has increased to the point that a section of the route in Pennsylvania is now covered by cirrus clouds about 90 percent of the time—which was not the case before the jet age. A similar situation has been noted along the New York-to-London jet route where cloud cover is up 10 percent over the 1950s and is still increasing.

While some scientists say the greenhouse effect in the cities is causing our average temperatures to increase, others disagree and say that the fact that there is greater cloud formation at high altitudes will give us a trend toward cooler temperatures around the world. They estimate that average annual temperatures have dropped by one-third to one-half a degree during the past three decades which sounds like little until you stop to consider that the last ice age came about through a temperature drop of only four or five degrees. Despite this controversy over the results, the fact is that air pollution is affecting our world climate and we do not know exactly how seriously.

The long dispute over building the American version of the SST found many of its critics concerned with high-altitude pollution and its eventual effects. The climatic phenomenon which it may create may offer no reason to cease experimentation but this is an area of concern for the future and one in which we should move very slowly.

DETECTION

Since subsequent chapters of this book will cover the method of detection and measurement of atmospheric pollutants, only a few words will be said at this point. Obviously we have detecting mechanisms of the effect of air pollution in human beings and from signs in nature which have been described before. However, these afford neither accurate quantitative measurements nor can they accurately discern between various types of pollution. A normal man can tell when he smells a hydrocarbon vapor but there are very few of us who would be able to detect sulfur dioxide or carbon monoxide with our nose alone. We can see the effects of pollution on our own health and plant life, but even experts in the field cannot determine whether one pollutant or a combination of pollutants has caused the symptoms which they record. It is therefore vitally necessary that we develop not only qualitative but quantitative analytical equipment which will not only label the pollutant but determine its level and its source.

Particulate matter is perhaps the easiest to determine on a quantitative basis through the use of air samplers. These can collect such matter which can later be weighed and then related to a volumetric standard. Sulfur dioxide and carbon monoxide are detectable on a quantitative basis with modern instrumentation, and equipment has recently been developed for accurate determination of oxides of nitrogen; however, the reliability of such equipment for use on long-term sampling operations has not been proven.

Hydrocarbon emissions are a somewhat different story. These can be determined qualitatively and quantitatively only through elaborate instrumentation such as gas chromatographs which require time and skill to operate. Nevertheless, there is perhaps more research work being accomplished in the area of pollutant measurement at the present time than in any other phase of the pollution problem. While measurement is necessary to the proper solution of the overall problem, there is no reason to await better detection devices before beginning to reduce some of the major sources of air pollutants which we all know and recognize. Reduced expenditure on detection at the present time and greater expenditures on the reduction of carbon monoxide and sulfur dioxide levels should have a much better payout for the future.

CORRECT POLLUTION PROBLEMS—WHY?

The information included in Chapter 1 is merely an abstract of other chapters in this book. We are trying to set the scene to give you a picture of the total air pollution problem and its effect on our lives.

While the reasons for correcting air pollution should be obvious, it is amazing how many industries don't or won't recognize their role in this task. They should have three major motivating forces enlisting their support in the battle against air pollution. The first is economic, the second is social, and the third is medical, and all three are interconnected.

Economic reasons for correcting air pollution problems can be appreciated from the statistics. It was stated before that agricultural losses in the United States alone exceed $500,000,000 a year due to air pollution. But consider some of the other major areas of economic loss in our society because of air pollution: (1) building property and machinery damage due to corrosion; (2) lost time because of the health effects of pollution on employees; (3) lost product which escapes from the manufacturing plant because of inadequate collection devices; (4) lost goodwill and therefore lost sales because of local reaction to air pollution problems.

These are four ways in which your plant, or any plant, can lose directly because of air pollution which you cause or tolerate in your plant environment. All of these can be tied into an economic loss which must ultimately increase the cost of your product and decrease your profits.

Industry has already assumed a large portion of the responsibility for the health costs of its employees. Is it not then foolish to pay for these benefits while creating an atmosphere which is detrimental to the health of these employees?

Industry has come to assume a role in the community because it finds that it is desirable for it to be socially acceptable; yet if it pollutes the air of the community, no amount of work on the part of its executives or people can make it an acceptable neighbor.

Industry spends millions of dollars a year on preventive maintenance for the physical appearance of its facilities and yet its own air pollution corrodes the machinery and discolors the paint for which that very industry paid.

Industry must begin to look at air pollution abatement as an economic plus instead of an economic minus. The problem has been that the first jump is such a big one. It's like a man going to the dentist for the first time in 40 years; the first job is bound to be a big one, but after this hurdle has been overcome, regular preventive action should prove no more costly than for any other program of preventive maintenance within the plant. Certainly the initial cost of air pollution prevention equipment, when none has ever been used, is going to be a burden. One of the reasons given by many plant managers for not starting on the program is that local and state regulations are not definitive, and they do not want to install a piece of equipment which will be obsolete within two years. This is not an entirely justifiable attitude, however, since in most cases equipment which should meet any future requirements can be selected.

Financing for this type of major capital outlay can usually be done within the community. Bankers like to lend money but will lend it much more freely when it involves a community improvement. In fact, they can take part of the credit for the solution. In the next few years we will see municipal monies used to help industry in this area, and many of the tax provisions which have been suggested to aid in pollution abatement will soon become fact.

As industry begins to put its own house in order the pressure for air pollution abatement will begin to become a matter of individual conscience, and without industry to blame for the problem, individuals will have to begin to "clean up their own houses." At this point we should begin to see some significant strides made in solving the total problem.

2 Effects of Air Pollution on the Environment

INTRODUCTION

Abundant evidence has established that air pollution damages vegetation, accelerates the deterioration of materials, soils property, affects climate, reduces visibility and solar radiation, aggravates public relations, adds to production costs, contributes to safety hazards, interferes with the comfortable enjoyment of life and property, and is a definite factor in human and animal morbidity and mortality. The perplexing social questions are: how do we attain acceptable levels of air pollution at minimum public and private expense? What are the trade-offs and what trade-offs are we willing to make, and what are the acceptable levels of air pollution? The private and public sectors have been struggling to answer such questions for many years. Engineers must first identify the problems, the major information sources, and where possible quantify the effects in economic terms and the mechanisms and/or theory of the cause-and-effect relationships. The purpose of this chapter is to assist such efforts.

The Socioeconomic Implications

It may well be that our economic systems have contributed to the perpetuation of air pollution. Because of the nature of the environmental problems, the motivations for control are generally external to the pressures of the marketplace. While in some cases the sources and the effects on the receptors are obvious, many of the effects of air pollution are insidious, difficult to quantify, and are caused by a multiplicity of sources. It is thus understandable that our legislative and judicial systems have and will continue to be faced with the responsibility of arbitrating and resolving the myriad conflicts involved.

Scores of judicial decisions could be cited to illustrate the complex legal questions involved.[1] Prior to the early 1930s the courts tended to balance equities in favor of the major pollutant sources. However, in the past decade there has been a swing in the other direction. In 1970 there were several landmark decisions at state and federal levels[2,3,4,5] which at least clarified the rights of the public and the responsibilities of the sources. The Supreme Court of New Jersey affirmed the judgment of the Appelate Division of the Superior Court (100 N. J. 366, 242A. 2d 21) which in essence found the State definition of air pollution to be a reasonable standard. The state of New Jersey definition is as follows:

> Air pollution is the presence in the outdoor atmosphere of one or more air contaminants in such quantities and duration as are, or tend to be, injurious to human health or welfare, animal or plant life, or would unreasonably interfere with the enjoyment of life and property. . . .

The legislative and executive branches of government have in many cases shown their awareness of the public desire for a cleaner environment and at the same time the need to balance equities. Legislation has been adopted which not only requires sources to control contaminant emissions, but also requires the executive branches of government to analyze in more detail the costs and benefits involved, determine the state of the technological abilities to control emissions, and establish through research more economic and effective ways of reducing or eliminating air contaminant emissions.[6,7,8,9]

Like many private businesses, governments, particularly the federal government programs, are adopting programming, planning, and budgeting systems to facilitate the making of management decisions influenced by cost benefit considerations. The Clean Air Act as amended in 1967 while recognizing "that the prevention and control of air pollution at its source is the primary responsibility of States and local governments . . ."[6] makes the executive branch of the federal government responsible for the issuance of air pollution criteria documents where "Such criteria shall accurately reflect the latest scientific knowledge useful in indicating the kind and extent of all identifiable effects on health and welfare which may be expected from the presence of an air pollution agent, or combination of agents in the ambient air, in varying quantities." Since there are about half a million contaminants that could conceivably be emitted into the atmosphere and the possible number of combinations and ranges of concentration are virtually infinite, this is a Herculean task.

Numerous articles, theses, and books have struggled with the purely economic aspects of air pollution and its control. Peckham[10] surveyed the literature on economics of air pollution and cited more than 385 references. But with all the conceptual and fact-finding studies, we are far from producing a comprehensive ledger quantifying the cost of control versus the costs of unabated air pollution. On the "Costs of Control" side of the ledger there are some rather comprehensive figures.[7,11] This topic has been quite thoroughly reviewed in Senate Document No. 91-65, entitled "The Cost of Clean Air," a March 1970 report to the Congress of the United States.

While this chapter deals mainly with the adverse effects of air pollution, we recognize that the cost of control may also have some adverse public effects, particularly on the price of goods and services. The conclusions of the aforementioned Senate report which estimated the costs for controlling 4 major forms of pollution (particulates, sulfur oxides, hydrocarbons, and carbon monoxide) for 21 source categories in 100 metropolitan areas, are worthy of note:

> In general, the annual cost to control these four pollutants is a very very small percentage of the value of shipments in each of these industries—usually less than 1 percent. To control air pollution from solid waste disposal operations, about $0.39 per ton of waste will be required; the cost for steam-electric powerplants is estimated at 1.3 mils per kilowatt-hour.

While there is still a great deal of effort needed to determine the total cost of control and to consider the impact in specific geographic areas, balancing the ledger on the benefit side in dollars and cents is the more difficult task. How does one measure the value of an improved quality of life? What trade-offs are individuals in specific areas willing to make? How safe is safe? How do the air pollution hazards compare with the risks to which we voluntarily expose ourselves each day? These and many similar questions have been considered by ecologists, economists, engineers, epidemiologists, lawyers, physicians, political scientists, legislators, sociologists, etc. A thorough review of the socioeconomic and sociotechnological problems associated with air pollution and its control, may not, however, be of significant assistance to the engineer or plant manager faced with the problem of designing an abatement system. Suffice to say that the problem is complicated by many social factors, that public involvement in the decision-making process is nessary and that it will take our government agencies, public institutions, and private resources and talent working together to find and implement the optimal solutions. Both the public and private sectors can benefit:

Public Benefits

1. Improved health
2. Reduced safety hazards
3. Reduced health risks to man and animal
4. More comfortable enjoyment of life and property
5. Reduced property damage
6. Increased property values
7. Less vegetation damage

Private Benefits

1. Lower employee absenteeism
2. Reduced risk of civil damage suits
3. Better employee relations
4. Better public relations
5. Reduced maintenance costs
6. Increased property values
7. Product recovery
8. New markets for new products relating to air pollution control
9. Reduced product contamination or damage

EFFECTS ON MATERIALS

This section is limited to a discussion of the effects various forms of air pollution have on inanimate objects and the socioeconomic implications thereof. Peckham, of the Division of Economic Effects Research of the National Air Pollution Control Administration, and his co-workers, have compiled rather complete literature surveys on the subject.[10,12,13] These and the "Criteria Documents"[14,15,16,17,18] should be referred to for a more comprehensive summary of this subject.

Socioeconomic Implications

Assuming, for the sake of simplicity, that the public would be willing to accept the health risks associated with living in a polluted environment, as they apparently accept voluntarily many other risks like driving an automobile, flying, skiing, hunting and the like,[19] would they voluntarily accept the damage to property if they had an adequate definition of the costs involved? Assuming many would not, what could they do? It would appear that they have only about three options. They could move, in the hope of finding a less polluted area. They could institute a suit for damages, or they could prevail upon their various public officials to take corrective action on their behalf. All three have been tried.

The first option, that of moving, can in itself be quite a hardship. Many of those living in highly polluted areas cannot afford the luxury of moving to improve the quality of their life, but neither can they afford to institute a civil suit for damages when faced with their many other economic problems.

Some individuals have successfully sued for damages caused by air pollution. One New York State case that was widely reported by the news media involved a civil suit against a cement plant.[3] The plaintiffs successfully sued for damages and then sued for an injunction. The decision in the case was quite interesting in that the injunctive relief was denied on the basis that the equities had been balanced by the damages awarded. To take this course of action, various public groups would have to have both proof of personal damages as well as proof that the responsible sources had been properly identified. To balance equities in this manner is not only very time-consuming, but would probably be the most expensive approach for both the public and private interests involved.

Legislation has been passed in Michigan which might set a unique precedent. It gives citizens the right to file suit against public and private sources when there is sufficient proof to show that the sources are damaging the environment.[20] Under this legislation it will apparently not be necessary for the plaintiffs to establish that they are being directly affected by the emissions from the source. Because health damage is extremely difficult to prove, it is quite likely that initial suits under this law will be concerned with the evidence of damage to materials and lower forms of life.

The third public option, which intuitively seems the most appropriate, is to support or stimulate public programs that measure the costs and benefits as well as evaluate the technical feasibility of control, and to support the adoption of legislation and regulations and the enforcement thereof to prevent and abate air pollution problems accordingly. This has been the major objective of the many concerned publics.

The preceding discussion is perhaps oversimplified. There are numerous problems that have to be resolved. There is the matter of logistics. The major problems are concentrated in "hot spots" and in many cases those most affected by the pollution depend upon the source for their livelihood. Market competition places a burden on the industries, and the industries' cooperation and assistance is vital if pollution problems are to be resolved. All kinds of data and information are needed to make rational management decisions but the surfacing of such information is complicated not only by the complex systems that would be necessary to obtain it, but also by the costs of some of the proposed data collection systems, and the competitive privileged nature of some of the data. The National Air Pollution Control Administration has a Division of Economic Effects Research which is charged with the responsibility of investigating this matter. NAPCA has also negotiated several contracts to acquire more data. But there are enough data to stimulate concern and continued study.

One of the earliest and most commonly cited studies of the economic significance of air pollution was conducted by the Mellon Institute during 1913.[21] Primarily the costs of soiling and wasted fuels (from incomplete combustion) were considered. It was concluded that the average annual per-capita costs due to such effects in the Pittsburgh area was about $20. Subsequent updating of this figure on the basis of the commodity price index apparently led to the now common range of estimates between $60 and $70 per capita per year.[22,23] Kneese[24] has examined the problems associated with these extrapolations and has presented a more comprehensive basis for analyzing the economic effects. Many articles and even complete books have been published on the concepts for economic analysis of air pollution damage.[25,26,27,28,29,30]

The remaining portion of this section is a capsule summary of the effects on materials by soiling, general particulates, sulfur oxides, photochemical oxidants, and acid gases.

Soiling

One only needs to walk through some of our urban and industrialized areas to see some of the soiling damage caused by air pollution. Soot covered buildings, dirty windows and windshields, soiled textiles, clothes that need to be laundered after being hung out on a line to dry, facial grime, corroded metal surfaces, lawn furniture and park benches too dirty to use, and many other instances are readily observable. Michelson and Tourin[194] made a comparative study of the costs that could be attributed to air pollution (primarily soiling) in Steubenville and Uniontown, Ohio. Interestingly enough they found that the increased maintenance and cleaning costs which they correlated with the atmospheric particulate loadings in Steubenville were about $84 per capita per year more than those costs in the similar but cleaner community of Uniontown. The Michaelson and Tourin data were also plotted out in a Senate report[31] so as to graphically illustrate the relation-

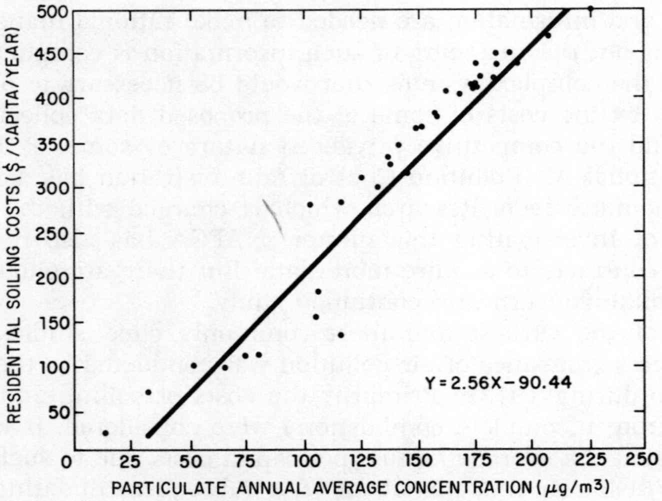

Fig. 2-1 Annual average particulate concentration versus residential soiling costs.

ship between soiling costs and the particulate levels in the atmosphere (see Fig. 2-1).

General Particulates

Obviously not all the potential types of damage that could be related to particulates were considered in the aforementioned studies. It is difficult to assign an economic value to damaged forms of art or the aesthetic value of a clean environment. The corrosion and erosion damage are significantly related to the presence of other pollutants and weather conditions. Particulates can cause corrosion through galvanic action. But coupled with acid or alkaline particulates and moisture the corrosion rates are greatly accelerated. The soiling and corrosion effects not only add to the cost of maintenance of property but depreciate property values.[32] After a comprehensive examination of the methodology for evaluating the relationship between property values and air pollution, Ridker[25] acknowledged the limitations of the available data but supported the conclusion that property values are adversely affected by air pollution.

Sulfur Oxides

Since sulfur compounds in solution are so chemically reactive both as oxidizing as well as reducing agents, it is not surprising that many forms of material damage can be related to atmospheric levels of sulfur compounds. Paints have been adversely affected.[33,34] Building and construction materials including masonry have been slowly destroyed.[35] Electrical contacts in expensive instrumentation have been ruined.[36] Textiles have in some cases not only faded,[37] or weakened,[38] but "dissolved."[39,193] Valuable art objects, papers, and books have undergone deterioration. While it is true that a simple chemistry course could demonstrate the potential damages that could be caused by the presence of sulfur compounds in the air, it is more important to note that such damage is occurring. We know the general mechanics involved, but we can only estimate the total damage costs. But isolated studies have shown the costs could be astronomical in our urban areas.[40,41]

Photochemical Oxidants

The effects of photochemical oxidants on material have been well summarized by Jaffe,[42] and in the *Air Quality Criteria for Photochemical Oxidants*.[17] Photochemical smog has been found to contain a rather complex and varied mixture of oxidizing and reducing agents. Thus the common analytical techniques have measured total oxidants or net oxidation capacity referred to a specific reference such as a buffered potassium iodide solution.

The principal oxidant, ozone, reacts rapidly with a variety of organic materials.[43] This ability has led to its beneficial use in the treatment of water and wastewater. Rubber, when under stress is particularly susceptible to chemical attack by ozone. Because the reaction is relatively fast and the results quite easily observed, rubber cracking has been used as one of the techniques for qualitatively measuring the presence of ozone. Reportedly, concentrations as low as 0.01 ppm ozone may cause such damage.[44]

Ozone and other oxidants such as NO_x have been shown to have adverse effects on dyed fabrics,[45,46] textiles, synthetic fibers, and other organic materials.[17] An exhaustive annotated bibliography abstract on economic effects,[26] and this one was not specific for NO_x. The economic significance of materials damage caused by ozone was investigated on a limited scale by Garet *et al.*[48] As with compiled by NAPCA on nitrogen oxides (NO_x)[47] includes only one many other economic studies, extrapolating data for national application was impossible.

Other Acid Gases

While the information on the detrimental effects of other acid gases on materials is limited, such effects can be observed.

Experience around swimming pools, water and wastewater treatment plants has shown that acid gases, particularly those with high redox potentials such as chlorine and its compounds, can be quite corrosive. Low concentrations in water have been found to react quite rapidly with organic and nitrogenous materials as well as with metals.

Because of the almost infinite number of variables involved, it is unlikely that we will be able to identify the full extent of materials damage due to air pollution before taking steps to limit the emissions of those substances which might be reasonably expected to have adverse effects. The absorptive capacities of particulates, the synergistic and catalytic reactions that conceivably may occur, the profound influence of weather conditions and the complex time concentration relationships each must be analyzed before we can extrapolate the results of the numerous studies that have been done. These and the limits of our testing abilities make what might seem a relatively easy subject to understand a very difficult problem to define in economic terms.

HUMAN HEALTH EFFECTS

While there may be levels of air pollution that will be acceptable to the public as a trade-off for other benefits (assuming there are significant benefits to accepting air pollution), many key public groups have shown that they are unwilling to compromise human health. By now most of the interested public groups are well aware of the acute toxicological effects of air pollution evidenced during periods of adverse meteorological conditions around the world. Such disasters as those itemized in Table 2-1 not only demonstrated the lethal potential of air pollution, but stimulated many public fears and questions concerning long-term chronic exposure to air pollution. The definition of the latter problem is complicated by a host of factors.

Acute Effects

The air pollution episodes have established that high levels of pollutants have caused the premature death of thousands of people. Meteorologists have shown that the unfavorable weather conditions that contributed to such disasters are likely to occur quite often.[49] Dr. Ralph Larsen, after studying the particulate and sulfur oxide levels and resulting death rates during the London and New York episodes, graphically illustrated the correlation between deaths and pollutant concentration (Fig. 2-2). While ad-

TABLE 2-1[52]

Year and Month	Location	Reported Excess Deaths	Reported Illness	Reference
1873 Dec. 9–11	London	?[a]	?	Logan (1953), Scott (1953)
1880 Jan. 26–29	London	?	?	Logan (1953)
1892 Dec. 28–30	London	?	?	Logan (1953)
1930 December	Meuse Valley, Belgium	63	6,000	M. Firket (1931), J. Firket (1936)
1948 October	Donora, Pa	20	6,000	PHS (1949)
1948 Nov. 26–Dec. 1	London	700–800	?	Logan (1948, 1953)
1950 Nov. 21	Poza Rica, Mexico	22	320	McCabe & Clayton (1952)
1952 Dec. 5–9	London	4,000	?	Ministry of Health (1954)
1953 November	New York	?	?	Greenberg, *et al.* (1962)
1956 Jan. 3–6	London	1,000	?	Martin (1961)
1957 Dec. 2–5	London	700–800	?	Bradley, *et al.* (1958)
1959 Jan. 26–31	London	200–250	?	Martin & Bradley (1960)
1962 Dec. 5–10	London	700	?	Scott (1963)
1963 Jan. 7–22	London	700	?	Scott (1963)
1963 Jan. 9–Feb. 12	New York	200–400	?	Greenberg, *et al.* (1967)
1966 Nov. 23–25	New York	?	?	Glasser, *et al.* (1967)
1966 Nov. 24–30	New York City	168	?	Fensterstock & Frankhauser (1968)

[a]Figures not available.

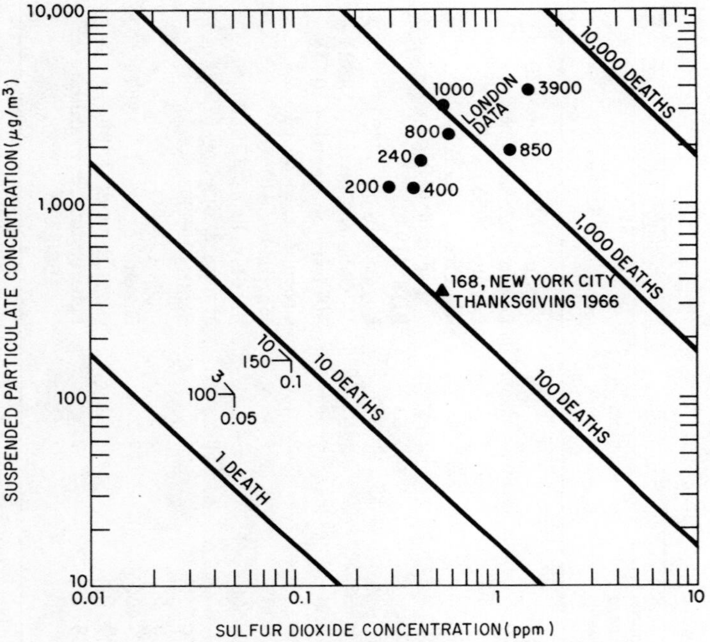

Fig. 2-2 Number of excess deaths per episode expected in London or New York (population of about 10 million) when the maximum 24-hour-average concentrations of sulfur dioxide and suspended particulate reach various values.[50] *(Courtesy J.A.P.C.A.)*

mittedly many other factors may have been involved, the statistical exercise provides a concise estimate of the potential mortality risks involved.[50] Larsen's calculations suggest the possibility that deaths might be more closely related to the product of sulfur dioxide and particulate concentrations rather than to the concentration of either pollutant alone (Fig. 2-3).

Hodgson[51] analyzed 2.5 years of New York City data for acute effects of air pollution at nonepisodal levels and concluded that "slight or moderate increases in concentration of air pollution during a month can be expected to result in increased mortality from heart and respiratory diseases amounting to several hundred deaths. And this is occurring for levels of pollution that are not considered unusually high and for changes in levels that may indeed pass unnoticed."

Critics of this approach might emphasize that many other unmeasured factors could have influenced the results. Perhaps oxides of nitrogen or adverse weather conditions, etc., may have been significant. But whether one considers the particulates and sulfur

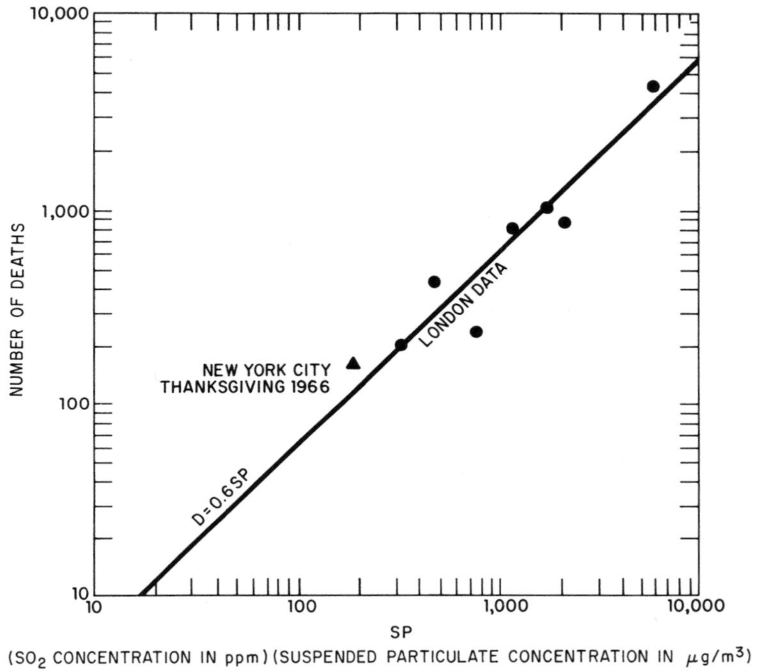

Fig. 2-3 Number of deaths in London or New York air pollution episodes as a function of the product of sulfur dioxide and suspended particulate concentrations.[50] *(Courtesy J.A.P.C.A.)*

oxides as the etiological agents or as indicators of the general air pollution levels, the correlation is still alarming to the public.

Chronic Effects

While the acute effects of air pollution are alarming, the chronic effects are of more concern. Epidemiological and clinical studies have related various air pollutants to everything from the common cold to various forms of cancer. Epidemiological and clinical studies too numerous to review in one section have repeatedly correlated air pollution and respiratory diseases.[55–66] Again many of the studies can be subject to questions. Human susceptibility to air pollutants varies widely. The number of possible air contaminants and concentration ranges thereof is practically infinite. The means of testing and expressing pollutant concentrations and even the medical terminology has varied. Clinical tests could not possibly expose human receptors to all possible pollutant mixtures and concentrations. In the epidemiological investigations it was and is still quite impractical to measure all the contaminants present during the entire exposure period. Thus, perhaps the best we can do is to determine possible correlations between chronic exposures to air pollution and health effects.

One of the most limiting factors in epidemiological studies is the relatively limited air quality data. In most cases the data is limited to various forms of measurements of particulates and sulfur compounds. Thus, it is understandable that the first two "criteria documents" released by NAPCA[14,15] dealt with these more prevalent pollutants. Since these documents represent a rather thorough review of the literature and should be referred to by any interested party, we will not attempt to paraphrase them, but rather include in tabular form the major conclusions supported by these documents and their numerous references (Table 2-2).

Perhaps the most famous form of air pollution is the photochemical smog prevalent in California. The smog, which results from a series of interactions in the atmosphere involving certain forms of hydrocarbons, oxides of nitrogen, ozone, and sunlight, is well known for its effects on visibility, vegetation, and human physiological responses, particularly eye irritation. The second series of "criteria documents"[16,17,18] released by NAPCA deal mainly with motor vehicle emissions (hydrocarbons, carbon monoxide, and photochemical oxidants). Table 2-3 summarizes the major health effects reported therein.[67] While certain hydrocarbons are known to cause physiological irritation, the NAPCA criteria document[18] states:

> Studies conducted thus far of the effects of ambient air concentrations of gaseous hydrocarbons have not demonstrated direct adverse effects from this class of pollution on human health. However, it has been demonstrated that ambient levels of photochemical oxidant which do have adverse effects on health, are a direct function of gaseous hydrocarbon concentrations; and when promulgating air quality standards for hydrocarbons, their contribution to the formation of oxidant should be taken into account.

Jaffe[81] also compiled a summary of the biological effects of photochemical pollutants on man and animals. He concluded that "Photochemical oxidants, such as frequently found in urban community atmospheres . . . cause repeated and continuing biological impact on man and animals . . ."

At the time of this writing the "Air Quality Criteria for Oxides of Nitrogen" has not been released by NAPCA. It may well be that epidemiological information gathered in the last couple of years substantially adds to the bank of information. Mueller and Hitchcock[82] summarized the toxicological effects of low concentrations of nitrogen dioxide in 1969 as shown on Table 2-4.

Health Effects of Specific Industrial Emissions

To this point we have been dealing with the more prevalent pollutants which may come from a wide variety of sources. While it would be impossible to discuss the potential health effects of all possible contaminants, some of the information on the more toxic contaminants is included.

The most comprehensive documents on specific industrial pollutants have been prepared by and for industrial hygienists.[83,84,85] It must be emphasized that the threshold limit values and other occupational health standards have been established based usually upon an 8-hour exposure of generally healthy persons. Extrapolating such standards to levels of chronic exposure by the general public is seldom possible. While in comparison there is relatively limited epidemiological data, the differences in levels of the air contaminants acceptable for occupational health and public health differ by several orders of magnitude (Table 2-5).

In addition to the "Criteria Documents" NAPCA has released a series of 27 reports compiled for them by Litton Systems, Inc., covering the following 30 pollutants:[87]

Aeroallergens (pollens)
Aldehydes (includes acrolein and formaldehyde)
Ammonia
Arsenic and its compounds
Asbestos
Barium and its compounds
Beryllium and its compounds
Biological aerosols (microorganisms)
Boron and its compounds
Cadmium and its compounds
Chromium and its compounds (includes chromic acid)
Chlorine gas
Ethylene
Hydrochloric acid
Hydrogen sulfide
Iron and its compounds
Manganese and its compounds
Mercury and its compounds
Nickel and its compounds
Odorous compounds
Organic carcinogens
Pesticides
Phosphorus and its compounds
Radioactive substances
Selenium and its compounds
Vanadium and its compounds
Zinc and its compounds

Not all the above have established public health significance, but they should be a part of the plant engineer's library. The following summary includes a review of industrial emissions having general air pollution significance where other than occupational health data has been analyzed.

TABLE 2-2

Pollutant	Conc. Level Producing Adverse Health Effects	Adverse Health Effects	Reference
Particulates and Sulfur Oxides	1) 80–100 $\mu g/m^3$ particulates (annual geometric mean) with sulfation levels of about 0.3 mg/cm^2-mo.	1) Increased death rates for persons over 50 years of age.	68
	2) 130 $\mu g/m^3$ (0.046 ppm) of SO_2 (annual mean) accompanied by particulate concentrations of 130 $\mu g/m^3$.	2) Increased frequency and severity of respiratory diseases in schoolchildren.	69
	3) 190 $\mu g/m^3$ (0.068 ppm) of SO_2 (annual mean) accompanied by particulate concentrations of about 177 $\mu g/m^3$.	3) Increased frequency and severity of respiratory diseases in schoolchildren.	70
	4) 105–265 $\mu g/m^3$ (0.037 to 0.092 ppm) of SO_2 (annual mean) accompanied by particulate concentrations of 185 $\mu g/m^3$.	4) Increased frequency of respiratory symptoms and lung disease.	71
	5) 140–260 $\mu g/m^3$ (0.05–0.09 ppm) of SO_2 (24-hour average).	5) Increased illness rate of older persons with severe chronic bronchitis.	72
	6) 300–500 $\mu g/m^3$ (0.11–0.19 ppm) of SO_2 (24-hour mean) with low particulate levels.	6) Increased hospital admissions for respiratory disease and absenteeism from work of older persons.	73
	7) 300 $\mu g/m^3$ particulates for 24 hours accompaned by SO_2 concentrations of 630 $\mu g/m^3$ (0.22 ppm).	7) Chronic bronchitis patients suffer acute worsening of symptoms.	74

TABLE 2-3

Pollutant	Conc. Level Producting Adverse Health Effects	Adverse Health Effects	Reference
Carbon Monoxide (CO)	1) 58 mg/m^3 (50 ppm) for 90 minutes (similar effects) upon exposure to 10 to 17 mg/m^3 (10 to 15 ppm) for 8 or more hours.	1) Impaired time interval discrimination.	75
	2) Effects upon equivalent exposure to 35 mg/m^3 (30 ppm) for 8 or more hours.	2) Impaired performance in psychomotor tests.	76
	3) Effects upon equivalent exposure to 35 mg/m^3 (30 ppm) for 8 or more hours.	3) Increase in visual threshold.	77
Photochemical Oxidants	1) In excess of 130 $\mu g/m^3$ (0.07 ppm).	1) Impairment of performance by student athletes.	78
	2) 490 $\mu g/m^3$ (0.25 ppm) maximum daily value. (This value would be expected to be associated with a maximum hourly average concentration as low as 300 $\mu g/m^3$ (0.15 ppm).	2) Aggravation of asthma attacks.	79
	3) 200 $\mu g/m^3$ (0.1 ppm) maximum daily value.	3) Eye irritation.	80

TABLE 2-4 NITROGEN DIOXIDE—TOXICOLOGIC EFFECTS AT LOW CONCENTRATIONS[82]

Species	*Conc. ppm*[a]	*Duration of Exposure*	*Effects*	*Comment*	*Reference*
Man	5	10 min	Increased airway resistance in 5/5 subjects. Mean increase 92%.	Response was delayed. Maximal increase 30 min after exposure. Time taken to return to normal not investigated.	Abe (1967)
Rabbit	0.25	6-hr day/hr 6 days	Alteration in structure of lung collagen (by electron microscopy).	Change still apparent in animal sacrificed 7 days after exposure.	Buell *et al.* (1966)
Rabbit	1	1 hr	Alteration in lung collagen and elastin (by spectroscopy) of animals sacrificed immediately after exposure. Reduced effect observed in animal sacrificed 24 hr after exposure.	Chemical changes suggest denaturation of structural protein only partially reversed.	
Rats	0.5 or 1	6 hr 1 hr	Morphologic change in lung mast cells and release of granular substance in animals sacrificed immediately after exposure.	Effect was partially reversed in 24- to 27-hr post-exposure changes are indicative of inflammation.	Thomas *et al.* (1967)
Rats	1	4 hr	Lipoperoxidation of lung lipids.	Effect delayed. Maximum response occurred 24-48 hr after exposure.	Thomas *et al.* (1968)

Rats	2	Continuously for natural lifetime	Pathological changes in bronchiolar epithelium (by electron microscopy), in normal cellular activity (blebbing).	Changes indicative of a preemphysematous condition. Exposure to 0.8 ppm indicates similar process.	Freeman *et al.* (1968)
Rats	1	4 hr			Thomas (1968)
Mice	0.5	6- to 24 hr/ day for 3 wk	Distended alveoli, loss of cilia, changes in bronchiolar epithelium.	Indicative of emphasematous condition.	Blair *et al.* (1969)
		Continuous for 3 mo	Increased susceptibility to bacterial pneumonia.		Ehrlich & Henry (1968)

[a]By volume.

TABLE 2-5 COMPARISON OF INDUSTRIAL THRESHOLD LIMIT VALUES (TLV) WITH MAXIMAL AIR POLLUTION VALUES[86]

Substance	Industrial Threshold Limits (TLV)[84]		Maximal Community Air Pollution Levels	Place and Date Where Observed
	ppm	mg/m^3		
Gases				
Acrolein	0.1	0.25	0.011 ppm	Los Angeles, California, 1960
Carbon monoxide	50	55	360 ppm	London, England
Fluoride (as F)	—	2.5	0.56 mg/m^3	Ural Alum (USSR)
Formaldehyde	5	6	1.87 ppm[a]	Pasadena, California, 1957
Heptane (*n*-heptane)	500	2000	4.66 ppm[b]	Los Angeles, California, 1957
Hexane	500	1800	—	—
Hydrogen sulfide	10	15	0.9 ppm	Santa Clara, California, 1949–54
Nitrogen dioxide	5	9	1.3 ppm	Los Angeles, California, 1962
Ozone	0.1	0.2	0.90 ppm	Los Angeles, California, 1955
Sulfur dioxide	5	13	3.16 ppm	Chicago, Illinois, 1937
Particulates				
Arsenic	—	0.5	0.069 mg/m^3	Zemianske Kostol, Czechoslovakia, 1964
Beryllium[c]	—	0.002	0.0011 mg/m^3	Pennsylvania, 1958
Lead	—	0.2	0.042 mg/m^3	Los Angeles, California, 1949–54
Sulfuric acid	—	1	—	—
Vanadium (V_2O_5)	—	—	0.0007 mg/m^3	Mihama, Japan, 1964
dust	—	0.5	—	—
fume	—	0.1	—	—

[a] Aldehydes as formaldehyde.
[b] Total hydrocarbons.
[c] Neighborhood, i.e., ambient levels (TLV) also have been set uniquely for this substance at 0.00001 mg/m^3 as a 30-day average.

Asbestos. Recently several air pollution control programs have taken action to reduce or eliminate asbestos emissions, particularly those from spraying operations. The numerous investigations by Selikoff[88,89] have provided the basis for such control action. The "Preliminary Air Pollution Survey of Asbestos" includes more than 238 references. The report indicates that asbestos bodies are causal agents and commonly found in the lungs of persons having asbestosis, pleural calcification, pleural plaques, lung cancer, and pleural and peritoneal mesotheliomas. Most of the information cited dealt with occupational hazards. One study by Utidjian *et al.*[90] did study the distribution of asbestos bodies in lungs in the Pittsburgh area. Although it was not concluded in the above survey report, it would appear that chronic public exposure to any level of asbestos is a public health hazard.

Beryllium. The state of New York has established an air quality objective for beryllium of 0.01 micrograms per cubic meter (24-hr average).[91] Chapter VII of the New Jersey air pollution control code, among others, severely limits the emission rates of particulate beryllium and its compounds.[92] Such compounds are extremely toxic, but are "commonly found as an atmospheric pollutant within the confines and in the proximity of industrial plants producing or using beryllium substances."[93] Studies conducted in the state of Pennsylvania[94,95] found cases of chronic poisoning and relatively high concentrations of beryllium. While it appeared that the cause may well have been airborne beryllium, it was apparently impossible to rule out other vehicles of beryllium transmission.

Fluorides. Hodge and Smith[96] thoroughly reviewed the literature on this subject including litigation proceedings, and concluded that species of vegetation are far more susceptible to effects of fluoride than is man and thus air quality standards established for the protection of vegetation should adequately protect man. The authors also conclude that "If the air quality standard is set at less than 10 ppb (parts per billion) of fluoride, and the total intake by man from this source is calculated as less than 0.2 mg/day, the increase in intake might not be detectable even by good analytical techniques."

As a means of reference the ambient air quality objectives established by the state of New York range from 1 to 4 ppb as HF (24-hr average).

Lead. Lead in the atmosphere has become quite a controversial subject. Lead is found in many forms. It is in food and water as well as air; therefore, it is difficult to clearly relate blood levels of lead to the form of ingestion.[97,98] The public concern over lead in gasoline has generated a variety of arguments. It may well be that airborne lead may not only be inhaled, but deposited on objects in congested areas and thus indirectly enter the bloodstream through abrasions suffered by children at play. This, of course, is conjecture. John R. Goldsmith presented a case for establishing a long-term average standard of 2 $\mu g/m^3$ where the pollution is largely from motor vehicles.[99] However, Stopps, in a discussion of Dr. Goldsmith's article concluded ". . . since I must reject the premises upon which Dr. Goldsmith bases his air quality standard, I must also reject the number he offers."[100] A California report on lead in the atmosphere[101] provides a comprehensive review of the biological effects of lead on man but the only unequivocal conclusion that can be drawn is that lead in the atmosphere is not doing anyone any good.

Mercury. Stahl[102] indicated that mercury vapor and many mercury compounds have been found in the atmosphere in sufficient concentration and duration as to cause human health effects. A host of serious hazards were suggested, but the epidemiological information was inadequate to establish a "safe" ambient air quality standard.

Vanadium. Athanassiadis[103] shows that in some of the major metropolitan areas of the country the average concentration of vanadium has reached 0.9 $\mu g/m^3$. The report concludes that "exposure to vanadium or its compounds at low concentrations through inhalation has produced observable adverse effects on the human organism. Chronic exposure to environmental air concentrations has been statistically associated with the incidence of cardiovascular disease and certain types of cancer."

Health Effects Research is a continuing responsibility of NAPCA. It is quite apparent that as we expand our air-sampling networks and research into this area we will find that many of the contaminants in the atmosphere are causing or contributing to health deterioration of "sensitive" groups. We will continually be challenged by the conflicts between our right of life, liberty, and pursuit of happiness and our economic needs and the human activities associated therewith that in one way or another contribute to environmental pollution.

Safety Hazards

Air pollution can cause or contribute to a variety of safety hazards. Some are quite obvious, e.g., the hazards associated with reduced visibility,[104] but all are difficult to quantify in terms of the risk.

The effects of air pollution on visibility and the economic and ecological significance of such effects are discussed in more detail in a following section.

Concentrations of and exposure periods to carbon monoxide have been related to human blood levels of carboxyhemoglobin[105] and these have been related to impairments of human judgment and coordination, as summarized in Table 2-3. Attempts are presently being made to develop mathematical models for predicting urban concentrations of carbon monoxide in the hope that such models may guide highway designers. From the knowledge we have on the effects of carbon monoxide and the potential for very high concentrations in confined areas, it appears almost self-evident that where unimpaired human judgment and reactions are essential for safety, a concentration of more than 30 ppm for 8 hours or higher concentrations for shorter exposure periods are hazardous. A good example is the possible relationship between carbon monoxide and traffic accidents.[106,107]

Recently an investigation of a fatal bus crash implicated air pollution as a possible cause of hazardous road conditions.[108] It was theorized that chemical reactions on and with the concrete road surface rendered it slippery. Air pollution officials suggested that an incinerator in the vicinity may have been burning plastic material producing acid emissions. These hydrochloric acid emissions were thought to have reacted with the concrete to produce a "soapy" film. While it appears that the accident may well have been caused by air pollution, it will probably be extremely difficult to prove it beyond a "reasonable doubt."

Nuisances

Air contaminants that are detectable by the sense of sight, touch, smell, and/or taste can be a nuisance in many ways even if they do not result in direct adverse economic or health effects. With the possible exception of smoke, where emissions are frequently limited to a particular optical density, few air contaminant emissions are legally a nuisance per se. Presently most nuisance regulations require proof that the contaminant emissions interfere with the comfortable enjoyment of life and property before they are legally accepted. As the body of legal precedent develops, additional types of emissions will probably be legally considered nuisances per se.

Courts in New Jersey have found that airborne odors which interfere with the comfortable enjoyment of life and property are pollutants and constitute a violation of the regulations prohibiting air pollution.[4] Chapter 11 of the New Jersey Air Pollution Control Code prohibits the emission from incinerators of odors that are detectable in an area of human use or occupancy.[109] The state is also considering the formation of a code which would prohibit the emission of any odors that are detectable off the premises owned by the source.[110] But, until such codoes are adopted and tested in the courts, control officials will have to be guided by complaints.

Much has been written about the detection and measurement of odors,[111,112] but man still remains the best "stinkometer." Sullivan[113] has made a comprehensive survey of the literature. He has summarized the results of threshold odor investigations, testing techniques, as well as control measures. It can be concluded that while human perception and response to odor differs as to receptor, short- and long-term conditioning to the odor, and physiological condition, such perception generally varies exponentially with the concentration of the odorant. However, interactions with other contaminants in the effluent and/or the atmosphere may mask or intensify the odor.

Because odors tend to affect the sense of taste, it is understandable that the presence of odorous materials in the air affects appetites. The presence of settled particulates on railings, benches, automobiles, clothes, and swimming pools is also quite an insult to the senses of sight and touch. Balancing equities in such cases is virtually impossible, but the experience in New Jersey indicates that when a sufficient number of persons (about a dozen) can show that such conditions unreasonably interfere with the comfortable enjoyment of life and property, the courts generally determine equity to be on the side of the plaintiff.

EFFECTS OF AIR POLLUTION UPON ANIMALS

Although the evidence is not as extensive or as well documented as it is for humans it is apparent that many of the acute air pollution episodes which have been disastrous to man have been but little less accommodating to his domestic animals. The animal casualties recorded during the "classic" air pollution disasters of the Meuse Valley, Donora, Poza Rica, and London have helped verify the severity of the conditions experienced by human inhabitants of the communities.

The Meuse Valley, Belgium, episode of December 1930 resulted in the sickening and deaths of many cattle.[114]

In Donora, Pennsylvania, in October 1948, when 6,000 persons—43 percent of the population—were made ill during an acute air pollution episode, a sizable number of animal victims were noted.[115] A Public Health Service survey disclosed that 15.5 percent of the dogs, 20 percent of the canaries, 11 percent of the fowls, and 77 percents of the cats became ill during the smog days. Twenty-four animal deaths were attributed to this episode; however, cattle, sheep, horses, and swine were not significantly affected.

In November 1950 a predawn industrial "accident" permitted a half-hour release of large amounts of hydrogen sulfide gas over the small town of Poza Rica, Mexico.[53] Three-hundred and twenty people were hospitalized and twenty-two died. About 50 percent of the chickens, cattle, pigs, geese, ducks, dogs, and cats died as a result of the exposure. In addition all the canaries in the area were killed.

During the London, England,[116] disaster of December 1952, which caused 4,000 excess human deaths, 52 cattle out of 351 on the ground floor at the Smithfield Cattle Show became seriously ill. Five of these died and nine others had to be slaughtered later. It appears that in an earlier smog episode, which occurred at the same time as the Smithfield Cattle Show, one-third of the animals had to be removed to the cleaner air of the surrounding countryside; but despite this move, a number of animals eventually had to be slaughtered.

In addition to being subjected to suffering the ill effects of breathing polluted air, animals can become ill by eating fodder contaminated by air pollutants. Although the ambient air concentration of a pollutant may be too low to cause any adverse effect through inhalation, an accumulation and concentration of airborne contaminants by vegetation and forage may cause the subsequent poisoning of animals when they eat the contiminated vegetation. Fluorides, arsenic, molybdenum, lead, and zinc have been identified as the responsible agents in several such occurrences.[117,118,119,31] Carelessness or misuse of pesticides including those containing mercury and/or lead have also accounted for a number of domestic animal losses.[120] Of all of these air pollutants which have caused widespread damage and have proved most troublesome to control, fluorine probably should be given first place in this list of bad actors.[117]

MacKintosh[121] reports that chronic lead poisoning has been observed frequently in horses that have been grazing on forage near smelters, lead mines, and in orchards that have been sprayed with insecticides containing lead. Stokinger[118] has reported that cattle and horses were poisoned by inhaling air contaminated by zinc and lead within a half-mile of zinc and lead smelters. In the vicinity of a coke oven the herbage was found to contain 25 to 46 ppm of lead. This was sufficient to cause poisoning of cattle and sheep.[120] Lead is a cumulative poison. As a result, the continuous ingestion of very small daily doses will ultimately be as effective as one toxic dose. Depending on the amount of lead that may be deposited from dusts or sprays, poisoning and death can take many months if slightly contaminated hay is fed; or it can occur within 24 hours in animals feeding in or near orchards which have been heavily sprayed.[121]

Street[120] cites a personal communication from Radeleff in 1963 stating that pesticide accidents account for 0.5 to 1.0 percent of all domestic animal losses from disease. These acute poisoning accidents usually were due to carelessness or misuse of the pesticides and involved the more toxic organophosphates in addition to endrin and dieldrin. As in humans, chronic effects have been difficult to identify in domestic animals.

In June 1962, a gold mine and smelter in one of the western states were reopened; after they were in operation for approximately 5 months it became apparent that the emission control equipment was inadequate. Both sulfur dioxide and arsenic trioxide were emitted in visible quantities into the atmosphere. Air samples taken at the plant showed 60 to 13,000 $\mu g/m^3$ of arsenic. June grass near the community school showed an arsenic concentration of 925 $\mu g/g$. The community pet population was reduced from over 24 to 1. The surviving collie had a large ulcer in the mouth and another on the right forepaw.[122]

Extensive poisoning of cattle, horses, and sheep occurred in the vicinity of a copper smelter located at Anaconda, Montana, in 1902 due to the ingestion of arsenic trioxide, which had contaminated the forage crops.[123,124] When a flock of 3500 sheep was brought from an area 28 miles away to graze 15 miles from the smelter, 625 of the sheep died. Grass and moss from the area in which the animals had fed was found to contain 52 and 405 ppm of arsenic trioxide, respectively. Moreover, horses in an area remote from the smelter died after eating hay grown in a location on which smelter fumes could have fallen. The arsenic trioxide content of the hay was 285 ppm. Large quantities of both arsenic and copper were found on vegetation up to 15 miles from the smelter. Both the type of symptom and data from analysis of tissues failed to implicate arsenic as the poisoning agent, whereas evidence concerning animals fed on graded doses of arsenic verified experimentally that arsenic was responsible.

Sullivan[125] states that a study of the effects of airborne arsenic from a nearby smelter on animals in Saxony Forest was reported by Prell. Red deer, foxes, and horses were all affected. Bees in the area also had a high mortality rate.

Stokinger and Coffin[119] indicate that fluoride air pollutants have probably caused more widespread damage to livestock in the United States than any of the other pollutants. Sources of airborne fluorides include the manufacture of phosphate fertilizer, aluminum production, fluorinated hydrocarbons and plastics, and the production of uranium and other heavy metals.

Irrespective of the nature of the fluoride, the effects from eating contaminated forage are the same—an abnormal calcification of bone and tooth structures termed fluorosis, owing to a large increase in fluoride in these structures. Milk production is also decreased. Animals lose weight, acquire a stiff posture, become lame, and the hair coat becomes rough. Phillips[117] reports that cattle and sheep are the most susceptible to fluoride toxicosis of all farm animals. Swine occupy a second rank, although fluorosis of swine in the United States has not been a major problem. Horses appear to be quite resistant to fluoride poisoning, and authentic cases of fluorosis of horses in the United States are rare. Poultry are probably the most resistant to fluorine of all farm animals and present no problem so far as fluorosis is concerned.

The courts have made numerous monetary awards to livestock farmers for losses caused by the ingestion of fodder containing excessive amounts of fluorides. A number of these are summarized in Table 2-8 of this chapter.

The case of Rocky Mountain Phosphates, Inc., superphosphate plant, versus the Garrison, Montana, area ranches is a good example of this problem.[31]

The Montana State Board of Health received complaints of adverse health and welfare effects resulting from the air pollutant emissions of the plant almost immediately after operations began in 1963. Measured concentrations of fluoride in grass forage near the plant have ranged as high as 10 mg/g (10,000 ppm)—far in excess of the maximum .035 mg/g (35 ppm) allowed by Montana. Symptoms of fluorosis in cattle and other livestock had not been observed before the plant began operations; however, after the opening of the plant, many ranchers declared that fluoride-induced injury to cattle caused substantial economic losses, especially in milk and calf productivity. In March 1968, the Montana State Supreme Court upheld a decision of a lower court to award $113,283 in livestock damage payments to seven Garrison area ranchers.

To protect farm animals from fluorides Suttie[126] has suggested that the yearly average fluoride content of forage be limited to 40 ppm and there should not be more than 80 ppm for any one month nor more than 60 ppm for more than two consecutive months.

Since animals frequently obtain a majority of their food from one growing area for long periods of time they are much more prone to fall victim to local sources of air pollution contaminants of vegetation. While ingestion of the above-named pollutants might not kill the animals it could result in other significant defects such as decreased reproductivity, growth, or outputs of milk, eggs, and wool.

An excellent review of the literature concerning the effects of air pollutants on animals has been compiled by Lillie.[192] His summary, part of which follows, indicates that we still have much to learn about the effects up on animals of many of our current air pollutants.

> The symptomology, tolerance limits, and alleviators have been well-defined for cattle and sheep afflicted with fluorosis, a disease that is worldwide. Arsenic was much more of a health hazard during the first 40 years of the 20th century than since then. Atmospheric lead, as differentiated from lead in pesticides and in materials such as paints and storage batteries, is recognized as a potential hazard. This is illustrated by the high lead content of vegetation and soils along heavily travelled roads. Nevertheless, no information could be found on the effect of high lead concentration on farm animals grazing along highways.
>
> Other air pollutants commonly associated with domestic animals are: ammonia with poultry and swine; carbon monoxide with all domestic animals; dust with poultry and rabbits; hydrogen sulfide with swine and poultry; sulfur dioxide with cattle, sheep, and swine; and nitrogen oxides with all animals. Little or no information was found on automobile exhausts, smoke, beryllium, cadmium, hydrocarbons, manganese, mercury, molybdenum, ozone, vanadium, and zinc affecting domestic animals in this country.

VEGETATION INJURY

Air pollution damage to vegetation has been recognized for at least a century.[127] However, this knowledge was often disregarded or poorly disseminated as evidenced by such major contemporary episodes as San Francisco, California;[128] Trail, British Columbia;[129] Sudbury, Ontario;[130] and Ducktown, Tennessee[131] where plant life was almost entirely exterminated over large areas.

The complete destruction of vegetation associated with the early history of SO_2 pollution from smelters is no longer a major problem. However, other problem sources of SO_2 have developed and are becoming more widespread.[149] Other gaseous pollutants that are important phytotoxicants include fluorides, ozone, PAN (peroxyacetyl nitrate), and ethylene. Gaseous pollutants which have been known to cause lesser vegetation damage include chlorine, herbicides, hydrogen chloride, ammonia, and mercury.[133,150]

Particulates have also caused phytotoxicity, generally by physically covering the leaves of the plant thereby possibly inhibiting the plant's respiration through the stomata and/or reducing photosynthesis by the absorption of light. Injury due to the direct effect of high or low pH's on cell constituents has also occurred. Fluorides in particulate form are less damaging to vegetation than gaseous types. The fluoride from particulates apparently has great difficulty penetrating the leaf tissue in a physiologically active form. However, fluorosis in animals has been reported due to the ingestion of vegetation covered with particles containing fluorides. Cement-kiln dust deposits have been frequenty reported as adversely affecting plants when naturally deposited on a moist leaf surface. Other aerosols which have been reported as injurious to plants include soot, magnesium oxide, foundry dusts, and sulfuric acid types.

The gaseous pollutants enter with the air via the stomata during the normal plant respiration cycle. The pollutants destroy plant chlorophyll, disrupt the photosynthetic process, and consequently reduce food production. The damage may vary from a difficult-to-detect reduction in growth rate to complete extermination. Between these extremes such things as leaf tissue collapse, chlorosis, and growth alterations will be noted. As Hindawi[132] points out the leaf is the primary indicator of the injurious effects of air pollution. Its structure plays a most important role in building carbohydrates and other vital plant products and foods. Therefore, injury to the leaf will affect the vitality of the total plant.

When vegetation is injured by an air pollutant, symptoms characteristic of the specific pollutant usually develop. Table 2-6, adapted from Hindawi, describes these briefly. Since the pollutant itself normally undergoes chemical change soon after it contacts plant tissue, symptoms are often the only remaining evidence of pollutant assault. Thus, symptoms provide the major basis for diagnosis. Observations of injury on sensitive plant species have provided a means of monitoring pollutant emission from a source, and observations of plant injury have also been a valuable tool for determining the geographical distribution of a pollutant over a large area.[151] In addition to the type of injury other keys to the pollutant are type and part of leaf affected, species and varieties of plant affected, and the appearance of the injury.

Table 2-7 lists those plants which are more sensitive to air pollution injury from the various major pollutants. However, when the air pollution fumigations are excessive, even resistant species of vegetation can exhibit injury or even death. An excellent description of air pollution injury to vegetation is contained in the Jackobson and Hill.[133]

TABLE 2-6 SUMMARY OF POLLUTANTS, SOURCES, SYMPTOMS, VEGETATION AFFECTED, INJURY THRESHOLDS, AND CHEMICAL ANALYSES[132]

					INJURY THRESHOLD[a]				
Pollutants	*Source*	*Symptom*	*Type of Leaf Affected*	*Part of Leaf Affected*	*ppm*	*μg/m³*	*Sustained Exposure*	*Reference*	*Chemical Analysis for Pollutants in Plants*
Ozone (O_3)	Photochemical reaction of hydrocarbon and nitrogen oxides from fuel combustion, refuse burning, and evaporation from petroleum products and organic solvents.	Fleck, stipple, bleaching, bleached spotting, pigmentation, growth suppression, and early abscission. Tips of conifer needles become brown and necrotic	Old, progressing to young	Palisade	0.03	70	4 hr	135	None
Peroxyacetyl nitrate (PAN)	Same sources as ozone.	Glazing, silvering or bronzing on lower surface leaves.	Young	Spongy cells	0.01	250	6 hr	136	None
Sulfur dioxide (SO_2)	Combustion of fossil fuels, refining of petroleum, smelting of ores containing S, manufacture of H_2SO_4, burning of S-containing refuse, papermaking, and burning coal refuse banks.	Bleached spots, bleached areas between veins, bleached margin, chlorosis, growth suppression, early abscission, and reduction in yield.	Middle-aged	Mesophyll cells	0.3	800	8 hr	137	[b]

Hydrogen fluoride (HF)	Phosphate rock processing aluminum industry, iron smelting, brick and ceramic works, and fiber-glass manufacturing.	Tip and margin burn, chlorosis, dwarfing leaf abscission, and lower yield.	Young, fully phyll	Epidermis and meso- expanded	0.1 (ppb)	0.2	5 wk	138 and 152	Distillation and titration
Chlorine (Cl_2)	Leaks in chlorine storage tanks; hydrochloric acid mist, chlorine and chlorine products manufacture; incineration of chlorine-containing plastics.	Bleaching between veins, tip and margin burn, and leaf abscission.	Mature	Epidermis and mesophyll	0.10	300	2 hr	139	[b]
Ethylene (C_2H_4)	Incomplete combustion of coal, gas, and oil for heating, and automobile and truck exhaust.	Sepal withering, leaf abnormalities, flower dropping, and failure of flower to open properly.	(Flower)	All	0.05	60	6 hr	140	None

[a]Metric equivalent based on 25°C and 760 mm mercury.
[b]Chemical analysis often is not reliable for diagnosing chloride or sulfate accumulation in leaf tissue because undamaged plants often contain higher concentrations of these pollutants than are found in damaged plants.

TABLE 2-7 PLANTS RELATIVELY SENSITIVE TO INJURY FROM VARIOUS AIR POLLUTANTS[133]

SULFUR DIOXIDE				
Crops	*Garden Flowers*	*Trees*	*Garden Plants*	*Weeds*
Alfalfa	Aster	Apple	Bean	Bindweed
Barley	Bachelor's button	Birch	Beet, table	Buckwheat
Bean, field	Cosmos	Catalpa	Broccoli	Careless weed
Clover	Four o'clock	Elm, American	Brussel sprouts	Curley dock
Cotton	Morning glory	Larch	Carrot	Fleabane
Oats	Sweet pea	Mulberry	Endive	Lettuce, prickly
Rye	Verbena	Pear	Lettuce	Mallow
Safflower	Violet	Pine, eastern white	Okra	Plantain
Soybean	Zinnia	Pine, ponderosa	Pepper (bell, chili)	Ragweed
Wheat		Poplar, lombardy	Pumpkin	Sunflower
			Radish	Velvet-weed
			Rhubarb	
			Spinach	
			Squash	
			Sweet Potato	
			Swiss Chard	
			Turnip	

TABLE 2-7 (Continued)

OZONE: *Crops*	OZONE: *Trees, Shrubs, Ornamentals*	FLUORIDE	PEROXYACETYL NITRATE (PAN)	ETHYLENE
Alfalfa	Alder	Apricot, Chinese and royal	Bean, pinto	Bean, Black Valentine
Barley	Apple, crab	Blueberry	Chard, Swiss	Carnation
Bean	Aspen, quaking	Boxelder	Chickweed	Cotton
Clover, red	Boxelder	Corn, sweet	Dahlia	Cowpea
Corn, sweet	Bridalwreath	Fir, Douglass	Grass, annual blue	Cucumber
Grass, bent	Carnation	Gladiolus	Lettuce	Marigold, African
Grass, brome	Catalpa	Grape, European	Mustard	Orchid
Grass, crab	Chrysanthemum	Grape, Oregon	Nettle, little-leaf	Pea, cream
Grass, orchard	Grape	Larch, western	Oat	Peach
Muskmelon	Honeylocust	Peach (fruit)	Petunia	Philodendron
Oat	Lilac	Pine, eastern white, lodgepole, scotch, Mugho	Tomato	Privet
Onion	Maple, silver	Pine, ponderosa		Rose
Peanut	Oak, gambel	Plum, Bradshaw		Sweet potato
Potato	Petunia	Prune, Italian		Tomato
Radish	Pine, eastern white	Spruce, blue		
Rye	Pine, ponderosa	Tulip		
Spinach	Privet			
Tobacco	Snowberry			
Tomato	Sycamore			
Wheat	Weeping willow			

Hindawi[132] reports that the cost of agricultural losses due to air pollution in the United States was recently estimated to be in the neighborhood of $500 million annually. Losses to agronomic species in California alone are estimated to amount to $132 million. No estimates have been made of the real economic loss caused by suppression of growth, delayed maturity, reduction in yield, and the attendant increase in the cost of crop production. Losses to landscape plantings have generally not been calculated. In fact, they have seldom been recognized. Exacting records of the vegetation economic losses due to air pollution are not numerous; however, some are summarized in Table 2-8. It has been difficult to obtain exact economic figures because of the many factors such as variable market price, degree of possible plant recovery, hidden injury, differentiation from other causes of injury, and lack of adequate follow up and people trained to recognize plant injury.

What is probably the most extensive statewide survey of vegetation damage caused by air pollution[134] ever conducted was initiated in 1969 to determine the losses on food, fiber, and ornamental crops in Pennsylvania. County agricultural agents and Pennsylvania State University extension specialists, all experienced biologists, provided the specialized manpower needed for complete coverage of the state. A 3-day short course was held for the participants and a handbook containing illustrations of typical damage and guidelines in diagnosing injury was distributed to all participants. Each report of injury was investigated by a post-doctoral scholar in plant pathology.

Ninety-two investigations were made in 28 counties. The reporters knew of air pollution damage which was not reported or which they were unable to follow up in detail. These were not included. Greatest economic losses were observed on vegetable, fruit, and agronomic crops, followed by lawns, ornamentals, greenhouse flowers, and forest trees.

The suspected major pollutants in decreasing order of importance were: oxidants, oxides of sulfur, lead, hydrogen chloride, particulates, herbicides, and ethylene.

Direct economic losses exclusive of profits were estimated to be in excess of $3.5 million. Indirect economic losses were estimated to account for an additional $8 million. Indirect losses include grower profits, crop substitutions, relocation costs, and reforestation of denuded land. In addition, other indirect losses were observed, but could not be quantified. These included changes in vegetation type, erosion, stream silting, decrease in water retention capacities of watersheds, aesthetic losses, and farm abandonment.

Similar surveys by other states would undoubtedly produce more

TABLE 2-8 SUMMARY OF REPORTED VEGETATION LOSSES DUE TO AIR POLLUTION

$132 millon per year for all commercial fruits and vegetable losses in California.[141]

Yearly total cost of air pollution on vegetation is close to a billion dollars.[141]

Field and vegetable crop losses in California at about $8 million/year.[142]

Field and vegetable crop losses along the eastern seaboard of the United States of about $18 million/year.[142]

A 1969 survey of the state of Pennsylvania produced an estimated direct economic loss exclusive of profits of $3.5 million. Indirect economic losses were estimated to account for an additional $8 million.[134]

In California it has been variously estimated that the direct economic losses in agricultural production from photochemical air pollution approaches $10 to $12 million/year. The indirect loss due to growth retardation, delayed maturity, reduced yields, and impaired quality might be 10 times this.[143]

In March 1968, the Montana State Supreme Court upheld a decision of a lower court to award $113,282 in livestock damage payments to seven Powell County ranchers. This was in payment for the symptoms of fluorosis in cattle and other livestock caused by fluoride-containing particulate and gaseous pollutant emissions coming form a superphosphate plant. The livestock had grazed on forage made excessively high in fluoride by these emissions.[31]

In January 1968, Wm. G. Roe and Co. were awarded $55,887 by a U.S. District Court in Florida for the partial destruction of Roe's citrus crop for the 1963–64 season due to excessive fluoride emissions from a superphosphate plant located 2 miles away.[31]

In July 1965, W. N. Shurling and seven others were awarded $206,050.78 for fluoride damage to fruit and vegetable crops. A phosphate plant was the defendant.[31]

A primary aluminum smelter located in Multnomah County, Ore., has a long history of paying monetary "awards" for damages caused by fluoride emissions. They include:[31]

(a) $450,000 paid to surrounding property owners between 1942 and 1946.
(b) $38,823 in 1955 to a family of three for injuries to human health.

TABLE 2-8 (Continued)

(c) $91,500 awarded in 1956 for the loss of use of land for grazing and for deterioration of land from August 1951 to January 1956.
(d) $300,000 for damages to a 500-acre dairy farm as well as for loss of milk production.
(e) In 1962 the aluminum company purchased the above dairy farm.
(f) In 1968 the company made an out-of-court settlement for damages to a 1,500-acre cattle ranch that included cash and the purchase of the ranch.

Darley *et al.*[136] reported that during 1959 the combined damages incurred by three orchid growers in northern California due to ethylene damage amounted to $70,000.

James[145, 144] surveyed losses in the San Francisco Bay area. Data supplied up to 1964 by several cooperating orchid growers indicated an annual loss ranging from $60,000 to $100,000. Reduction in profit suffered by carnation growers in 1963 was estimated at $700,000.

Tulip bulb farmers in the Beverwijk, Netherlands, area receive from $10,000 to $200,000 per year for air pollution damage to relatively insensitive tulip varieties caused by blast furnaces and steelworks 3 miles away. The farmers are not permitted to grow types that are sensitive to the hydrogen fluoride air pollutant.[146]

Nearly 1.3 million trees in the San Bernardino National Forest were reported by the U.S. Forest Service in October 1969 to be dying from the effects of smog. The trees are distributed over 161,000 acres in the San Bernardino Mountains of Southern California. The environmental factors were said to be worsening, with the number of acres affected being four times greater than forestry experts had estimated in 1968.[147]

The U.S. Dept. of Agriculture has estimated that annual losses to agricultural crops from air pollution ranged from $150 million to $500 million for the period 1951 to 1960.[148] These estimates do not include an allowance for damage to noncommercial crops of vegetation in urban areas where plants are more likely to be exposed to high pollutant levels.

accurate information upon which to base our ambient air and emission standards.

Since air pollution is growing in intensity in many sections of the country, the losses from vegetation damage will undoubtedly increase unless additional control measures are instituted.

METEOROLOGICAL EFFECTS

Although our atmosphere may appear to be of tremendous size it is really a relatively thin skin covering this home called earth. Our "ocean of air" extends technically for perhaps 1,000 miles. However, the atmosphere thins so rapidly as one leaves the earth that after only 3½ miles over one-half the atmosphere, by weight, lies below. As we fly from city to city in today's jet aircraft, at the 30- to 35-thousand-feet level, we have 6–7 tenths of our atmosphere beneath us. At this level the sky above has lost much of its color and appears very dark due to the loss of aerosols and gases which scatter the blue light toward our line of sight.

If all of our atmosphere were concentrated to the density found at sea level we would run out of air completely before we reached the 30,000-foot level. For millions of years until the industrial revolution the atmosphere had probably not changed appreciably. During the last 200 years man has begun to poison his atmosphere at an increasing rate—factory chimneys, exhaust pipes, atomic explosions, high-flying jet aircraft, etc. continue to encumber our atmosphere with unnatural pollutants. These impurities may not yet threaten our lives, but we should be on guard. Man's very existence is imperiled and we cannot take the threat too lightly.

Effects of Global Air Pollution

The major factor influencing the climate of the world is the earth's heat balance. The amounts of heat received, retained, and reflected by the earth and its atmosphere determine the state of this delicate ecological balance upon which the health and well-being of the earth's peoples depend. Weather phenomena such as wind, storms, clouds, rain and snow and the energy changes associated with them are in a great part dependent upon this heat budget. Man influences this heat budget through the addition of carbon dioxide (CO_2) and aerosols to the atmosphere.

CO_2 is nearly transparent to visible light but it is a strong absorber and back radiator of infrared radiation, particularly in the wavelengths from 12 to 18 microns. Consequently, an increase of atmospheric CO_2 could act, much like the glass in a greenhouse, to raise the temperature of the lower atmosphere. Manabe and Wetherald[153] analyzed the temperature effect of CO_2 at a constant relative humidity and an average amount of cloudiness. Their results indicated that for a change in CO_2 from 300 to 600 ppm, an increase in temperature of 4.25°F is estimated. The average atmospheric background concentration is now estimated to be 320 ppm. In summarizing their CO_2 modeling experiments, Manabe and Wetherald point out that:

1. The higher the CO_2 concentration, the warmer is the equilibrium temperature of the earth's surface and the troposphere (the lower atmosphere in which all "weather" occurs).
2. The higher the CO_2 concentration the colder is the equilibrium temperature of the stratosphere.

The estimated average CO_2 concentration in the atmosphere was about 290 ppm at the beginning of this century.[154] The 30 ppm increase since that time would produce an estimated temperature increase of about 0.5°F. The current average annual increase in atmospheric CO_2 content seems to be about 0.7 ppm.[155] This is taking place despite the uptake of CO_2 by the biosphere and the oceans. Photosynthesis removes CO_2 from the atmosphere and forms plant carbon which is deposited as litter and humus. The major sink for the carbon from atmospheric CO_2 is the settling action of dead marine organisms which have taken up the dissolved CO_2 during their life at the surface of the oceans.

The sources of atmospheric CO_2 are the oxidation of carbonaceous materials by both plants and animals, decay of organic materials, and the combustion of organic fuels. Until recently there was a steady-state balance between the release of CO_2 and its consumption by photosynthesis in plants and by the formation of calcium carbonate.[156] This balance has been disturbed by the emission into the atmosphere of additional CO_2 from the increased combustion of ancient carbonaceous fuels.

It is estimated that on a relative basis CO_2 emissions in the year 2000 from organic-fuel-burning will be almost three times higher than those in 1965. It is estimated that the total CO_2 emissions over the period 1965–2000 will be great enough to increase atmospheric CO_2 to about 370 ppm by the year 2000 for an average atmospheric temperature increase of 0.9°F.[155]

Atmospheric turbidity is also an important factor in the heat balance of the earth's atmospheric system, and it may be having a greater influence upon our world climate today than the increase in CO_2 content of our atmosphere.[157] The observed increase in turbidity over the past few decades may be playing a vital role in the reported decrease in worldwide air temperature since the 1940s by increasing the planetary albedo or reflectivity.[158,159] Humphreys[160] has shown that the interception of outgoing radiation by fine atmospheric dust is wholly negligible in comparison with the interception of incoming solar radiation.

Washington, D.C., during the period 1903 to 1966, has experienced a possible decrease of nearly 3 percent in the total available solar energy at ground level and a possible increase in the average annual number of $0.1–1\mu$ radius aerosol particles over the city of $2.8 \times 10^7/cm^2$.[157] The net effect of this apparent secular increase in turbidity (which from the Davos, Switzerland, and other evidence[161] appears likely to be worldwide) is probably to increase the mean albedo of the planet and reduce the mean temperature of the earth-atmosphere system. This effect would be the opposite of that caused by an increase in CO_2. The argument has been made that the large-scale cooling trend observed in the northern hemisphere since about 1955 is due to the disturbance of the radiation balance by fine particles and this effect has already reversed any warming trend due to CO_2 increase.

Other factors which may mask the "greenhouse" effect of CO_2 in the atmosphere are the water vapor and cloud contents there. Möller[162] points out that a fairly significant 10 percent increase in CO_2 can be counterbalanced by a 3 percent in atmospheric water vapor or a 1 percent change in total average cloudiness.

Although apparently we are unsure as to whether the "cooling-off" or the "warming-up" major air pollutants are winning supremacy, there should be no doubt that the potential damage to our total environment could be severe. Whether you choose the CO_2 greenhouse warming theory described by Revelle[156] and others, or the newer atmospheric turbidity cooling theory professed by McCormick and Ludwig,[159] the prospect for the future must be of serious concern. We must not just win local battles while we lose the war. Not only must we concern ourselves with the local injurious effects of pollutants in high concentrations but small total quantity, but we must consider the total ecological effects of the grossly abundant pollutants—CO_2 and inert submicron particles.

Air Pollution Effects upon urban Climates

Urban climates are influenced by many factors not found in the surrounding countryside. Among these factors which influence the city climate to make it different than its rural environs are:[163]

1. The difference between surface materials.
2. The far greater variety of shapes and orientations of city structures than the features of the natural landscapes of the rural areas.
3. The prodigious generation of heat, particularly in winter when heating systems are in operation.
4. The rapid removal of precipitation from the city's environment.
5. The heavy load of solid, liquid, and gaseous contaminants contained in the air over the urban areas.

Landsberg[164] describes some of the climatic changes produced by cities as follows:

Radiation:	
Total on horizontal surface	15–20 percent less
Ultraviolet, winter	30 percent less
Ultraviolet, summer	5 percent less
Cloudiness:	
Clouds	5–10 percent more
Fog, winter	100 percent more
Fog, summer	30 percent more
Precipitation:	
Amounts	5–10 percent more

Effects on Fog

Neuberger and Gutnick[165] have carried out a series of laboratory experiments in which fog characteristics were studied as a function of condensation nuclei concentration. From these experiments they concluded:

1. Fog density at time of initial formation increases rapidly as the nuclei concentration increases from clean (1000/ml) to moderately polluted (70,000/ml), but a further increase in nuclei does not increase fog density.

2. Fog duration increases continuously as the nuclei concentration increases. This is attributed to a smaller initial droplet size and to a lessened rate of coalesence for the higher nuclei concentrations since in the chamber fog is dissipated by gravitational settling.

Longer lasting fogs in urban areas are related to the overabundance of condensation nuclei, which produces a fog with smaller-sized drops. The chemistry of the condensation nuclei and the droplets may also play an important role.

Increasing fogs reduce insolation, hamper transportation, and inhibit atmospheric ventilation of an area.

Effects on Precipitation

Robinson[166] indicates that it is more difficult to determine a relationship between air pollution and precipitation than it is to relate fog formation to pollution. First, the rain mechanism is not one that can be readily reproduced in the laboratory and studied as a function of changing parameters. Second, the mechanism of rain formation is very complex and a city can influence this mechanism in significant ways which are quite unrelated to its air pollution. According to Landsberg[164] most urban factors tend to increase precipitation. He enumerated these causes without emphasis on the order as follows:

1. Water vapor addition from combustion processes and factories.
2. Thermal updrafts from local heating.
3. Updrafts from increased friction turbulence.
4. Added condensation nuclei leading more readily to cloud formation.
5. Added nuclei which might possibly act as freezing centers for subcooled cloud particles.

Landsberg cited as evidence the gradual increase of rainfall over Tulsa, Okla., with its growth from a village to a city in 5 decades and the concomitant increase in particle concentration.

Air Quality Criteria for Particulate Matter[157] notes that R. H. Frederick in a recent analysis, showed a definite minimum Sunday rainfall for a 10-year period in Louisville, Pittsburgh, and Buffalo. Precipitation occurred in these cities less often on Sundays than on other days of the week, and the average rainfall was less for Sundays than for weekdays. A strong city influence has also been suggested in the snow patterns in Toronto. *Air Quality* also describes an interesting and significant increase in precipitation that has been observed at La Porte, Ind., since 1925. La Porte is 30 miles east of the large complex of heavy industries in the metropolitan Chicago area. Chagnon[167] compared precipitation at La Porte, Valparaiso, and South Bend, Ind., with a 5-year moving average of the number of smoky and hazy days in Chicago (Fig. 2-4). The temporal distribution of the smoke-haze days after 1930 is rather similar to the La Porte curve. A notable increase in smoke-haze days began in 1935, becoming more marked after 1940, coincident with the sharp increase in the La Porte precipitation curve. The reduction in the frequency of smoke-haze days after the peak reached in 1947 also generally matches the decline of the La Porte curve since 1947.

Stout[168] has shown that the shape of the time-series curve for La Porte precipitation also generally matched a time-series curve

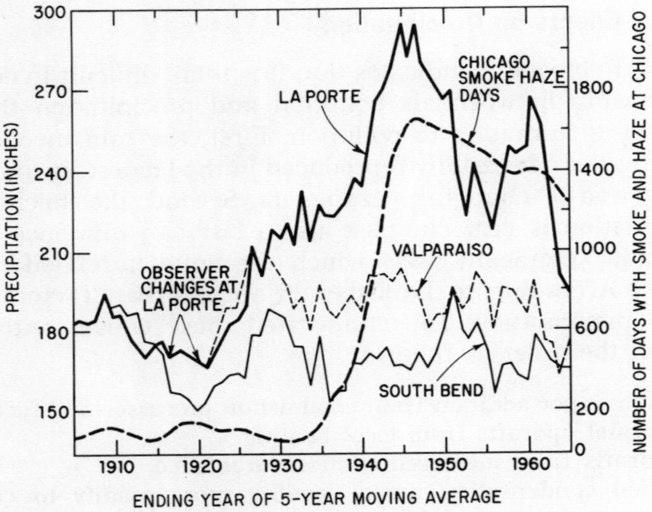

Fig. 2-4 Precipitation values at selected indiana stations and smoke-haze days at Chicago. (This Figure shows the way in which precipitation trends at La Porte follow the haze changes in Chicago. The results are plotted as five-year moving averages.) (From Ref. 167)

for annual steel production in the Chicago industrial complex. Between 1905 and 1965, peaks in steel production, which occurred when production in most other industries was also high, were all associated with high points in the La Porte precipitation curve.

The effect of industrial pollution on precipitation has also been studied by Telford.[169] He found that smoke from steel furnaces was a prolific source of freezing nuclei, increasing counts by a factor of 50 over those in nearby clean air. He concluded that there should be increased rainfall downwind of such installation.

In summary we can say that while the sum of evidence currently available is not at all conclusive it appears that the precipitation patterns around cities studied indicates that small rainfall increases are usually found downwind of the cities, and that cloud seeding by aerosol air pollutants is at least part of the cause of this increase.

Insolation Effects

The blanket of particulates over most large cities causes the solar energy that reaches an urban complex to be significantly less than that observed in rural areas. The particles are most effective as attenuators of radiation when the sun angle is low, since the path length of the radiation passing through the particulate material is dependent on sun elevation. Thus, for a given amount of particulates, solar radiation will be reduced by the largest fraction at high-latitude cities and during the winter.

Studies indicate a fair approximation of the association between atmospheric aerosol concentration and relative solar radiation levels, as shown in Table 2-9.

TABLE 2-9 APPROXIMATE ASSOCIATION BETWEEN ATMOSPHERIC AEROSOL CONCENTRATIONS AND RELATIVE SOLAR RADIATION LEVELS[157]

Aerosol Concentration $\mu g/m^3$	SOLAR RADIATION, % OF VALUE FOR 100 $\mu g/m^3$	
	Total	*Ultraviolet*
50	105	104
100	100	100
200	95	92
400	90	77

Most radiation scattered by atmospheric dust is directed forward and thus attenuation of total solar radiation is primarily due to absorption.

Peterson[170] notes that the introduction of smoke controls in London during the mid-1950s has afforded an opportunity to check the radiation smoke relation. Monteith[171] summarized data on particulate concentration and solar energy at Kingsway (central London) for the years 1957 to 1963. During this time smoke density decreased by 10$\mu g/m^3$ while total solar radiation increased by about 1 percent. The average smoke concentrations of 80 $\mu g/m^3$ at Kew (inner suburbs) represents an energy decrease of about 8 percent, and in the center of town, where smoke concentrations average 200 to 300 $\mu g/m^3$, the income of solar radiation is about 20 to 30 percent less than that in nearby rural areas. Similarly, Jenkins[172] reported that the frequency of bright sunshine in London also increased in recent years after implementation of the air pollution control laws. During the period 1958 to 1967, the average number of hours of bright sunshine from November through January was 50 percent greater than that observed from 1931 to 1960.

Measurements of ultraviolet radiations in downtown Los Angeles and on Mt. Wilson[173] showed its dependence upon the clearness of the atmosphere. Attenuation of ultraviolet radiation by the atmosphere between the 5,700-foot level at Mt. Wilson and the 350-foot level in Los Angeles averaged 14 percent on no-smog days; when smog was present peak attenuation was as much as 58 percent with a mean value of about 38 percent or almost 3 times that measured on the no-smog day.

In general, the total solar radiation received is inversely related to the concentration of smoke and suspended particles; thus, solar radiation measurements may be used, in the absence of clouds, as a crude index of particulate air pollution.

The radiation most affected is that of the ultraviolet with the infrared least affected. This is of importance due to the bactericidal effect of ultraviolet radiation. Reductions in the intensity of solar radiation or changes in its spectral distributions have significance for the photosynthesis of vegetation, the distribution of plants and animals on the earth, the weathering of natural and man-made materials, and man's aesthetic enjoyment and physical well-being.

The "inadvertent" modification of the atmosphere by man-made air pollutants is a dangerous business. We must keep track of changes in the meteorology of the atmosphere so that we do not disasterously alter the world's weather systems.

VISIBILITY

The deterioration of visibility is probably one of the most common effects of air pollution noted by the citizen. Therefore, it is something that he will tend to complain about, and in turn he will complain about those emissions which he can see.

Reduced visibility creates an economic burden upon many communities. Transportation and general well-being are adversely affected. Reduced visibility slows landing and takeoff frequencies at airports. Additional hazards to safety are also imposed which could result in deaths, personal injury, and/or property damage. Automobile traffic may be slowed due to impairment of atmospheric visibility. When the motorist's vision is limited accidents, bodily injury, deaths, and property damage increase. Shipping in our harbors suffers from similar impairments.

Impaired visibility makes it necessary to enlarge our transportation facilities in order to accommodate normal traffic demands. The money for such construction will come ultimately from the pocket of the citizen as will that for the increased insurance rates which result from more transportation accidents.

It may be necessary to increase the use of electricity to compensate for reduced lighting, and a dirty atmosphere is not conducive to producing a general feeling of well being in man.

Peckham[13] lists many of the ecological and economic effects of visibility reduction by air pollution as reported in the literature. Many of these, plus others, are summarized in Table 2-10.

People want and need safe and dependable transport. They want to be able to travel without extraordinary risks or delay. Furthermore, most people also want access to pleasing scenery in bright clear weather. Air pollution can often defeat these wants by depressing visibility, blocking sunshine, and intensifying fog. This seems clear from the evidence, but what is not so clear, however, is the monetary magnitude of the injuries suffered.

Relationship between Visibility and Particle Concentration

Visibility in the atmosphere is reduced by two optical effects which gas molecules and aerosols (solid particles and liquid droplets) have upon visible radiation. One is the attenuation by the molecules and aerosols of light passing from object to observer. It is the result both of absorption of light and of the scattering of light out of the incident beam. The other optical effect that degrades the contrast between object and background is the illumination of the intervening air which results when light is scattered into the line of sight by the molecules and particles within that line. This results in dark objects becoming progressively lighter in shade as they become more distant.

Visibility observations frequently show a strong directional variation which is dependent upon the angle of the sun. Lower visibilities will normally be observed in the direction of the sun than away from it, even in a uniformly mixed atmosphere. This is due to a greater scattering of light by small particles in the polluted air in a forward direction—toward an observer looking toward the sun—then back toward the illuminating source. This is somewhat similar to the scattering of the oncoming car light beams by a dirty windshield. Blue light (the shorter visible wavelengths) will be scattered to a greater degree than red light (the longer ones). This is the reason that our sky appears blue and our sunsets red or orange.

The observed visibility is dependent both upon the amount and nature of the particulate matter in the atmosphere and upon the volume of air into which the material is mixed. This volume of air is dependent upon the height of the inversion layer and velocity of the wind. The lower the inversion base and the lower the wind velocity the less air is available for diluting the pollutants. However, high winds may decrease visibility in local areas due to the entrainment of surface dust.

Since it is mostly the aerosols which cause a reduction in visibility, relative humidity can have an important effect on overall

TABLE 2-10 ECOLOGIC AND ECONOMIC EFFECTS OF REDUCED VISIBILITY BY AIR POLLUTION

Effect	*Date*	*Description*	*Reference*
Increased fog	Late 1800s	Increase in number of foggy days/year in London from 50.8 for 1871–5 to 74.2 for 1886–90.	(174)
	1936–37	London experienced 40 days of fog while the suburbs had only 15.	(174)
Reduced insolation	1926	A difference of 55% was found between the sunlight of Manchester and that of its suburbs.	(175)
	1929	Rural areas 10 miles outside Baltimore received 50% more UV radiation than the urban center.	(176)
	1945	Chicago smoke, together with fog and partial cloudiness, deprived the city of approximately 44% of its possible hours of sunshine.	(177)
	1954	Isolation of UV radiation was reduced 90% in some areas due to absorption of light by water, ozone, carbon monoxide, and hydrocarbons.	(178)
	1965	A mean reduction of the insolation of ultraviolet light of 38% was found in moderate to heavy smog as compared to that on Mt. Wilson at 5,700 feet. Peaks reached 58%.	(173)
Vehicular traffic	1962	The reduction of visibility by air pollution resulted in 15 cars colliding near Wallsend; and 12 buses, petrol tankers, and cars were wrecked in two accidents near Cheshire.	(181)
	1962	The 1962 air pollution disaster in London caused visibility reductions to 30 yards in many parts of the city halting all London bus service and the ambulances of the Royal Automobile Club rescue squad.	(182)
	1963	"Any source of dense smoke close to a modern high-speed highway can pose a serious threat to travelers. In at least two recent instances—one in Pennsylvania close to a smoldering culm pile, another in Louisiana near a burning dump—the sudden application of brakes by a single motorist led to a chain reaction which involved the wrecking of a number of vehicles and injury to many of their occupants."	(180)

Vehicular traffic	1967	N. J. Turnpike Authority showed that air pollution contributes to visibility restrictions that endanger the safety of traffic.	(179)
	1970	Officials of the N. J. Turnpike Authority seek to prevent garbage dumping near turnpike because the garbage dumps "do burn," despite regulations forbidding open burning, producing serious driving hazards. The turnpike engineer said that unless something is done the northern section of the turnpike will be plagued by dense and dangerous smog conditions when smoke from burning garbage combines with fog and high humidity.	(183)
Aircraft	1963	The Civil Aeronautics Board in a review of one-third of the 1962 aircraft accidents in the U.S. showed six in which "obstruction to vision" (by smoke, haze, sand, and dust) were listed as a cause.	(180)
	1965	An Eastern Airlines pilot reports that the risk of near-collisions due to smoke was so great in 1965 that "you've had a dull trip when you don't experience at least one on every trip sequence as an airline pilot."	(184)
	1967	At the Kansas City Municipal Airport smoke plumes have been found to obscure runways and even to envelop the control tower. At the nearby Fairfax Airport smoke from burning dumps, a petroleum refinery and a manufacturing plant often interfered with airport operations.	(185)
	1969	Another Eastern pilot, testifying before the Virginia Air Pollution Control Board, described the many delays and cancellations caused by air pollution. He further indicated air pollution as the cause of at least two midair collisions during 1967: one over Hendersonville, N. C., in which 82 persons died and another near Urbana, Ohio, in which 26 persons died.	(186)
Other transportation	1962	London Air Pollution Disaster—shipping was virtually at a standstill in the Port of London with 60 ships fog-bound between Gravesend and the Nore. The Clyde was also closed to shipping.	(187)
	1962	Two trains crashed together at Gillingham, Kent, killing one and injuring three.	(181)

visibility. The size of the hygroscopic particles in the atmosphere is altered by the moisture conditions under high, but still not saturated, humidity conditions. Hygroscopic particles pick up water and increase in size. As they increase in size they become more effective in reducing visibility. When the relative humidity exceeds approximately 70 percent many types of particles exhibit delinquent behavior and grow into fog droplets, thereby greatly decreasing visibility. Most aerosols exhibit little deliquescence below 70 percent relative humidity. However, sodium hydroxide, calcium chloride, sulfuric acid, magnesium chloride, and sodium iodide all undergo phase change at humidities below 50 percent.

Carbon, tar, metal, and other opaque particles including those which tend to occur as agglomerates also contribute to reduced visibility. Transparent crystals impair visibility by scattering the light. These include crystalline compounds of iron, aluminum, silicon, and calcium which may exist as sulfates, nitrates, chlorides, and fluorides.

Although most visibility reduction is caused by aerosols, nitrogen dioxide (NO_2), a yellow-brown visible gas, also reduces visibility. The brown haze that is frequently seen over our metropolitan areas during the summer is attributed mainly to NO_2 which is strongly absorbent over the blue-green area of the visible spectrum. In Southern California, up to 30 percent of the visibility reduction may result from light absorption by NO_2.[188]

Air Quality Criteria for Particulate Matter[189] presents an excellent mathematical description of the relationship between atmospheric particulate concentration and visibility. This relationship follows in summary form.

Relationship between Atmospheric Particulate Loadings and Visibility

Middleton[190] describes the simple attenuation of a light beam along its path as follows. The intensity, I, decreases by an increment dI over the increment of path length dx. The relationship between intensity and distance is :

$$\frac{dI}{I} = -b dx \qquad (2\text{-}1)$$

or in integrated form:

$$I = I_0 e^{-bx} \qquad (2\text{-}2)$$

where b is the extinction coefficient assumed to be constant over x, and I_0 represents the intensity of light at $x = 0$.

The extinction coefficient, b, is the sum of the scattering coefficients of the air molecules and of the aerosols and the light absorption by the gases and aerosols. Of these, the scattering due to the aerosols is assumed to dominate in haze.[166] In scattering, the light energy is spatially redistributed, and by being removed from the line of viewing does result in visibility extinction.

Assuming 2 percent contrast as the threshold for the "average" human eye, Eq. 2-2 can be written:[166,190]

$$L_v = \frac{3.9}{b_{scat}} \tag{2-3}$$

where L_v = visual range in meters

b = extinction coefficient per meter along the path of sight

The extinction coefficient, b, has been found to be directly dependent upon the amount of atmospheric aerosol concentration, provided the relative humidity is below 70 percent, the aerosol is not particularly hygroscopic, and it is "aged" enough to contain a size distribution close to standard typical pattern. Such an aged pattern may develop within one hour for an aerosol from a combustion source. Because of this, and because the visible light scattering is caused primarily by particles from about 0.1–1μ radius, the scattering can be empirically related to the particulate concentration as follows:

$$L_v \simeq \frac{A \times 10^3}{G'} \tag{2-4}$$

where L_v = equivalent visual range

$A = 1.2 \frac{2.4}{0.6}$ for L_v expressed in kilometers

$= 0.75 \frac{1.5}{0.38}$ for L_v expressed in miles

G' = particle concentration ($\mu g/m^3$)

Deviations from Eq. 2-4 would be expected to occur when the relative humidity exceeds 70 percent, since many particles exhibit deliquescent behavior and grow into fog droplets. This relationship may not hold for photochemical smog, since it is not known whether its size distribution conforms to the necessary pattern of an "aged" aerosol.

Equation 2-4 provides a convenient means for estimating the expected visibility for different levels of particulate concentrations, under the conditions stated. From Eq. 2-3 and 2-4, Table 2-11, and

TABLE 2-11 RELATIONSHIP BETWEEN EQUIVALENT VISUAL RANGE AND PARTICLE CONCENTRATION[191]

Mass Concentration $\mu g/m^3$	*Scattering Coefficient (due to aerosol)* b_{scat}/m	*Equivalent Visual Range (in km)* L_v	*Equivalent Visual Range (in miles)* L_v	*Area Description*
10 20 5	0.3×10^{-4}	120.0	75.00	Remote rural
30 60 15	1.0×10^{-4}	40.0	25.00	Rural
75 150 37.5	2.4×10^{-4}	16.0	10.00	California ambient air standard
100 200 50	3.3×10^{-4}	12.0	7.50	Urban
300 600 150	10.0×10^{-4}	4.0	2.50	Urban-industrial
1000 2,000 500	33.0×10^{-4}	1.2	0.75	Very heavily polluted

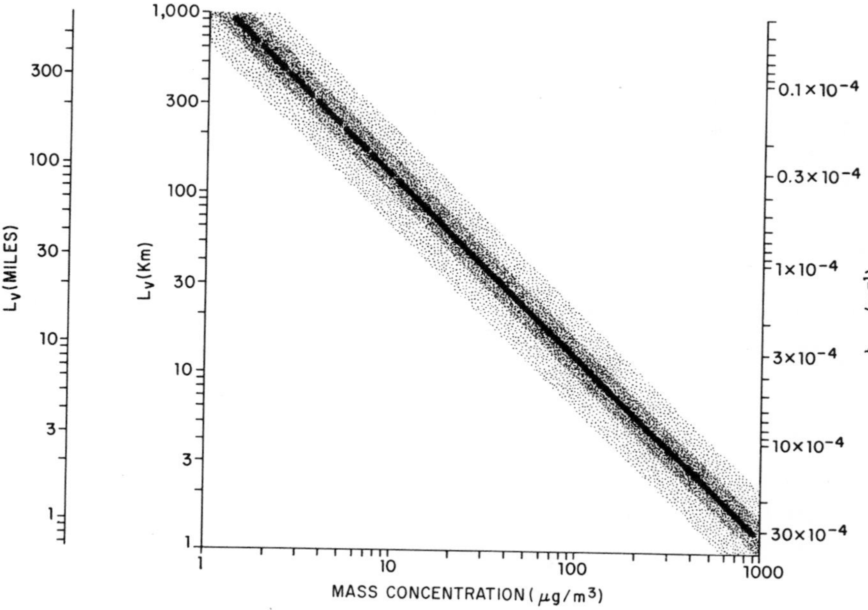

Fig. 2-5 Relation between visual range and mass concentration.[191] *(Courtesy J.A.P.C.A.)*

Fig. 2-5 relationships are derived which show the inverse proportionality between visual range and mass aerosol concentration.

To summarize, the obscuration of visibility per unit weight of suspended material is directly proportional to the scattering coefficient and the concentration of particles in the atmosphere, while it is inversely proportional to the specific gravity and radii of the particles for those 0.1μ or greater in diameter. The particles with the highest scattering efficiency are those with a size range of about $0.1–1.0\mu$.

SUMMARY AND CONCLUSIONS

Air pollution problems result from deleterious effects caused by the air contaminants rather than the mere presence of those contaminants in the air. Numerous adverse effects have been documented; many others are probably still unknown. Since the number and relative concentrations of potential pollutants are practically infinite, lack of data continues to be a persistant problem. On the other hand, the data gathered thus far and the financial penalties assessed indicate a need for caution in the use of dilution as solution to pollution.

It is apparent that the public will become less tolerant of environmental degradation. Economic and population growth are often cited as a cause of environmental problems. While there may be merit to this argument, it would appear self-evident that pollution reduction rather than economic stagnation or population control is the more feasible course of action at this time.

REFERENCES

1. "Air Pollution Control," *Law and Contemporary Problems,* School of Law, Duke University, 33:2 (Spring 1968).
2. Walsh, R. T., "The Bishop Case—An Interstate Odor Problem," *Proceedings Mid-Atlantic States Section—A.P.C.A. Semi-Annual Technical Conference on Odors: Their Detection, Measurement and Control,* Rutgers University, New Brunswick, N. J., May 13, 1970, p. 1.
3. Better, James W., "Odor as an Air Pollution Nuisance—New York State," *Proceedings: Mid-Atlantic States Section—A.P.C.A. Semi-Annual Technical Conference on Odors: Their Detection, Measurement and Control,* Rutgers University, New Brunswick, N. J., May 13, 1970, p. 11.
4. "New Jersey State Department of Health *v* Owen-Corning Fibreglas Corp." 100 N. J. Super. Er. 366,242, A. 2d 21(1968) from ed. 73 N. J. 248,250 A 2d 11.
5. "Consolidation Coal Co. *et al. v* Roscoe P. Kandle, M.D., Commissioner *et al.*," argued before the Supreme Court of New Jersey April 22, 1969; Decided May 8, 1969; Reported: 54 N. J. 11.
6. *Clean Air Act* as amended through the "Air Quality Act of 1967," Public Law 90-148, 90th Congress, Nov. 21, 1967.
7. *The Cost of Clean Air,* Senate Document No. 91-65; Washington, D.C., March 1970.
8. *Economic Impact of Air Pollution Controls on the Secondary Nonferrous Metals Industry,* U.S. Dept. of Commerce, Washington, D.C., 1969.
9. Wilson, R. D., and Minnotte, D. W., "A Cost-Benefit Approach to Air Pollution Control," *J.A.P.C.A.*, 19:5 (May 1969), p. 303.
10. Peckham, B. W., *Recent Literature on the Economics of Air Pollution,* National Air Pollution Control Administration, Raleigh, N. C., Oct. 1969 (mimeo).
11. *New York State Air Pollution Control Cost Survey,* February 1969, N. Y. State Dept. of Health, Albany, N. Y.
12. Jones, A. C., "Studies to Determine the Costs of Soiling Due to Air Pollution: An Evaluation," *Economics of Air and Water Pollution,* Virginia Polytechnic Institute, Blacksburg, Va., Oct. 1969, p. 146.
13. Peckham, B. W., "Odors, Visability, and Art: Some Aspects of Air Pollution Damage," *Economics of Air and Water Pollution,* Virginia Polytechnic Institute, Blacksburg, Va., Oct. 1969, p. 157.

14. *Air Quality Criteria for Sulfur Oxides,* NAPCA Pub. No. AP-50, USDHEW, Washington, D.C., Jan. 1969.
15. *Air Quality Criteria for Particulate Matter,* NAPCA Pub. No. AP-49, USDHEW, Washington, D.C., Jan. 1969.
16. *Air Quality Criteria for Carbon Monoxide,* NAPCA Pub. No. AP-62, USDHEW, Washington, D.C., March 1970.
17. *Air Quality Criteria for Photochemical Oxidants,* NAPCA Pub. No. AP-63, USDHEW, Washington, D.C., March 1970.
18. *Air Quality Criteria for Hydrocarbons,* NAPCA Pub. No. AP-64, USDHEW, Washington, D.C., March 1970.
19. Starr, C., "Social Benefit Versus Technological Risk," *Proceedings of Symposium on Human Ecology,* PHS Pub. No. 1929, Washington, D.C., 1968, p. 24.
20. "Michigan Law Permitting Citizen Suits Over Pollution Becomes Effective Oct. 1," *Air/Water Poll. Rep.,* **8**:31, Silver Springs, Md., Aug. 3, 1970, p. 307.
21. O'Connor, J. J., Jr., "The Economic Cost of the Smoke Nuisance to Pittsburgh," Univ. of Pittsburgh, Mellon Institute of Industrial Research and School of Specific Industries, Pittsburgh, Pa., *Smoke Investigation Bulletin No. 4,* 1913.
22. Schmidt, A. W. *The Pittsburgh Program in Retrospect: The Economic Evaluation,* ASME Paper No. 59 PBC-3, Pittsburgh Bicentennial Conference, Pittsburgh, Pa., 1959.
23. Michelson, I., and Tourin, B., "Comparative Method for Studying Costs of Air Pollution," *Public Health Reports,* **81**:6 (June 1966).
24. Kneese, A. Y., "How Much Is Air Pollution Costing Us in the United States?" *Proceedings of the Third National Conference on Air Pollution,* Washington, D. C., Dec. 12–14, 1966, p. 529.
25. Ridker, R. G., *Economic Costs of Air Pollution,* Frederick A. Praeger, New York, 1967.
26. Kohn, R. E., "Leaf Burning: An Economic Case Study," *Scientist and Citizen,* **9**:4, (April 1967), pp. 71–75.
27. O'Connor, J. F., and Citarella, J. F., "An Air Pollution Control Cost Study of the Steam-Electric Power Generating Industry," *J.A.P.C.A.,* **20**:6 (May 1970), p. 283.
28. Walker, W. R. (ed.), *Economics of Air and Water Pollution,* Bulletin No. 26, Water Resources Research Center, Virginia Polytechnic Institute, Blacksburg, Va., October 1969.
29. Goldman, M. I., *Controlling Air Pollution—The Economics of a Cleaner America,* Prentice-Hall, Englewood Cliffs, N. J., 1967, p. 82.
30. Michelson, I., *The Costs of Living in Polluted Air Versus the Costs of Controlling Air Pollution,* Report to the U.S. Public Health Service Conference on Air Pollution Abatement in the N. Y.–N. J. Area at New York, N. Y., January 11, 1967.
31. *National Emission Standards Study,* Senate Document No. 91-63, Govt. Printing Office, Washington, D.C., March 1970.
32. *Progress in the Prevention and Control of Air Pollution,* Senate

Document No. 91-64, Govt. Printing Office, Washington, D.C., March 1970.

33. Holbrow, G. L., "Atmospheric Pollution: Its Measurement and Some Effects on Paint," *J. Oil Color Chem. Assoc.*, **45** (Oct. 1962), pp. 701–718.
34. Yocom, J. E., and McCaldin, R. O., "Effects of Air Pollution on Materials and the Economy, in A. C. Stern (ed.), *Air Pollution*, Academic Press, New York, 1968, Vol. I, p. 617.
35. Anderson, G., "Europe's Crumbling Cathedrals and the 20th Century," *The Sunday Home News*, New Brunswick, N. J., Jan. 4, 1970, p. D-5.
36. Tice, E. A., "Effects of Air Pollution on the Atmospheric Corrosion Behavior of Some Metals and Alloys," *J. Air Pollution Control Assoc.*, **12**:12 (December 1962), p. 553.
37. Ajax, R. L., *et al.*, "The Effects of Air Pollltion on the Fading of Dyed Fabrics," *J. Air Pollution Control Assoc.*, **17:4** (April 1967), p. 220.
38. Salvin, V. S., "Effect of Air Pollutants on Dyed Fabrics," *J. Air Pollution Control Assoc.*, **13**:9 (Sept. 1963), p. 416.
39. *The Effects of Air Pollution,* PHS Pub. No. 1556, USDHEW, Washington, D.C., 1967, p. 14.
40. *The High Cost of Air: Economic Effects of Air Pollution,* USDHEW, PHS, Govt. Printing Office, Washington, D.C., 1958.
41. Beaver, H., *Committee on Air Pollution Report,* H. M. Stationery Office, London, 1954.
42. Jaffe, L. S., "The Effects of Photochemical Oxidants on Materials," *J.A.P.C.A.*, **17**:6 (June 1967), p. 375.
43. Bailey, P. S., "The Reactions of Ozone with Organic Compounds," *Chem. Rev.*, **58**:925 (1958).
44. Fisher, H. L., "Antioxidation and Antiozonation," *Chemistry of Natural and Synthetic Rubber,* Van Nostrand Reinhold, New York, 1957, p. 49.
45. Salvin, V. S., "Relation of Atmospheric Contaminants and Ozone to Light-fastness," *Am. Dyestuff Reptr.*, **53:1** (Jan. 6, 1964), pp. 33–41.
46. Ajax, R. L., *et al.,* "The Effects of Air Pollution on the Fading of Dyed Fabrics," *J. Air Pollution Control Assoc.*, **17**:4 (April 1967), pp. 220–224.
47. *Nitrogen Oxides: An Annotated Bibliography,* NAPCA Pub. No. AP-72, USDHEW, Washington, D.C., August 1970.
48. Garet, R., Drapkin, R. and Schaeffer, A., "The Economic Effect of Ozone on Home Maintenance Costs in the Los Angeles Basin," *Air Pollution Project: An Educational Experiment in Self-Directed Research,* Associated Students of the Calif. Inst. of Tech., Pasadena, 1968, p. 223–254.
49. Havens, A. V., "Weather—An Increasingly Significant Factor in Air Pollution Occurrence and Control," *News and Views,* Rutgers Univ. College of Agriculture, New Brunswick, N. J., Oct. 1963, p 1.

50. Larsen, R. I., "Relating Air Pollutant Effects to Concentration and Control," *J. Air Pollution Control Assoc.*, **20**:4 (April 1970), p. 214.
51. Hodgson, T. A., Jr., "Short-term Effects of Air Pollution on Mortality in New York City," *Environ. Sci. Technol.*, **4**:7 (July 1970), p. 589.
52. *Air Quality Criteria*, Staff Report, Subcommittee on Air and Water Pollution, Committee on Public Works, U.S. Senate, Govt. Printing Office, Washington, D.C., July 1968, p. 34.
53. McCabe, L. C., and Clayton, G. D., "Air Pollution by Hydrogen Sulfide in Poza Rica, Mexico," *Arch. Ind. Hyg. Occup. Med.*, **6** (1952), p. 199.
54. Fensterstork, J. C., and R. K. Fankhauser, *Thanksgiving 1966 Air Pollution Episode in the Eastern U. S.*, NAPCA Pub. No. AP-45, Durham, N. C., July 1968.
55. Epstein, S. S., "Carcinogenicity of Organic Extracts of Atmospheric Pollutants," *J.A.P.C.A.*, **17**:11 (Nov. 1967), p. 728.
56. Goldsmith, J. R., and Nadel, J. A., "Experimental Exposure of Human Subjects to Ozone," *J.A.P.C.A.*, **19**:5 (May 1969), p. 329.
57. Henry, M. C., Ehrlich, R., and Blair, W. H., "Effect of Nitrogen Dioxide on Resistance of Squirrel Monkeys to Klebsiella Pneumoniae Infection," *Arch. Environ. Health*, **18**:4 (April 1969), p. 580.
58. Holland, W. W., Reid, D. D., Seltser R., and Stone, R. W., "Respiratory Disease in England and the United States, Studies of Comparative Prevalence," *Arch. Environ. Health*, **10** (1965), pp. 338–345.
59. Ipsen, J., Ingenito, F. E., and Deane, M., "Episodic Morbidity and Mortality in Relation to Air Pollution," *Arch. Environ. Health*, **18**: 4 (April 1969), p. 458.
60. Ishikawa, S., Bowden, D. H., Fisher, V., and Wyatt, J. P., "The Emphysema Profile in Two Midwestern Cities in North America," *Arch. Environ. Health*, **18**:4 (April 1969), p. 660.
61. McCarrol, J., "Measurements of Morbidity and Mortality Related to Air Pollution, *J.A.P.C.A.*, **17**:4 (April 1967), p. 203.
62. Petrilli, R. L., Agnese, G, and Kanitz, S., "Epidemiology Studies of Air Pollution Effects in Genoa, Italy," *Arch. Environ. Health*, **12** (1966), pp. 733–740.
63. Rylander, R., "Alteration of Lung Defense Mechanisms Against Airborne Bacteria," *Arch. Environ. Health*, **18**:4 (April 1969), p. 551.
64. Sterling, T. D., Pollack, S. V., and Weinkan, J., "Measuring the Effect of Air Pollution on Urban Morbidity," *Arch. Environ. Health*, **18**:4 (April 1969), p. 485.
65. Verma, M. P., Schilling, F. J., and Becker, W. H., "Epidemiological Study of Illness Absences in Relation to Air Pollution," *Arch. Environ. Health*, **18**:4 (April 1969), p. 536.
66. Winkelstein, W., Jr., and Kantor, S., "Stomach Cancer," *Arch. Environ. Health*, **18**:4 (April 1969), p. 544.

67. Arbesman, P., Region II, USPHS, National Air Pollution Control Administration, USPHS, New York, Personal Communication, Oct. 1970.
68. Winkelstein, W., "The Relationship of Air Pollution and Economic Status to Total Mortality and Selected Respiratory System Mortality in Man," *Arch. Environ. Health,* **14** (1967), pp. 162–169.
69. Douglas, J. W. B., and Waller, R. E., "Air Pollution and Respiratory Infection in Children." *Brit. J. Prevent. Soc. Med.,* **20** (1966), pp. 1–8.
70. Lunn, J. E., Knowelden, J., and Handyside, A. J., "Patterns of Respiratory Illness in Sheffield Infant School Children." *Brit. J. Prevent. Soc. Med.,* **21** (1967), pp. 7–16.
71. Petrilli, R. L., Agnese, G., and Kanitz, S., "Epidemiology Studies of Air Pollution Effects in Genoa, Italy," *Arch. Environ. Health,* **12** (1966), pp. 733–740.
72. Carnow, B. W., Lepper, M. H., Shekelle, R. B., and Stamler, J., "The Chicago Air Pollution Study: SO_2 Levels and Acute Illness in Patients with Chronic Bronchopulmonary Disease," *Arch. Environ. Health,* **18** (1969), pp. 768–776.
73. Brasser, L. J., Joosting, P. E., and Von Zuilen, D., *Sulfur Dioxide —To What Level Is It Acceptable*? Research Institute for Public Health Engineering, Report G-300, Delft, Netherlands (July 1967).
74. Lawther, P. J., "Climate, Air Pollution and Chronic Bronchitis." *Proc. Roy. Soc. Med.,* **51** (1958), pp. 262–264.
75. Beard, R. R., and Wertheim, G. A, "Behavioral Impairment Associated with Small Doses of Carbon Monoxide," *Am J. Pub. Health,* **57** (Nov. 1967), pp. 2012–2022.
76. Schulte, J. H., "Effects of Mild Carbon Monoxide Intoxication," *Arch. Environ. Health,* **7**:5 (Nov. 1963), pp. 524–530.
77. McFarland, R. A., *et al.*, "The Effects of Carbon Monoxide and Altitude on Visual Thresholds," *Aviation Med.,* **51**:6 (Dec. 1944), pp. 381–394.
78. Wayne, W. S., Wehrle, P. F., and Carroll, R. E., "Oxidant Air Pollution and Athletic Performance," *J. Am. Med. Assoc., **199**:*12 (March 20, 1967), pp. 901–904.
79. Schoettlin, C. E., and Landau, E., "Air Pollution and Asthmatic Attacks in the Los Angeles Area," *Public Health Repts.,* **76** (1961), pp. 545–548.
80. Renzetti, N. A., and Gobram, V., *Studies of Eye Irritation Due to Los Angeles Smog 1954–1956,* Air Pollution Foundation, San Marino, Calif., July 1957.
81. Jaffe, L. S., "The Biological Effects of Photochemical Air Pollutants on Man and Animals, *Am. J. Pub. Health,* **57**:8 (Aug. 1967), p. 1269.
82. Mueller, P. K., and M. Hitchcock, "Air Quality Criteria—Toxicological Appraisal for Oxidants, Nitrogen Oxides, and Hydrocarbons," *J.A.P.C.A.*, **19**:9 (Sept. 1969), p. 670.
83. Sax, N. Irving, *et al, Dangerous Properties of Industrial Materials,* 2d ed., Van Nostrand Reinhold Co., New York, 1963.

84. *Threshold Limit Values of Airborne Contaminants*, American Conference of Governmental Industrial Hygienists, Cincinnati, Ohio, 1969.
85. Fasset, D. W., and Irish, D. D. (eds.), *Toxicology*, in F. A. Patty (ed.), *Industrial Hygiene Toxicology*, Vol. II, 2d rev. ed., Wiley Interscience New York, 1963.
86. Goldsmith, J. R., "Effects of Air Pollution on Human Health," in A. Stern (ed.) *Air Pollution*, 1968, p. 550.
87. Litton Systems, Inc., *Preliminary Air Pollution Survey of 30 Pollutants*, NAPCA Pub. Nos. APTD 69-23 to APTD 69-49, prepared for National Air Pollution Control Administration, USDHEW under Contract No. Ph22-68-25, Oct. 1969.
88. Selikoff, I. J., Churg, J., and Hammond, E. C., "Asbestos Exposure and Neoplasia, *J. Am. Med. Assoc.*, **188**:22 (1964).
89. Selikoff, I. J., and Hammond, E. C, "Community Effects of Non-Occupational Environmental Asbestos Exposure," *Am. J. Pub. Health*, **58**:9 (1968), p. 1658.
90. Utidjian, M. D., *et al.*, "Ferruginous Bodies in Human Lungs," *Arch. Environ. Health*, **17** (1968), p. 327.
91. *Ambient Air Quality Objectives—Classifications Systems*, Air Pollution Control Board, N. Y. State Dept. of Health, Albany, N. Y., Jan. 11, 1964.
92. *New Jersey Air Pollution Control Code*, Effective Oct. 1, 1964, Division of Environmental Quality, Dept. of Environmental Protection, Trenton, N. J., Chap. 7, "Solid Particles."
93. Durocher, N. L., *Preliminary Air Pollution Survey of Beryllium and Its Compounds—A Literature Review*, National Air Pollution Control Administration Publication No. APTD 69-29, Washington, D.C., Oct. 1969, p. 1.
94. Sussman, V. H., *et al.*, "An Air Pollution Study of a Community Surrounding a Beryllium Plant." *Am. Ind. Hyg. Assoc. J.*, **20**:6 (1959), p. 504.
95. Lieben, J., and Williams, R. R., "Respiratory Disease Associated with Beryllium Refining and Alloy Fabrication 1968 Follow-up," *J. Occup. Med.*, **11**:9 (Sept. 1969), p. 480.
96. Hodge, H. C., and Smith, F. A, "Air Quality Criteria for the Effects of Fluorides on Man," *J. Air Pollution Control Assoc.*, **20**:4 (April 1970), p. 226.
97. Atkins, P. R., "Lead in a Suburban Environment," *J.A.P.C.A.*, **19**:8 (Aug. 1969), p. 591.
98. "Air Quality and Lead—American Chemical Society Symposium," *Environ Sci. Technol.*, **4**:3 (March 1970) and 4 (April 1970).
99. Goldsmith, J. R., "Epidemiological Bases for Possible Air Quality Criteria for Lead," *J. Air Pollution Control Assoc.*, **19**:9 (Sept. 1969), p. 714.
100. Stopps, G. J., "Discussion—Epidemiological Bases for Possible Air Quality Criteria for Lead," *J. Air Pollution Control Assoc.*, **19**:9 (Sept. 1969), p. 71.
101. *Lead in the Environment and Its Effects on Humans*, State of

California, Dept. of Public Health, Berkeley, Calif., March 1967.
102. Stahl, Q. R., *Preliminary Air Pollution Survey of Mercury and Its Compounds—A Literature Survey,* NAPCA Pub. No. APTD 69-40, USDHEW, Raleigh, N. C., Oct. 1969.
103. Athanassiadis, Y. C., *Preliminary Air Pollution Survey of Vandium and Its Compounds—A Literature Review,* NAPCA Pub. No. APTD 69-48, USDHEW, Raleigh, N. C., Oct. 1969.
104. Ury, H. K., "Photochemical Air Pollution and Automobile Accidents in Los Angeles," *Arch. Environ. Health,* **17** (Sept. 1968), p. 334.
105. Goldsmith, J. R., "Effects of Air Pollution on Human Health," in A. C. Stern (ed.), *Air Pollution,* Academic Press, New York, 1968, Vol. I, p. 547.
106. Clayton, G. D., Cook, W. A., and Fredrick, W. G., "A Study of the Relationship of Street Level Carbon Monoxide Concentrations to Traffic Accidents," *Am. Ind. Hyg. Assoc. J.,* **21** (Feb. 1960), p. 46.
107. Clayton, G. D., *et al., Report of Research on Carbon Monoxide and Its Relation to Traffic Accidents,* U. S. Public Health Service Grant RG5676.
108. Stevenson, P., "Air Pollution Probed as Factor in Bus Crash," *The Morning Call,* Allentown, Pa., July 28, 1970, p. 5.
109. New Jersey Air Pollution Control Code, Effective June 12, 1968, Division of Environmental Quality, Dept. of Environmental Protection, Trenton, N. J., Chap. 11 "Incinerators."
110. Personal Communication, Division of Environmental Quality, N. J. Dept. of Environmental Protection, 1969.
111. Weisburd, M. I. (ed.), *Air Pollution Control Field Operations Manual,* USPHS Pub. No. 937, Washington, D.C., (1962).
112. *Proceedings: Mid-Atlantic States Section, Air Pollution Control Association Semi-Annual Technical Conference on Odors: Their Detection, Measurement and Control,* Rutgers Univ., New Brunswick, N. J., May 13, 1970.
113. Sullivan, R. J., *Preliminary Air Pollution Survey of Odorous Compounds,* NAPCA Pub. No. APTD 69-42, Washington, D.C., October 1969.
114. Firket, J., "Fog Along the Meuse Valley," *Trans. Faraday Soc.,* **32** (1936), p. 1192.
115. Schrenk, H. H., *et al., Air Pollution in Donora, Pa.,* "USPHS Bull. No. 306," Washington, D.C., (1949).
116. Scott, J. A., "Fog and Deaths in London, December, 1952," *Pub. Health Rept., 68* (1953), p. 474.
117. Phillips, P. H., "The Effects of Air Pollutants on Farm Animals," in P. Magill *et al., Air Pollution Handbook,* McGraw-Hill Book Co., New York, 1956.
118. Stokinger, H. E., "Effect of Air Pollutants on Wildlife," *Conn. Med.,* 27:8 (1963), p. 487.
119. Stokinger, H. E., and Coffin, D. L., "Biologic Effects of Air Pollutants," in A. C. Stern (ed.), *Air Pollution,* Academic Press, New York, 1968.

120. Street, J. C., "Ecological Systems: Domestic Animals," in C. O. Chichester (ed.), *Research in Pesticides*, Academic Press, New York, 1965.
121. MacKintosh, P. G., "Clinical Manifestations and Surgical Treatment of Lead Poisoning in the Horse," *J. Am. Vet. Med. Assoc.*, **74** (1929), p. 193.
122. Birmingham, D. J., Key, M. M., Holaday, D. A., and Perone, V. B., "An Outbreak of Arsenical Dermatosis in a Mining Community," *Arch. Dermatol.*, **91** (1965), p. 457.
123. Harkins, W. D., and Swain, R. E., "The Chronic Arsenical Poisoning of Herbivorous Animals," *J. Am. Chem. Soc.*, **30** (1908), p. 928.
124. Haywood, J. K., "Injury to Vegetation and Animal Life by Smelter Fumes," *J. Am. Chem. Soc.*, **29** (1907), p. 998.
125. Sullivan, R. J., *Preliminary Air Pollution Survey of Arsenic and Its Compounds—A Literature Review*, NAPCA Pub. No. APTD 69-26, USDHEW, Raleigh, N. C., 1969.
126. Suttie, J. W., "Air Quality Standards for the Protection of Farm Animals from Fluorides," *J. Air Poll. Contr. Assoc.*, **19:4** (April 1969).
127. Cameron, C. A., *Garden Chron.*, **1** (1874), p. 274.
128. Holmes, J. A., Franklin E. C., and Gould, R. A., *U.S. Bur. Mines Bull.* **98** (1915).
129. National Research Council of Canada, *Effects of Sulfur Dioxide on Vegetation*, NRC 815, Ottawa, Canada (1939).
130. Linzon, S. N., *The Influence of Smelter Fumes on the Growth of White Pine in the Sudbury Region*, "Contribution No. 439," Forest Biol. Div., Dept. of Agriculture, Ottawa, Canada (1958).
131. Seigworth, K. J., *Am. Forests*, **49** (1943), pp. 521 and 558.
132. Hindawi, I. J., *Air Pollution Injury to Vegetation*, NAPCA Pub. No. AP-71, USDHEW, Raleigh, N. C. (1970).
133. Jackobson, J. S., and Hill, A. V., (eds.), *Recognition of Air Pollution Injury to Vegetation: A Pictorial Atlas*, TR-7 Agricultural Committee, Air Pollution Control Association, Pittsburgh, Pa., 1970.
134. *Statewide Survey of Air Pollution Damage to Vegetation—1969*, CAES Pub. No. 148-70, Center for Air Environment Studies, Pennsylvania State Univ., University Park, Pa., 1970.
135. Heck, W. W., Dunning, J. A., and Hindawi, I. J., "Interaction of Environmental Factors on the Sensitivity of Plants to Air Pollution," *J. Air Poll. Contr. Assoc.*, **15** (Nov. 1965), p. 511.
136. Darley, E. F., *et al.*, "Plant Damage by Pollution Derived from Automobiles," *Arch. Environ. Health*, **6** (June 1963), p. 761.
137. Thomas, M. D., and Hendricks, R. H., "Effects of Air Pollutants on Plants," in P. L. Magill *et al.* (eds.), *Air Pollution Handbook*, McGraw-Hill Book Co., New York (1956).
138. Thomas, M. D., "Effects of Air Pollution on Plants," *World Health Organization Monograph Series No. 46*, Columbia Univ. Press, New York, 1961.
139. Brennan, E. G., Leone, I. A., and Daines, R. H., "Chlorine as a

Phytotoxic Air Pollutant," *Int. J. Air Water Poll.*, 9 (Dec. 1965), p. 791.
140. Brennan, E. G., Leone, I. A., and Daines, R. H., "Fluorine Toxicity in Tomato as Modified by Alternations with Nitrogen, Calcium, and Phosphorous Nutrition of the Plant," *Plant Physiol.* 25 (Oct. 1950), p. 736.
141. Brandt, C. S., and Heck, W. W., "Effects of Air Pollutants on Vegetation," in A. C. Stern (ed.), *Air Pollution*, Academic Press, New York, 1968, Vol. I, p. 401.
142. Middleton, J. T., *Ann. Rev. Plant Physiol.*, 12 (1961), p. 431.
143. Darley, E. F., "The Role of Photochemical Air Pollution on Vegetation," in *Proceedings of the First European Congress on the Influence of Air Pollution on Plants and Animals*, Wageningen, The Netherlands, 1968, p. 137.
144. James, H. A., *Flower Damage—A Case Study*, California Information Bulletin No. 8-63, San Francisco Bay Area Air Pollution Control District, 1963.
145. James, H. A., *Commercial Crop Losses in the Bay Area Attributed to Air Pollution*, San Francisco Bay Area Air Pollution Control District, 1964.
146. Pace, E., "Industrial Pollution a Problem for Some Dutch Tulip Growers," *New York Times* April 1, 1970.
147. "Air Pollution Killing 1.3-Million Trees in San Bernardino National Forest," *Air/Water Poll. Rept.*, Silver Spring, Md., Oct. 27, 1969, p. 368.
148. "Losses in Agriculture," *Agricultural Handbook No. 291*, USDA, Agricultural Research Service, Washington, D.C. (1965).
149. Leone, I. A., Brennan, E. G., and Daines, R. H., "Factors Influencing SO_2 Phytotoxicity in New Jersey," *Plant Disease Reporter*, 49:11 (Nov. 1965).
150. Daines, R. H., Leone, I. A., and Brennan, E. G., *Air Pollution as It Affects Agriculture in New Jersey*, Bull. No. 794, N. J. Agriculture Experiment Station, Rutgers Univ., New Brunswick, N. J., 1960.
151. Brennan, E. G., Leone, I. A., and Daines, R. H., "Characterization of the Plant Damage Problem by Air Pollutants in New Jersey," *Plant Disease Reporter*, 51:10 (Oct. 1967).
152. Leone, I. A., Dept. of Plant Biology, College of Agriculture and Environmental Science, Rutgers Univ., New Brunswick, N. J., personal communication, Sept. 11, 1970.
153. Manabe, S., and Wetherald, R. T., *J. Atmos. Sci.*, 24 (1967), p. 241.
154. Callendar, G. S., *Tellus*, 10 (1958), p. 243.
155. Robinson E., and Robbins, R. C., *Sources, Abundance and Fate of Gaseous Atmospheric Pollutants Supplement*, Stanford Research Institute, Stanford, Calif., June 1969.
156. Revelle, R., *et al.*, "Atmospheric Carbon Dioxide," *Restoring the Quality of Our Environment*, The White House, Nov. 1965.
157. "Effects of Atmospheric Particulate Matter on Solar Radiation and

Climate Near the Ground," *Air Quality Criteria for Particulate Matter,* NAPCA Pub No. AP-49, USDHEW, Washington, D.C., Jan. 1969.
158. McCormick, R. A., *Atmospheric Turbidity,* 60th Annual Meeting APCA, Cleveland, Ohio, June 1967.
159. McCormick, R. A., and Ludwig, J. H., "Climate Modification by Atmospheric Aerosols," *Science,* **156** (1967), p. 1358.
160. Humphreys, W. J., "Volcanic Dust and Other Factors in the Production of Climatic Changes and Their Possible Relation to Ice Ages," *Bull. Mount Weather Observatory,* **6** (1914), p. 1.
161. Peterson, J. T., and Bryson, R. A., "Atmospheric Aerosols: Increased Concentrations During the Last Decade," *Science,* **162** (1968), p. 120.
162. Möller, F., "On the Influence of Changes in the CO_2 Concentration in Air on the Radiation Balance of the Earth's Surface and on the Climate," *J. Geophys. Res.* **68**:13 (1963), p. 3877.
163. Lowry, W. P., "The Climate of Cities," *Scientific American,* **217**:2 (Aug. 1967), p. 15.
164. Landsberg, H. E., "City Air—Better or Worse," *Air Over Cities Symposium,* SEC Technical Report A 62-5, USDHEW, Nov. 1961.
165. Neuberger, H., and Gutnik, M., *Proceedings 1st National Air Pollution Symposium,* Stanford Research Institute, Pasadena, Calif., 1949.
166. Robinson, E., "Effect on the Physical Properties of the Atmosphere," in A. C. Stern (ed.), *Air Polllution,* 1968, p. 349.
167. Changnon, S. A., Jr., "The La Porte Weather Anomaly, Fact or Fiction," *Bull. Am. Meteorol. Soc.,* **49**:4 (1968).
168. Stout, G. E., "Some Observations of Cloud Initiation in Industrial Areas," *Air Over Cities Symposium,* SEC Technical Report A 62-5, USDHEW, Nov. 1961.
169. Telford, J. W., "Freezing Nuclei from Industrial Processes," *J. Meteorol.,* **17** (1960), p. 676.
170. Peterson, J. T., *The Climate of Cities: A Survey of Recent Literature,* NAPCA Pub. No. AP-59, USDHEW, Oct. 1969.
171. Monteith, J. L., "Local Differences in the Attenuation of Solar Radiation Over Britain," *Quart. J. Roy. Meteor. Soc.,* **92** (1966), p. 254.
172. Jenkins, I., "Increase in Averages of Sunshine in Central London," *Weather,* **24** (1969), p. 52.
173. Nader, J. S., *Pilot Study of Ultraviolet Radiation in Los Angeles, October 1965,* USPHS Pub. No. 999-AP-38, NAPCA, USDHEW, 1967.
174. Kratzer, P. A., *The Climate of Cities,* (Friedr. Vieweg and John Brauschweig trans.) American Meteorological Society, 1956, pp. 130, 132.
175. Clark, R. V., "Atmospheric Pollution and Health," *Med. Officer,* **38** (1927), p. 93.
176. Shreader, J. H., Coblentz, M. H., and Korff, F. A., "Effects of

Atmospheric Pollution Upon the Incidence of Solar Ultraviolet Light," *Am. J. Pub. Health,* 19 (1929), p. 717.
177. "Report on Sunshine, Smoke, and Dust," *J. Am., Med. Assoc.,* 133 (1947), p. 334.
178. Stair, R., and Gates, D. M., "Special Radiant Energy from the Sun Through Varying Degrees of Smog at Los Angeles," *Third National Air Pollution Symp.,* Pasadena, Calif., April 1955.
179. *Summary of Conference and Conclusions and Recommendations on Interstate Air Pollution, N. Y.-N. J. Metropolitan Area,* USDHEW, January 1967.
180. *A Study of Pollution—Air,* A Staff Report to the Committee on Public Works, U.S. Senate, Washington, D.C., Govt. Printing Office, 1963, p. 21.
181. "More Smog Deaths," *Manchester Guardian,* Manchester, England, Dec. 7, 1962.
182. "Fog Raises Demand for Hospital Beds," *The Times,* London, England, Dec. 7, 1962.
183. "N. J. Wants Garbage Off Turnpike," *The Home News,* New Brunswick, N. J., Aug. 27, 1970.
184. "Pilot Reports Smoke Poses Daily Collision Peril," *New York Times,* Dec. 7, 1965.
185. *Kansas City, Kansas—Kansas City, Missouri Air Pollution Abatement Activity: Visibility at Municipal and Fairfax Airports,* Technical Report, NAPCA, USDHEW, Jan. 1967, p. 47.
186. "Two Air Crashes Laid to Pollution," *Washington Post,* April 15, 1969.
187. "Fog Menace to Lungs," *The Times,* London, England, Dec. 6, 1962.
188. "Suspended Particles Seminar Highlights Methods Conference," *California Air Resources Board Bull.* 2:5 (March–April 1970).
189. "Effects of Atmospheric Particulate Matter on Visibility," *Air Quality Criteria for Particulate Matter,* NAPCA Pub. No. AP-49, USDHEW, Washington, D.C., Jan. 1969, p. 47.
190. Middleton, W. E. K., *Vision Through the Atmosphere,* Univ. of Toronto Press, Toronto, 1952.
191. Charlson, R. J., "Atmospheric Aerosol Research at the University of Washington," *J. Air Poll. Contr. Assoc.,* 18 (1968), p. 652.
192. Lillie, R. J., *Air Pollutants Affecting the Performance of Domestic Animals—A Literature Review,* Agriculture Handbook No. 380, Agricultural Research Service, USDA, Aug. 1970.
193. *Air Quality of 1967, Hearings Before the Committee on Interstate and Foreign Commerce, House of Representatives on H. R. 9509 and S. 780,* Serial No. 90-10, Washington, D.C., Govt. Printing Office, 41 1967, p. 41.
194. Michelson, I., and Tourin, B., "Comparative Method for Studying Costs of Air Pollution," *Pub. Health Rept.* 81:6, (June 1966), pp. 505–511.

3 Legislative and Regulatory Trends Regarding Air Pollution Control and Prevention

INTRODUCTION

Recent years exhibit, with increasing frequency, the environmental side effects of science and technology. Yet environmental degradation is not a new phenomenon; what is new is the degree and extent of the impact of modern man on the ecosystems he has attempted to dominate.

A descriptive picture[1] is presented in the analogy of our Western civilization as a great pendulum whose speed and direction of motion through time is determined by the forces of technology and economic determinism on the one hand and the opposing forces of human ecology on the other. . . . The forces of technology and economic determinism, fired subsequentially by the industrial age, the atomic age, and the space age, have pushed the pendulum at an ever-increasing and recently a dizzying speed. . . . We now have the knowledge, concepts, and attitudes, not only to reverse this motion, but to push the pendulum back to an equilibrium position which is favorable for human health and happiness. The unknown factor is man's will to do this.

It was the 1940s, however, before efforts were instituted in the United States to control and abate air pollution. During this period Los Angeles and California as well took steps to reduce the damage and discomfort caused by smog. It was 1955 before the first national air pollution control legislation was enacted.[2] This legislation provided the Public Health Service with limited authority to conduct research and provide technical assistance to the states and local governments.

By comparison, the English in 1273 enacted a law to prevent air pollution through the burning of soft coal. In 1854, Sir John Simon reported to the Court of the City of London his valid reasons for believing "that many persons of writable lungs find unquestionable inconvenience from these mechanical impurities [sulfur and fumes of sea coal] of the atmosphere."[3] In 1956 the English Clean Air Act was enacted as a direct consequence of the "air pollution episode" of December 1952, which caused an estimated 1,597 more deaths in the Administrative County of London than would have been expected during the same period from natural causes.[4]

LEGISLATIVE HISTORY

The United States followed in 1963 when federal statutes were broadened to include enforcement authority for the abatement of air pollution.[5] The Clean Air Act of 1963 also provided an expanded authority for research and technical assistance activities; and matching grants to state, regional, and local agencies for the creation or improvement of regulatory control programs.

This tardiness in response is characteristic of the development of public health programs in the United States. Until recently, the federal government has considered the protection of public health and welfare the responsibility of state and local government. Federal responsibility has been exercised only where interstate commerce was involved. For example, federal drinking standards apply only to water supplies used by the interstate carriers. There are many instances where state governments have tended to slough off responsibility for dealing with pollution problems.[6] The federal obstacle has been the legislative difficulty in defining federal, state, and local jurisdictional responsibilities. When dealing with the direct transportation of persons, goods, and services, in interstate commerce, or the effects of certain actions on such commerce, this is not too difficult. However, with environmental contamination, the problem was complicated by the difficulty of tracing causation: a direct connection between action (discharge), interstate movement, and effect.

For water there are well-stablished relationships between pollution and the transmission of communicable diseases. There are quantifiable direct ecomonic costs and damage to fish and wildlife also. The effects of air pollution are less obvious, particularly when chronic exposures are considered. In addition, there has been a tendency to consider the air as "free" and limitless, rather than a thin covering over the earth.

The Congress first considered environmental regulatory authority in 1948, with the Federal Water Pollution Control Act.[7] Concern centered on pollution originating in one state and adversely affecting the public health or welfare of another. The most frequently expressed concern was for states' rights and the federal initiated control and abatement was limited to interstate waters.

This issue was faced again with the Air Quality Act of 1967. This act provided a far-reaching federal enforcement role in air pollution control and abatement. Significantly absent from the discussion was concern for states' rights when public health protection is at stake. This national policy was reinforced in the Clean Air Amendments of 1970.

Clean Air Act of 1963

The intent of the Clean Air Act of 1963 was to encourage state, regional, and local programs to control and abate air pollution, whilst providing authority for the federal government to intervene in interstate problem areas. This approach was taken for two reasons: first, a desire to preserve and reinforce the federal system by supporting effective, viable action at all three levels of government (local, state, and federal); and, secondly, a recognition that the task of implementation and enforcement was so enormous that without effective programs at the state and local level the task could not be accomplished quickly and thoroughly.[8]

The preamble of the Clean Air Act was amended, giving recognition to urban areas, which sprawl across municipal county and state boundary lines, and to motor vehicles as a major contributor to air pollution. Federal enforcement authority was limited to two specific air pollution situations, and it was clearly intended to supplement the abatement powers of state and local governments.

First, for interstate problems where pollution originating in one state may be endangering the public health or welfare of persons in another state, the Secretary of Health, Education and Welfare was provided authority to intiate formal abatement proceedings. Second, for similar intrastate air pollution problems the secretary could invoke his abatement authority at the request of the governor involved.

The specific abatement procedures were modeled after those provided in the Federal Water Pollution Control Act Amendments of 1956.[9] A three-step procedure is involved: a conference with the cognizant official agencies, a public hearing, and finally court action. This procedure which could be terminated whenever the problem is resolved was modified with the Clean Air Amendments of 1970.

Two other provisions of the Clean Air Act of 1963 set the stage for subsequent legislation in 1967 and 1970. First, the act provided for the development of criteria of air quality as guides to state and local authorities when establishing ambient air quality standards and emission standards for a stationary source. Second, research authority was provided to develop methods to remove sulfur from fuels and control air pollution emissions. Both these features were reemphasized in the Air Quality Act of 1967.

Motor Vehicle Air Pollution Control Act of 1964

This discussion would not be complete without at least brief mention of automotive air pollution. In 1960 the Congress directed the Surgeon General to conduct a thorough study of motor vehicle exhaust as it affects human health through the pollution of air.[10] When the study[11] was completed the Air Pollution Control Act was amended in 1962,[12] requiring the Surgeon General to conduct studies relating to motor vehicle exhausts.

A formal process was provided in the Clean Air Act of 1963 for the continuing review of the motor vehicle pollution problem. A technical committee on motor vehicle pollution was formed with representatives from the Department of Health, Education and Welfare; the automotive industry; manufacturers of motor vehicle pollution control devices; and producers of motor fuels. The committee was to review progress toward the control of vehicular pollution and indicate specific areas in which additional research and development were needed.

The Clean Air Act of 1963 provided federal air pollution enforcement authority comparable to existing water pollution control authority. It was soon apparent, however, that a program modeled on the federal approach to water pollution control was not adequate. Air pollution control requires consideration not only of climate and meteorolgical effects but also moving and stationary sources of pollution. In 1964, extensive hearings were initiated by Senator Muskie's Subcommittee on Air and Water Pollution on the control of automotive emissions. A summary of these hearings states:[13]

> Automotive exhaust was cited (in the 1963 hearings) as responsible for some 50 percent of the national air pollution problems. It is, in many respects, the most important and critical source of air pollution, and it is, beyond question, increasing in seriousness despite pre-

liminary and isolated efforts to control it. One reason for the automobile's extreme importance in the overall national air pollution problem is in its omnipresence. Automobiles are found in every populated area, whether industrialized or recreational, urban or suburban. The Division of Air pollution of the Public Health Service has estimated that any place inhabited by 50,000 or more persons will have enough motor vehicles to create the potential for an air pollution problem.

Decisions in 1965 concerning the federal approach to controlling the automobile were later to affect the approach taken on stationary source control. Recognition was given in the Motor Vehicle Air Pollution Control Act[14] to three factors; first, to "technical and economic feasibility" in the setting of emission standards. It was anticipated that the establishment of standards would be a continuing process in which controls would be tightened as technical advances were made and economics improved.

Second, a decision was made to apply national standards to automotive exhausts. It was concluded that local control over vehicles moving in interstate commerce was impractical. Also, it was considered in the public interest to reduce the confusion that could result from up to fifty different state standards.

On the basis of the hearings, the Senate Subcommittee on Air and Water Pollution concluded[13] that it was technically and economically feasible to apply on a national basis the existing California standards for hydrocarbon and carbon monoxide emissions. It was widely accepted that this action would represent a major step toward the control and abatement of air pollution. At the time the automobile industry disputed the technical and economic feasibility of national standards but agreed to meet the national standards with the 1968 model automobiles.

During this period up to 1967, the federal program provided both planning and sustaining grants for the improvement in the effectiveness of state and local programs. The result was a notable improvement in the effectiveness of state and local programs for controlling and abating air pollution from stationary sources. Recognition of this improvement was reflected in the Air Quality Act of 1967.

Air Quality Act of 1967

The Air Quality Act of 1967[15] was perhaps the most controversial air pollution legislation enacted to date. The controversy, in part one of states' rights, was motivated by the regional nature of air pollution problems. Two alternatives were open to the Congress: first to provide for national emission standards for stationary sources, to be implemented by state and local agencies; or second, to provide for the establishment of regional ambient air quality standards and regional implementation schedules.

The congressional debate over national emission standards versus regional ambient air quality standards is considered to have begun when on December 13, 1966 Senator Muskie said:[16]

> With the exception of moving sources of pollution (for example, automobiles), I do not favor fixed national emission standards for individual sources of pollution. We do not need national ambient air quality criteria, applied as standards on a regional basis. The Federal Government is the logical entity to develop the criteria, with the cooperation of public and private groups. Those criteria must take into account health, aesthetics, conservation of natural resources and the protection of public and private property. The criteria must be modified, as our knowledge expands, to provide added protection against unforeseen pollution hazards. The ultimate goal should be to approach a level at which man will have to cope with little more than the "natural background level" of pollution.
>
> But setting criteria will not be enough.

Subsequently, on January 30, 1967, President Johnson proposed the establishment of national emission standards for "those industries that contribute heavily to air pollution."[17] His proposal would have authorized the Secretary of Health, Education and Welfare to

- Designate those industries in interstate commerce that are nationally significant sources of air pollution.
- Develop and publish industry-wide emission levels in consultation with the industry concerned.
- Provide each state to adopt equivalent levels—or stricter ones.
- Apply the federal levels to those states which do not adopt their own.

As proposed, national emission standards were emission limits that could not be exceeded regardless where an industrial plant was located. The stated objective was to provide uniformity and stability in emission limits thus reducing possible competitive advantages resulting in varying pollution control requirements.

This approach is similar to that in England's Alkali Act which requires that "the best practicable means" be used for preventing the discharge, either directly or indirectly, of noxious or offensive bases to the atmosphere and for rendering such gases where so discharged harmless and inoffensive. Absolute maximum emission levels are established for control equipment and seven years is considered a reasonable period for equipment obsolescence. For fifty-seven industry types each installation is required to obtain an annual license. The license is issued after proof that (1) the emission standards have been met, (2) the "best practicable means" of control have been employed, and (3) any emission that is being released has been rendered harmless and inoffensive.

Considerable attention was given to President Johnson's proposal for national emission standards. Administration representatives urged national emission standards on three grounds. First, they would eliminate the economic disadvantage of complying with varying local air pollution control requirements. Second, they would remove the temptation to industry to threaten to leave or avoid areas where such controls existed. Third, some industries are, by their nature, a danger to public health and welfare wherever they are located.

The judgment of the Senate Committee on Public Works, supported by the Congress, was that these arguments were offset by the following considerations:[18]

1. The administration itself did not propose uniform national emission standards, but rather minimal national standards. Clearly, therefore, there would be local variations which would not eliminate economic disadvantages.
2. Administration witnesses testified that the Public Health Service had made no findings with respect to industries which, in and of themselves, constituted a danger to public health and welfare.
3. National emission standards would eliminate some control options (relocation of pollution sources, fuel substitutes, and so forth) which may be essential in serious problem areas in the absence of effective technology.
4. Wise use of capital resources dictates that the first priority for the pollution control dollar be in those areas where the problem is most critical. National emission standards would give equal priority to critical areas and areas where no problem presently exists.
5. The difficulty in areas which have an air pollution problem is that the quality of the ambient air has deteriorated below a national level consistent with the protection of health and welfare. National emission standards would relate to the problem to the extent that "national" polluters happen to be located in the problem area. Other sources would not be touched by such standards. Such standards would not, therefore, represent a comprehensive attack on the problem of ambient air quality. The Subcommittee chose to deal with the problem of ambient air quality directly and to provide authority designed to improve it.

The resultant policy in the Air Quality Act attempted to implement a national air pollution control and abatement program within the context of traditional governmental institution arrangements. This was something new in the development of national policies. Its success depends upon the degree of commitment and cooperation that occurs between state and local governments, industry, and the private citizen. The ultimate alternative may well be national standards and national enforcement.

Recognizing that the Air Quality Act focused on air pollution problems in urban areas, the Congress adopted the view that further consideration of national emission standards for stationary sources was warranted. The act provided for a two-year study of the concept and the full range of its implications. This study served as the basis for proposals in 1970 for national emission standards for hazardous materials and for the application of the "best practicable means" of control in new plants.

The Air Quality Act provided a blueprint for a systematic effort to deal with air pollution on a regional basis, and for concurrent state and federal jurisdiction and enforcement requirements. In turn, the act established a framework for corollary state and local legislation.

The resultant public policy reflected the scientific nature of the air pollution problem and gave recognition to "technical and economic feasibility" when considering control methods. The procedure established called for the

- designation of "air quality control regions" based on jurisdictional boundaries, urban-industrial concentrations, and such other factors as climate, meteorology, and topography, for the purpose of setting ambient air quality standards.
- establishment and issuance of "air quality criteria" which reflect the latest scientific and medical knowledge of the identifiable health and welfare effects of varying concentrations of a contaminant, or class of contaminants, in the ambient air;
- development and issuance of information on recommended pollution control techniques, including their technical and economic feasibility;
- initiation of joint government-industry research to develop and demonstrate improved emission control technology to the degree required to prevent and abate air pollution.

Although the primary responsibility for control of air pollution was placed at the state level, a very strong federal fallback authority was provided. In addition to the enforcement authority contained in the Clean Air Act of 1963, authority was provided

1. to seek an immediate injunction to abate the emission of contaminants which present an imminent and substantial endangerment to health;

2. to designate "air quality control regions" for the purpose of implementing air quality standards, whenever and wherever deemed necessary to protect the public health and welfare;
3. in the absence of effective state action, to establish ambient air quality standards for such regions;
4. in the absence of effective state action, to enforce such standards;
5. in the absence of state action, to establish federal interstate air quality planning commissions.

During the two years following enactment of the Air Quality Act it was apparent that the federal program was not being implemented according to the anticipated time schedule. On December 10, 1969, Senator Muskie introduced the Air Quality Improvement Act, which provided an expanded regulatory authority governing motor vehicles.

On January 23, 1970, Senator Muskie also proposed accelerating the standard setting process covering stationary sources and extending the federal program to all areas where public health and welfare are at stake.[19]

To accomplish these objectives the proposed legislation

- required the immediate designation of all anticipated air quality control regions;
- extended federal enforcement authority to intrastate violations of air quality standards;
- permitted class suits to enforce standards;
- provided court authority to issue cease and desist or specific performance orders and to assess penalties for violations of the emission standards or compliance schedules;
- required that emission standards and compliance schedules be included in any air quality standards submitted for federal approval;
- encouraged the establishment of statewide air pollution control programs and provided authority for federal action in the absence of approved state programs;
- required that all new industries install the "best available" pollution control technology at the time of construction.

Legislation embodying these proposals, the National Air Quality Standards Act of 1970, was introduced by Senator Muskie on March 4, 1970.

During this same period the administration conducted its own evaluation of legislative needs. On February 10, 1970, President Nixon proposed the Clean Air Act Amendments of 1970.[20]

The proposal included new authority

- to set national ambient air quality standards;
- to establish national emission standards for classes of new facilities which are major contributors to air pollution;

- to establish national emission standards for hazardous materials;
- to seek court action when both interstate and intrastate air pollution levels fall below national standards, or there is local nonenforcement of emission standards or implementation schedules;
- to impose fines up to $10,000 per day for failure to meet established air quality standards or implementation schedules.

Clean Air Amendments of 1970

After extensive hearings and public debate, the Clean Air Amendments of 1970 were enacted on December 31, 1970. Recognizing the recent creation of the Environmental Protection Agency (EPA) all functions of the Secretary of Health, Education and Welfare were transferred to the administrator of the EPA. The 1970 amendments extended the geographical coverage of the federal program for the prevention, control, and abatement of air pollution from both stationary and moving sources. The Clean Air Amendments of 1970 provide new authority[21,22,23,24]

- requiring the inclusion of all areas within states in air quality control regions while preserving existing regions;
- requiring the establishment of statewide plans of implementation
- which are designed to achieve the primary or public health standard within three years;
- to promulgate performance standards for new major sources of atmospheric emissions;
- to promulgate national emission standards for pollutants found to be hazardous to public health;
- imposing fines and criminal penalties for violation of any air quality implementation plan, emission standard, performance standard, etc.;
- requiring emission monitoring, and record keeping and authorizing subpoenas, right of entry, and the issuance of abatement orders;
- providing for the control of automotive emissions, aircraft emissions, and the regulation of fuels.

New general authority was provided in a number of areas which can be expected to be incorporated in other environmental legislation. The Clean Air Amendments of 1970 provide authority

- to federally abate any air pollution that presents an imminent and substantial danger to health;
- to permit citizen suits to enforce violation of standards established pursuant to the act and mandatory functions to be performed by the administrator of EPA;
- to permit judicial review of standards, implementation plans, and other actions taken pursuant to the act;

- to prohibit the federal government from signing contracts with any company convicted of violating an air quality law;
- to enforce mandatory licensing of patents, trade, secrets, and know-how when the Attorney General determines such information is requisite to meeting standards established pursuant to the act.

Prospects for the Future

When the Senate passed the 1970 amendments there was provision for preconstruction review of the design and location of new sources and certification of compliance with new source performance standards. This provision was not incorporated in the final measure; however, such review requirements as certification can be anticipated under future state or federal laws.

The objective of an engineering plan review and evaluation is to assure that the construction of a major new facility will not conflict with existing implementation plans for air quality standards. Should additional control be required, appropriate actions could be taken during design or construction, increasing the probability of achieving effective, economic air pollution control. Certification of compliance would be provided at the time a new facility begins operation.

Under such requirements government and industry would be required to increase efforts to insure that new plants and equipment perform in accordance with design specifications and commitments made by engineers and equipment suppliers.

ABATEMENT AUTHORITY AND ACTIONS

As mentioned, the 1963 Clean Air Act provided the first federal regulatory authority to control and abate air pollution. At that time, only thirteen states had laws providing for statewide control of air pollution. Two other states had laws providing limited coverage. State laws left much to be desired. The Clean Air Act of 1963 and the Air Quality Act of 1967 have provided an impetus, however. Now most of the states have enacted legislation, although some of the statutes are inadequate to meet state requirements under the 1967 Air Quality Act.

We are witnessing a dynamic development of an entire body of law arising out of intense public reaction, political pressures, and the exigencies of health and the economic well-being of man. Occasionally the outcome broadens common law concepts. Frequently, the result means legal reform going to the very roots of the law, creating new principles, new torts, and new liabilities. Federal abatement actions have been instrumental in this evolutionary process.

Federal Abatement Actions

Under the abatement authority in the Clean Air Act the Secretary of Health, Education and Welfare could take action on his own initiative to abate and control interstate air pollution. At the request of a governor he could take action in intrastate problem areas. This authority was superseded in 1970 by the National Air Quality Standards Act.

As the capability of federal, state, or local authorities for enforcement has developed friction between them has sprung up. In most instances the friction means a simple jurisdictional squabble, but occasionally serious problems are created for industry, especially in areas with semiautonomous local agencies. For example, when the Chicago Metropolitan Sanitary District felt that their entrenched "empire" was threatened, they not only sought to inject themselves into local enforcement, but even attempted to launch water pollution suits[25] against companies operating out of the state.

Comparable zeal was reflected in the formation by the Illinois Attorney General of a task force on pollution control which is taking its crusade to the courts. The Deputy Attorney General was quoted as saying: "Our mission is to file suits and win cases. After many cases have been tried, there will be a doctrine established in meeting pollution control standards and the time required. We believe that a president of a corporation is subject to a jail sentence if he willfully disobeys a court order or the law." This tactic has never before been tried against officials of companies that violate pollution control laws.

Increasingly, state and local officials act in response to mounting public pressure. However, it is reasonable to assume that responses will never develop as far as they once did in England when Edward I ordered one of his subjects beheaded because he continued to burn foul-smelling high-sulfur coal instead of honest English oak.

Recently, United States attorneys were instructed to prosecute violations and seek maximum penalties. Under the 1963 Clean Air Act federal abatement authority was employed in ten areas:

1. Bishop, Maryland, and Selbyville, Delaware
2. Ticonderoga, New York, and Shoreham, Vermont
3. New York-New Jersey Metropolitan Area
4. Kansas City, Kansas-Kansas City, Missouri
5. Clarkston, Washington-Lewiston, Idaho
6. Parkersburg, West Virginia-Marietta, Ohio

7. Powel County (Garrison), Montana
8. National Capital Metropolitan Area (Washington, D.C.)
9. Ironton, Ohio-Ashland, Kentucky-Huntington, West Virginia
10. New Cumberland, West Virginia-Know Township, Ohio

The character of these actions ranged from a single local industrial plant to the complexities of the New York Metropolitan Area. Several outcomes served as a portent for the future. Of particular note are those for Bishop, Md.; Selbyville, Del.; metropolitan New York; and Parkersburg, W. Va.-Marietta, Ohio.

Bishop Processing Company. The first abatement conference under the provisions of the Clean Air Act was held in November 1965, at the request of the state of Delaware, in the rural community of Selbyville, Del. The problem was noxious and offensive odors arising from a rendering plant producing chicken feed and fertilizer in nearby Bishop, Md. The recommendations of this abatement action called for a vigorous cleanup program and installation of equipment to prevent and control odors.

On May 17, 1967, when the plant owner did not adequately comply with the recommendations, an abatement hearing was convened to receive testimony and evidence from all interested or affected parties. The Bishop Processing Co. council not only denied the charges but, on constitutional grounds, questioned the legality of the hearing board itself. The company's council was overruled and the company was asked to cease and desist not later than December 1, 1967, by installing and placing into operation the effective controls, systems, and devices recommended by the hearing board.

The company did not act within the compliance deadline. The Secretary of Health, Education and Welfare then asked the Attorney General to bring suit against the company. Subsequently, on March 7, 1968, the Justice Department filed a civil complaint in the United States District Court in Baltimore.

When the company's council offered a motion to dismiss the suit, the judge overruled the motion and awarded a consent decree on November 1, 1968, holding that the movement of air pollutants across a state line constitutes interstate commerce and "is subject to the power granted to Congress by the Constitution to regulate such commerce."[26]

The Bishop Processing Co. agreed to the terms of the consent decree which provided that if Delaware health officials found pollutants from the Bishop plant reached Delaware, a court order could be issued to close the processing plant immediately. Subsequently, a surveillance program was mounted in Selbyville and in March 1969 affidavits were filed with the court claiming that malodorous air pollution was still blowing from the Bishop's operation into Delaware.

The court requested additional evidence. This additional evidence of the occurrence of odors was submitted, and the court issued an injunction to take effect February 16, 1970. The injunction was appealed by the company before the Fourth Circuit Court of Appeals and the U.S. Supreme Court. Both courts upheld the cease and desist order.

Although this action took five years, it serves as a precedent in strengthening the hands of pollution control officials throughout the country. In awarding this decision the judge ruled that the commerce power may be exercised to achieve socially desirable objectives even in the absence of economic considerations. In rendering this opinion, he also affirmed the constitutionality of the Clean Air Act.

New York Metropolitan Area. The first phase of the New York-New Jersey abatement conference took place in January 1967. Besides New York City, three other New York Counties and nine New Jersey counties were involved. The first conference dealt with sulfur oxides and carbon monoxide. A second conference in January 1968 dealt with particulate matter.

Since then, both States have promulgated stringent regulations governing carbon monoxide and sulfur dioxide. New Jersey took the first state action to limit the sulfur content of fuel oil and coal. The regulation required the sulfur content of bituminous coal to be 1 percent by May 1968 and 0.2 percent by October 1971. This action aimed to achieve a 40 percent reduction in sulfur oxide emissions.

The significance lay in the extent to which the regulatory agencies were willing to go in restricting the sulfur content of fuel oil and coal to achieve air pollution control objectives. The effect may well be a major shift from reliance on coal—a domestic fuel—to low-sulfur oil—primarily imported.

Parkersburg, W. Va.—Marietta, Ohio. In March 1967, an abatement conference was convened to deal with interstate particulate and sulfur oxide air pollution in the area of Parkersburg, W. Va.-Marietta, Ohio. A second conference was held in October 1969 to permit the participants to present new data. At the second conference odorous chlorine compounds and eye irritants were considered also.

Recommendations regarding chlorine and particulates were issued on March 19, 1970. The conclusions were two: first, that the combustion of fuel, particularly coal, accounted for two-thirds of the total emissions of particulates in the area; second, that atmospheric reactions between chlorine and various chemicals were found to produce objectionable odor and lachrymators. Specific recommendations were the following:

(1) Existing regulations on open burning should be strictly enforced and additional prohibitions established. The burning of refuse should be permitted only in incinerators emitting less than 0.3 grain of particulate matter per cubic foot of gas, or of an opacity less than 1 on the Ringlemann Chart.

(2) Particulate emissions from fuel-burning equipment should be limited to 0.56 lb/hr million Btu input for sources producing from 1 to 10 million Btu/hr and 0.11 lb/hr/million Btu for sources of 10 billion or more Btu/hr. Sources with an input between 10 million and 10 billion Btu/hr should be limited to particulate emissions of no more than the amount determined by linear interpolation between these values. Visible emissions from such sources should be limited to an opacity less than 2 on the Ringelmann Chart. New plants should meet these limits at construction; existing plants should fully comply within three years.

(3) For industrial processes, emissions of particulate matter were specific, along with a visible emissions standard of an opacity less than 2 on the Ringelmann Chart. A three-year compliance schedule was provided.

(4) A six months' compliance schedule was provided for controlling chlorine emissions to no more than 3 lb/hr. Also, the emission concentrations were in no case to exceed 1.5 ppm by volume.

(5) At six-month intervals, principal polluters in the area were required to report data on their emissions, control efforts, and any explanations for delay.

Sulfur oxide recommendations were made on April 27, 1970. The primary recommendation was for the reduction of sulfur oxide emissions from a private power generating plant in Marietta, Ohio. This facility was found to emit 86 percent of all sulfur oxides in the abatement area. Also the point of release was found to be too low in proportion to surrounding buildings.

Specific recommendations called for a 40 percent reduction within 6 months, and a 70 percent reduction within 2 years. The facility also was required to take immediate steps to eliminate the "downwash" of its emissions; and at 6-month intervals to report data on emissions, control efforts, and any explanations for delay.

The most significant of these recommendations is the requirements for data on emissions.

Two additional abatement actions worthy of brief mention here were those for the national capital metropolitan area and Powell County, Mont. The District of Columbia action resulted in a July 1969 limitation on the sulfur content of fuels to 1 percent.

The Garrison, Mont., action is of interest because of the possibility for future court action. An abatement conference was held in August 1967 to consider fluoride emissions from a plant which processes phosphate rock to produce an animal food supplement. The single recommendation was that the plant be closed until proper abatement devices were installed. The plant closed within the month following the conference; however, in October 1967 operations were resumed.

In late 1969, the fluoride content of grasses in the area increased dramatically above the state standard for forage fed to livestock. In January 1970, the state Board of Health requested the plant to cease operation under the terms of a prior agreement. The plant then reopened on March 3, 1970, without the approval of the board or the state Attorney General.

The Attorney General's Office asked the state Supreme Court either to close the plant until the board could evaluate changes made to reduce pollution or to overrule the District Court order preventing the Attorney General from taking immediate legal action without a hearing. The state Supreme Court ruled the argument presented was "insufficient" to grant the request, thereby permitting the plant to continue operation.

The state Board of Health is continuing surveillance, and if state standards are exceeded, the board plans to invoke injunction procedures against the plant.

Federal Injunctive Authority

The Air Quality Act of 1967 provided a new federal enforcement provision. The act authorized the secretary to seek immediate court action to stop emissions when there is "imminent and substantial endangerment to the health of persons." This authority also must be contained in federally approved state standards and plans of implementation.

Unlike with abatement actions, there is no limitation on the court; accordingly, an injunction can be issued for a hazardous material or during an air pollution episode without consideration of technological or economic feasibility. This authority is broad enough to allow the secretary to stop the movement of traffic or shut down industries.

The National Air Pollution Administration defined emergency control levels for sulfur oxides and oxidants. Their position was that these levels should be prevented by whatever federal, state, and local emergency control actions were necessary. The ambient air quality level for sulfur oxides is 0.4 ppm (24-hr) and for oxidants it is 0.7 ppm (1-hr). Exposure to these levels presents an acute threat to health.[27]

State and local control agencies are being encouraged to develop appropriate air-monitoring systems and emergency procedures to implement the requisite action. Detailed knowledge is being acquired on the nature and location of pollution sources, local meteorological conditions, and air pollution levels. Detailed plans are being developed, also, tailored to the local needs. This authority was first invoked in Birmingham, Alabama in November 1971 when the EPA administrator obtained a court order to halt production for 23 major industries.

Private or Class Suits

Many control agencies are openly encouraging private and class suits as a means of seeking solutions to local and individual problems. Private or civil law suits brought against pollution under common law as a tort action, often a negligency or nuisance suit, are based on common law liability and are independent of governmental programs. Usually the private action is not superseded by statutory authority. As such, the value of private actions may well be to bring to the forefront issues that should be solved by legislation. On the other hand, tort actions create a body of common law dealing with environmental pollution control.[28]

By encouraging private and class suits against industry, government control agencies have opened a Pandora's box of problems for industry. Such suits are being urged whenever individuals believe that pollution control legislation is inadequate, that it is not being enforced, or that, through special waivers and exemptions, maximum pressure is not applied against industry to obtain abatement or elimination of pollution. In cases involving odors, such suits are said to be preferred, since odors are difficult to control by the use of standards.

Significantly, private or class suits are encouraged when pollution is deemed objectionable, even though emissions may not be illegal, and may meet all the prescribed standards of federal and state legislation. Two recent examples of such suits may be cited. One was a class action[29] for $39 billion claim against virtually all industry and all municipal corporations in Los Angeles County. This action was dismissed on the court's own motion.

The other was a suit[30] by two Chicago aldermen for $3 billion against all leading automobile, truck, and tractor manufacturers. It charged that they conspired to delay the research, development, and installation of air pollution control devices on their vehicles. This suit followed one in Los Angeles in which the Justice Department obtained a consent decree.[31]

In encouraging private and class action suits, control agencies are advocating:[32]

1. The development of a new tort of a "damage to health from air pollution." This would present a more effective and more readily recognized remedy than common law suits for nuisance or trespass. Government attorneys believe that precedents for such a tort already exist, and that technical information published by federal agencies could serve as expert testimony to prove such damage to health.
2. The broadening of the concepts of actual and proximate causes of damage, so that a plaintiff no longer has to prove that the damage was caused by a particular substance emitted from a specific installation. The contention is that the mere existence of an approximate concentration of pollutants required to cause the damage should be adequate, and that action should be against all polluters in the area as joint tort-feasors. This may have great appeal in highly industrialized areas where persons instituting such suits would have the option of seeking damages from the polluter of their choice.
3. Under this approach, compliance with existing standards and regulations would not in itself be sufficient and acceptable defense, especially when an industry has notice, either actual or inferred (constructive), that its emissions may be harmful. Such notice could be presumed from government documents, general technical knowledge, and published medical data.
4. In the area of "due care," enforcement agencies advocate that a court should require a source of pollution to meet the highest level of control practicable. They contend that failure to use the better systems could constitute negligence.

The thesis is that the courts should look only at the injury, devise remedies, and structure legal theories to fit the facts. The belief is that this would result in a more flexible application of the laws of nuisance, negligence, and trespass, and would help introduce new knowledge about tort liability for air pollution.

This is a relatively new problem with which industry will have to cope in the near future along with dealing with regulatory agencies.

Citizen Suits

Considerable attention was devoted by the 91st Congress to the question of whether federal laws should provide for citizen suits or class actions on environmental issues. Concern originated from restrained government initiative in seeking the control and abatement of air pollution.

After extensive debate it was the opinion of the Senate Public Works Committee[22] that citizen participation in the enforcement of air pollution standards and regulations is consistent with the underlying principles of the Clean Air Act. The identifiable standards provide manageable, indentifiable, precise benchmarks for enforcement.

It was believed that authorizing citizens to bring suits for violation of standards should motivate regulatory agencies to meet their enforcement and abatement responsibilities. Therefore, the Congress, in the 1970 amendments[24] provided citizen participation in the enforcement of air pollution standards and regulations.[22,23] The rights of persons (or classes of persons) to seek enforcement or other relief under any statute or common law is not affected.

Provision is made for suits against violators, including the United States, (1) to enforce an emission standard or other limitation established pursuant to federal law, or (2) to enforce an order issued with respect to a standard or limitation.

When violation of an order or a hazardous emission standard is alleged immediate action can be brought in the district in which the source is located. The right is preserved for the federal administrator to intervene.

In all other citizen suits, prior to commencing any action in the District Courts, the plaintiff must provide 60-days' notice to the violator, the state, and administrator. Suits against the administrator of the EPA are restricted to alleged failure to perform his mandatory functions under the act (e.g., promulgation of air quality criteria or new source performance standards).

Because concern was expressed regarding frivolous and harassing actions, provision was made that the courts may award costs of litigation, including reasonable attorney and expert witness fees. In addition, the court is provided discretionary authority to require filing of bond if a temporary restraining order or preliminary injunction is sought.[23] These safeguards should have the effect of discouraging frivolous or harassing actions.

The Congress has provided a public policy to the effect that citizens in bringing legitimate enforcement actions are performing a public service.

Penalties

Federal penalities for violation of the law were strengthened with the Clean Air Amendments of 1970. Conviction for a knowing violation is now subject to a penalty of $25,000 per day or imprisonment for one year, or both. For a second violation the penalty increases to $50,000 per day of violation or imprisonment for two years, or both.

STATE AND LOCAL STANDARDS

State and local air pollution control efforts in the United States are generally modeled after those of the pioneer—California. This was true to such an extent one could almost say that the way California goes, so goes the nation—although what is good for California may not necessarily be the best for other regions of the country.

Many state and local statutes and administrative rules were enacted hastily in response to public demands and outcrys for "clean air." Consequently, many are not in keeping with well thought out judicial and administrative processes. At the time, air pollution control legislation was in its infancy. The general approach was to establish emission standards or regulations applicable to selected source types based upon the nature of the local air pollution problem. Recent compilations[33,34] indicate that some forty-six states, under the impetus of federal program grants, have enacted air pollution control law, although some are too limited.[35] A summary of state statutory and regulatory authority is provided in Table 3-1.

Emission Standards

In general, state and local emission standards were developed before there were viable procedures for computing allowable emission limits from desired ambient air quality levels. The first emission standards were for particulates. Later emission standards were expanded to include gaseous emissions, individual chemical elements or compounds, and mixtures of specific substances.

During this evolutionary period the concept of application of the "best practicable means" for control was developed. Because of the economic or cost implications of "practicable," many political jurisdictions left the definition flexible and judgmental, resisting pressures to give numerical values to permissible emissions. In the United States the practice generally has been to select and apply the percentage reduction deemed practicable for all the units of a typical uncontroled source category.

The deficiencies in an approach based upon a percentage reduction are three. First, there is no assurance that application of the best practicable means of control will achieve either the desired ambient air quality standard or, at minimum, protect public health.

Second, since the emission standard is to be applied uniformly, it tends to be established at the lowest common denominator. This is done under the guise of not presenting an "undue economic hardship" on existing installations.

Third, percentage reductions in emissions do not reflect plant size. A large discharge of contaminants implies a large production capacity or a large number of production units. In general, the larger the installation the smaller is the economic impact of air pollution control equipment or processes. Although the total cost is larger for the larger production plant, the cost as a percentage of capital equipment costs is smaller. Thus, for the same percentage expenditure, the larger plant can usually acquire more efficient equipment with its associated added cost.

Also, a large installation produces a greater absolute quantity of pollution than does a smaller one, imposing a greater burden upon the atmosphere for the same control efficiency as that of a small installation. The concept of a "sliding scale" emission standard then emerges.

Effective sliding-scale emission standards specify not only the quantity of pollutant per unit volume of stack gas, but also the volume of stack gas per unit time or weight time. Otherwise, an increased flow rate in the stack could be induced through forced introduction of air and thereby reduce the concentration of pollutant in the stack gas without reducing the amount discharged in a unit time.

In general, state and local emission standards or regulations take effect immediately upon adoption. New facilities built after the date of adoption generally are required to be designed and equipped to comply with the applicable standards. Sources already in existence generally are given varying periods of time to achieve compliance, depending largely on the economic and technological problems associated with compliance. Where emission standards

TABLE 3-1 SUMMARY OF STATE AIR POLLUTION CONTROL LEGISLATION

State	Air Pollution Control Agency	AUTHORITY		AUTH. TO ESTABLISH			ENFORCEMENT CAPABILITY								
		Statutory	Regulatory	Control Regions	Ambient Air Standards	Emission Standards	Emergency	Permit	Fine	Inspection	Criminal	Injunctive	Orders	Local Option	Tax Incentive
Alabama	Alabama Department of Public Health Montgomery, Ala. 36104	Yes	Yes	Yes	Yes	Yes	E	P	F	I	C	I	O	Yes	Yes
Alaska	Alaska Department of Environmental Conservation Juneau, Alaska 99801	Yes	Yes	Ques.[b]	Yes	App.[a]	E	P	F	I	C	I	O		
Arizona	Division of Air Pollution Control Phoenix, Ariz. 85017	Lim.		Yes	Ques.	Ques.	E			I		I	O	Yes	
Arkansas	Arkansas Pollution Control Commission Little Rock, Ark. 72202	Yes	Yes	Yes	Yes	App.	E	P		I	C	?[c]	O		
California	California Air Resources Board Sacramento, Calif. 95814	Yes	Yes	Yes	Yes	Yes	E				C	I	O	Yes	Yes
Colorado	Colorado Department of Health Denver, Colo. 80220	Yes		Yes	Sta.[d]	Sta.	E		F	I	C	I	O	Yes	
Connecticut	Department of Environmental Protection Hartford, Conn. 06115	Yes	Yes	Yes	App.	App.	E		F	I		I	O	Yes	Yes
Delaware	Delaware Department of Natural Resources and Environmental Control Dover, Del. 19901	Yes	Yes	Yes	App.	App.	E	P	F	I		I	O		

District of Columbia	Department of Human Resources Washington, D.C. 20002	Yes		No	Yes	Yes	E		F		C	I	O		
Florida	Department of Pollution Control Jacksonville, Fla. 32304	Yes	Yes	Yes	Yes	App.	E	P	F	I	C	I	O	Yes	Yes
Georgia	Georgia Department of Public Health Atlanta, Ga. 30303	Yes	Yes	Yes	Yes	Yes	E			I	C		O	Yes	Yes
Hawaii	Division of Environmental Health Honolulu, Hawaii 96801	Yes	Yes	Yes	Yes	App.		P	F	I		I	O	Yes	
Idaho	Idaho State Health Department Boise, Idaho 83701	Yes		App.	App.	App.			F	I		I	O	Yes	Yes
Illinois	Illinois Environmental Protection Agency Springfield, Ill. 62706	Yes	Yes	Yes	App.	App.		P	F	I		I	O	Yes	Yes
Indiana	Indiana State Board of Health Indianapolis, Ind. 46206	Yes		No	App.	Lim.[e]				I			O	Yes	Yes
Iowa	Iowa State Department of Health Des Moines, Iowa 50319	Yes	Yes	Yes	Yes	Yes	E	P	F	I		I	O	Yes	
Kansas	Kansas State Department of Health Topeka, Kans. 66612		Yes	Yes	Yes	Yes	E	P	F	I		I	O	Yes	
Kentucky	Air Pollution Control Commission Frankfort, Ky. 40601	Yes	Yes	App.	Yes	Yes	E	P	F	I	C		O	Yes	
Louisiana	Louisiana State Department of Health New Orleans, La. 70160	Yes	Yes	Yes	App.	App.			F	I		I	O		

TABLE 3-1 (Continued)

State	Air Pollution Control Agency	AUTHORITY		AUTH. TO ESTABLISH			ENFORCEMENT CAPABILITY								
		Statutory	Regulatory	Control Regions	Ambient Air Standards	Emission Standards	Emergency	Permit	Fine	Inspection	Criminal	Injunctive	Orders	Local Option	Tax Incentive
Maine	Department of Environmental Protection Augusta, Me. 04330	Yes	lim	lim	lim	lim	E	P	F	I			O	Yes	Yes
Maryland	Maryland State Department of Health and Mental Hygiene Baltimore, Md. 21218	Yes	Yes	Yes	Yes	Yes	E						O	Yes	
Massachusetts	Department of Public Health Boston, Mass. 02133	Yes	Yes	Yes	App.	App.			F	I		I	O	Yes	Yes
Michigan	Michigan Department of Public Health Lansing, Mich. 48914	Yes	Yes	Yes	App.	App.	E							Yes	Yes
Minnesota	Minnesota Pollution Control Agency Minneapolis, Minn. 55440	Yes		Yes	Yes	Yes			F			I	O	Yes	Yes
Mississippi	Mississippi Air and Water Pollution Control Commission Jackson, Miss. 39205	Yes		Ques.	Yes	Yes	E	P	F	I	C	I	O	Yes	
Missouri	Missouri Air Conservation Commission Jefferson City, Mo. 65101	Yes	Yes	Yes	Yes	Yes	E		F	I		I	O	Yes	

Montana	Montana State Department of Health Helena, Mont. 59601	Yes	Yes	Yes	Yes	Yes	E	P	F	I	C	I	O		Yes
Nebraska	State Department of Environmental Control	Yes		Yes	Yes	Yes	E	P	F	I	C	I	O	Yes	Yes
Nevada	Bureau of Environmental Health Nevada Division of Health Reno, Nev. 89701	Yes		Ques.	App.	App.				I	C	I	O	Yes	
New Hampshire	Air Pollution Control Agency Concord, N. H. 03301	Yes	Yes	Ques.	App.	App.	E		F	I		I	O		Yes
New Jersey	Department of Environmental Protection Trenton, N. J. 08625	Yes	Yes	Ques.	App.	App.	E	P	F	I		I		Yes	Yes
New Mexico	Environmental Improvement Agency Santa Fe, N. M. 87501	Yes		Yes	App.	Yes	E					I		Yes	
New York	New York State Department of Environmental Conservation Albany, N. Y. 12205	Yes	Yes	Yes	Yes	Yes	E		F	I		I	O	Yes	Yes
N. Carolina	Department of Water and Air Resources Raleigh, N. C. 27603	Yes		App.	Yes	Yes	E	P	F	I	C	I	O	Yes	Yes
N. Dakota	North Dakota State Department of Health Bismarck, N. D. 58501	Lim.													
Ohio	Ohio Department of Health Columbus, Ohio 43216	Yes	Yes	Yes	Yes	Yes		P	F	I	?	I		Yes	Yes

TABLE 3-1 (Continued)

State	Air Pollution Control Agency	AUTHORITY		AUTH. TO ESTABLISH			ENFORCEMENT CAPABILITY								
		Statutory	Regulatory	Control Regions	Ambient Air Standards	Emission Standards	Emergency	Permit	Fine	Inspection	Criminal	Injunctive	Orders	Local Option	Tax Incentive
Oklahoma	Oklahoma State Department of Health Oklahoma City, Okla. 73105	Yes		Yes	App.	Yes					C			Yes	
Oregon	Department of Environmental Quality Portland, Ore. 97201	Yes	Yes	Yes	Yes	Yes	E			I	C	I		Yes	Yes
Pennsylvania	Department of Environmental Resources Harrisburg, Pa. 17120	Yes	Yes	Yes	App.	Yes			F		C	I		Yes	
Rhode Island	Division of Air Pollution Control Providence, R. I. 02903	Yes	Yes	Yes	Yes	App.	E	P	F	I	C	I	O		Yes
S. Carolina	South Carolina Pollution Control Authority Columbia, S. C. 29201	Yes	Yes	No	App.	App.		P	F	I	C	I	O	Yes	Yes
S. Dakota	South Dakota State Department of Health Pierre, S. D. 57501	Yes	Yes	No	No	No	?	P	F			I	?	Yes	
Tennessee	Tennessee Department of Public Health Nashville, Tenn. 37219	Yes		Ques.	Yes	Yes	E	P		I	C	I	O	Yes	
Texas	Texas State Department of Health Austin, Tex. 78756	Yes	Yes	Yes	App.	App.			F	I		I	O	Yes	

Utah	Utah State Division of Health Salt Lake City, Utah 84113	Yes		Yes	Yes	Yes			F	I	C	I	O	Yes	
Vermont	Agency of Environmental Conservation Barre, Vt. 05641	Yes		Yes	Yes	Yes	E		F	I			O		
Virginia	Air Pollution Control Board Richmond, Va. 23219	Yes	Yes	Yes	App.	App.			F	I		?		Yes	
Washington	Washington State Department of Ecology Seattle, Wash. 98504	Yes		Yes	Yes	Yes	E		F	I		I	O	Yes	Yes
W. Virginia	Air Pollution Control Commission Charleston, W. Va. 25311	Yes	Yes	Ques.	Ques.	App.	E		F	I		I	O	Yes	Yes
Wisconsin	Department of Natural Resources Madison, Wisc. 53705	Yes		Ques.	Ques.	Ques.	E			I			O	Yes	Yes
Wyoming	Department of Health and Social Services Cheyenne, Wyo. 82001	Yes	Yes	Ques.	Yes	Ques.									Yes
Guam	Sanitation Unit Agana, Guam 96910														
Puerto Rico	Air Pollution Control Section San Juan, P. R. 00908	Yes													
Virgin Islands	Bureau of Environmental Section St. Thomas, V. I. 00802	Yes		Yes	App.	Yes	E	P		I	C	I			

[a]Apparent [b]Question [c]Question [d]Specified in statute [e]Limited

and regulations include specific compliance schedules, the time allowed is generally from six months to three years. However, most state and local agencies can grant variances from compliance schedules.

Recent trends reveal a rapid replacement of concentration-emission standards with standards limiting total mass-emission rates on a schedule that requires increasing control with increasing size of source.

Control of process industries represents a real challenge in the design of equitable emission standards. The process-weight-rate concept, developed on the Pacific Coast is rapidly becoming the standard for this varied category of sources. The potential-emission-rate concept, developed more recently in the East, shows real promise for certain source types. Recent developments in control regulations include:

- Required elimination of all visible emissions.
- Emission standards for sulfur oxides from fuel combustion similar to those now used for particulate matter.
- Process-weight-rate and potential-emission-rate regulations for specific industry types for both particulate and gaseous pollutants.
- Mass-emission-rate standards that require application of modern fly-ash collectors to incinerators.
- Emission standards that limit the mass rate of emission of odors measurable by source sampling.

The approach being stimulated by the federal program is to establish emission standards on a regional basis after developing emission data on the number of sources in the region and their respective contributions to total regional emissions. This approach is discussed in detail subsequently.

Visible Emissions

The Ringelmann Chart, introduced in 1890 to regulate black smoke plumes, has been widely accepted. In 1947, the Los Angeles Air Pollution Control District (LAAPCD) originated the "equivalent opacity" concept which extended the Ringelmann Chart for application to visible plumes of any color which obscure the view of the observer to the same degree as black smoke. In 1948 the concept was incorporated in the California Health and Safety Code.

The concept has now spread throughout the nation to almost all major urban areas. The legality of the concept has received frequent challenge although upheld in the courts.[36]

Equivalent opacity regulations are especially useful for the surveillance of source installations without having to sample the source. Despite its usefulness, a number of technical questions have arisen concerning the validity of equivalent opacity.[37]

First, the visibility of a plume is largely determined by particle size rather than total weight. The optimum particle size for scattering light and producing a dark plume is 0.1–1.0μ. A high-collection efficiency may allow a visible plume although many small submicron particles are emitted. Yet, such particles remain suspended in the atmosphere for long periods of time and, during inversions, accumulate to cause severe visibility reduction and soiling of buildings and materials. These small particles are also inhaled by man and can be retained in the lower respiratory tract.

Second, concern is expressed regarding the reproducibility of equivalent opacity readings of emission plumes. Common objections are that opacity varies with the position of the observer relative to the sun, atmospheric lighting, and background. These sources of error also apply to observation of black plumes, but even the strongest opponents of pollution control have accepted the desirability of controlling smoke emissions. Observers can be taught to compensate for these variables and reproduce readings within 10 percent of actual plume transmittance.[38]

Third, there is a question as to the methods for complying with equivalent opacity compared with smoke regulations. When smoke is the offending agent, control can be achieved by improved combustion efficiency. When plume visibility is due to the emission of fine fly ash from fuel combustion or to fumes from metallurgical processes, control must then be achieved by use of collection equipment. Collection of submicron particles requires highly efficient devices such as baghouses, high-energy scrubbers, and high-efficiency electrostatic precipitators. Collection sufficient for compliance with mass-emission-rate standards may not be sufficient for compliance with equivalent opacity standards.

Mass concentration can be related to plume transmittance for specific particle sizes and types, and plume thickness. Conner and Hodkinson[39] demonstrated a close correlation between plume transmittance and mass concentration for oil particles by calculation and measurement. Other relationships have been published for different types of particles and sources.[40,41]

Equipment manufacturers make use of such existing data, however limited, to design control equipment to opacity requirements. This practice has, of necessity, depended primarily on the vendor's experience with similar installations on specific source types rather than on theoretical relationships. Correlation of particle size and concentration data with plume visibility for additional sources is needed to aid designers in eliminating offensively visible plumes. Many new industrial plants install equipment for purposes of eliminating all visible plumes, even if not required to do so.

Most visible emission standards are based on either No. 2 Ringelmann or its equivalent opacity (40 percent). Currently the trend is to require all incinerators and new sources to meet Ringelmann 1 (20 percent). Improved equipment designs have made it possible for many new sources to eliminate all visible discharges. Such an emission standard, Ringelmann 0, has been proposed by New Jersey.

Because mass-emission standards are unrelated to particle size, they are not always effective in eliminating visible plumes. The use of standards involving visible emissions is the only practical means for controlling submicron particles until measurement techniques and emission standards that limit the number of discharged particles according to size are developed.

Particulate Emission Standards

Some areas have actually replaced the Ringelmann standard by particulate-emission standards. Generally, preference is for "process weight" standards, because particulate concentration standards can be circumvented by diluting the effluent gas stream with excess air.

Under state statute, 28 States have adopted particulate emission standards for industrial processes and fuel combustion sources. Thirty have adopted emission or design standards for incinerators.

At the local and regional level, 97 agencies have adopted particulate emission standards for industrial process sources, 120 for fuel combustion sources, and 107 for incinerator design and operation.

The most widely accepted particulate concentration standard for fuel-burning equipment is that recommended by the American Society for Mechanical Engineers in 1949, and revised in 1966. The original 1949 model code,[42] limits emissions to 0.85 lb of dust per thousand pounds of flue gas, corrected to 50 percent excess air. The collection-efficiency requirements vary from about 50 to 85 percent, depending on the type of equipment used to burn coal with 10 percent ash and 13,000 Btu/lb.

The 1966 model code[43] limits the mass-emission rate of particulate matter rather than the in-stack concentration used in the 1949 model. A November 1968 revision recommended that particulate emissions vary with stack height, plant capacity, and air quality objectives (Fig. 3-1). This new model requires a varying degree

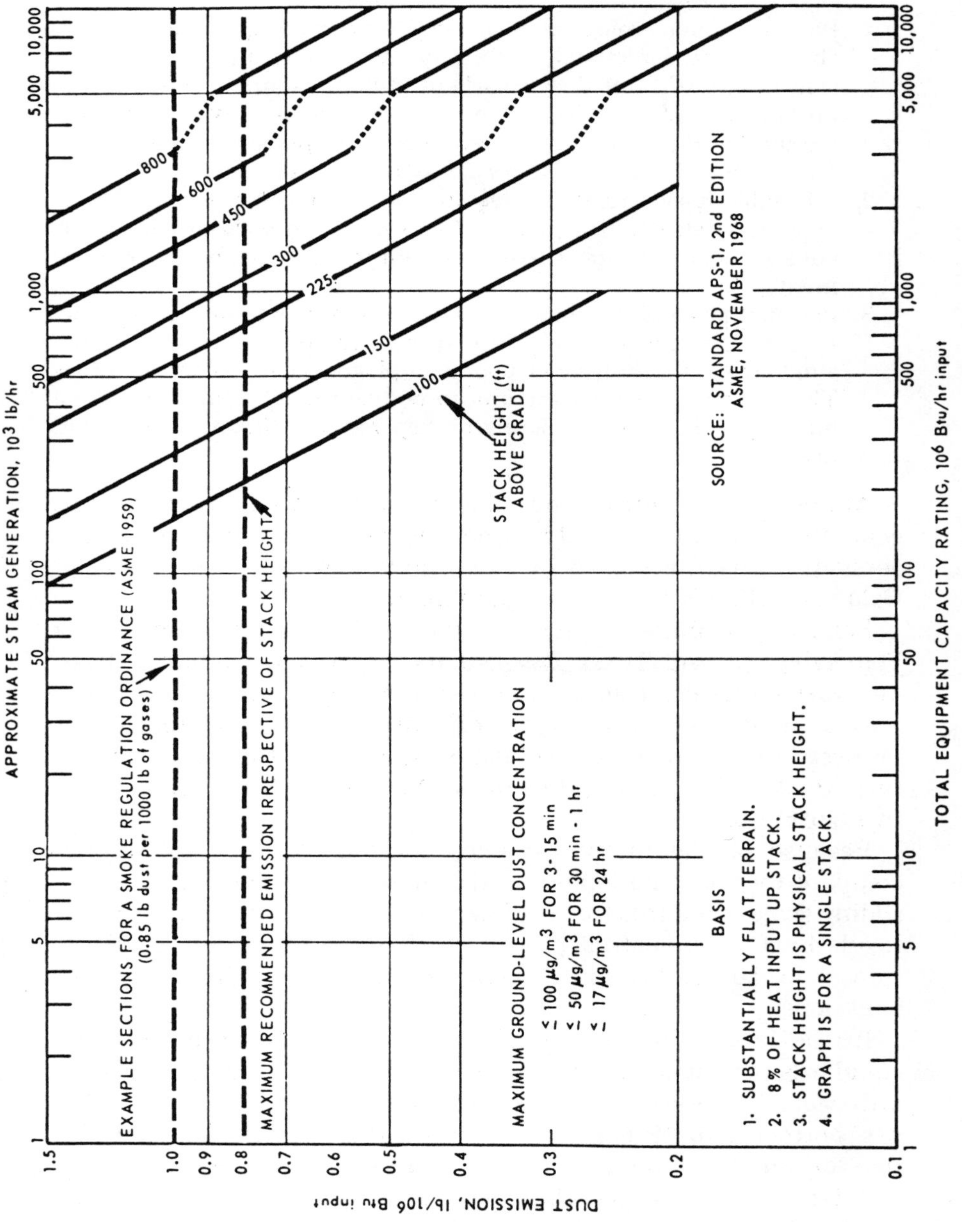

Fig. 3-1 Emission limits for selected stack heights.

of control dependent on plant size and stack height. The standard has received limited acceptance due to the following deficiencies:

1. Based on meteorological dispersion equations applicable only to single-source emissions located on essentially flat terrains, the standard is of limited value where multiple sources exist in close proximity (urban areas) or in mountainous or hilly regions. Also maximum allowable ground level concentrations are based on the "critical wind speed" without consideration for inversion.
2. Allowable mass-emission rates are determined by stack height. Increased stack height can be used to meet the standard in lieu of emission control, and there is no minimum stack-height requirement.
3. Because particulate emission standards are determined by stack height they are generally less restrictive for new plants with a single stack than for older plants employing many short stacks.
4. The standard does not attempt to reduce the total quantity of pollutants discharged but merely to disperse the effluent over a wider area.

Although the control requirements of the ASME Standard are generally lenient compared to other modern regulations, a modified version has been accepted by a few state and local agencies (e.g., Delaware, Illinois, and South Carolina).

Following a similar approach to the ASME Standard, New Jersey, Pennsylvania, and Texas have adopted emission standards based on expected levels of air quality. Using dispersion equations, allowance is made for stack height (effective), distance to the nearest property line, and, in some instances, particle size and pollutant toxicity. A typical standard from the New Jersey Standard is shown in Figure 3-2.

Because of the many variables considered, it is not easy to compare these regulations with the more conventional types (e.g., sliding-scale standards for fuel-burning equipment or process weight regulations for industrial processes). The basic concept is dispersion or broad distribution of pollutants rather than limitations on emissions.

In the late 1940s Los Angeles adopted concentration emission standards for particulate emissions from incinerators. These standards were expressed as grains per standard cubic foot of dry flue gas, corrected to 12 percent carbon dioxide, without the contribution of auxiliary fuel. In 1960, the LAAPCD published design standards for multiple chamber incinerators,[44] having banned single chamber incinerators in 1957. Other areas have banned single chamber incinerators along with open burning. The federal government adopted similar standards in 1966 for smaller incin-

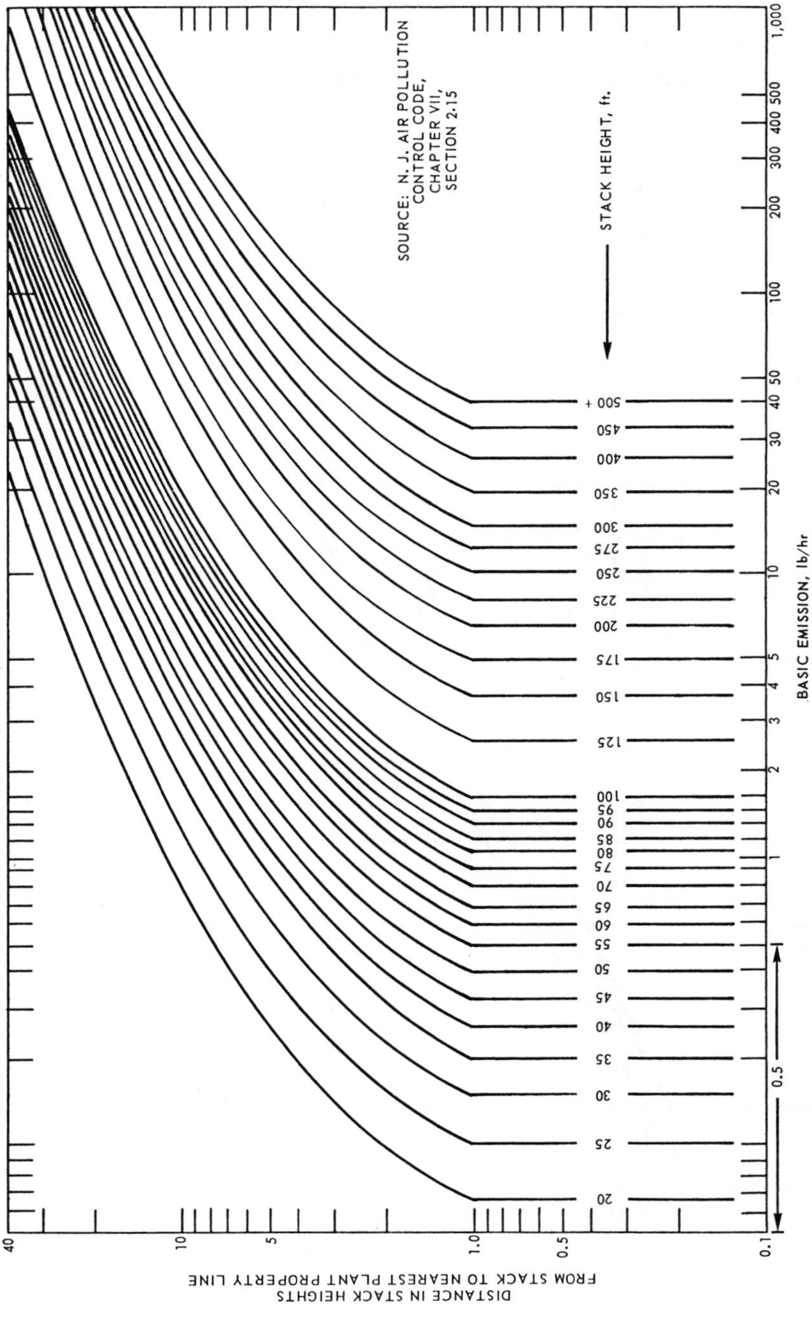

Fig. 3-2 New Jersey Standard for coarse solid particles.

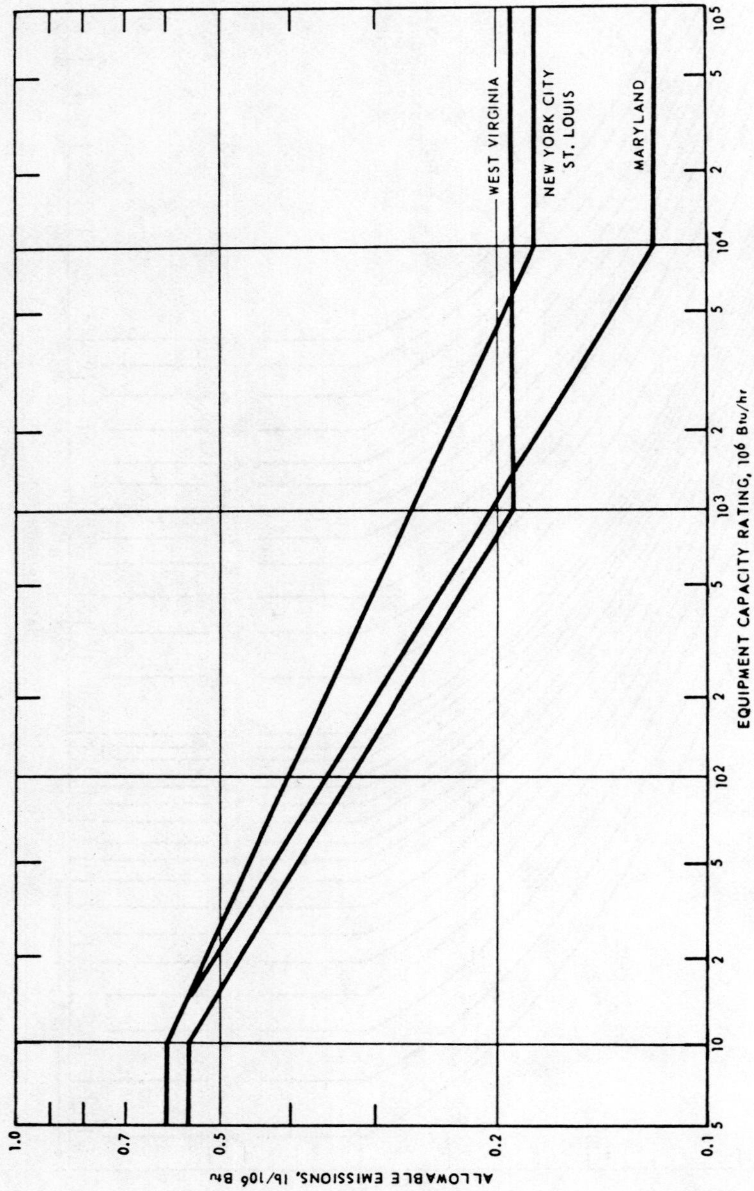

Fig. 3-3 Particulate matter standards for fuel-burning equipment.

erators (less than 200 lb/hr) at federal installations and required all larger government units to meet a more restrictive standard. Subsequently the federal government developed an interum guide for good practice for incinerators operated at federal facilities.[45] Based on the Los Angeles design standards, the purpose of the guide is to recommend incinerator designs that will operate in compliance with emission standards contained in the Code of Federal Regulations. Many state and local air pollution control agencies have adopted these two standards for fuel-burning installations.

Many concentration standards for fuel-burning equipment are rapidly being replaced with "sliding-scale" particulate emission standards such as those adopted by New York City in 1964. Three of the more restrictive of these are shown in Fig. 3-3. The sliding-scale standards indirectly restrict mass-emission rate. Thus, as plant size increases, total allowable particulate emissions also increase, but at a continually decreasing rate. Also, the sliding-scale standard eliminates the need to standardize flue gas volume.

TABLE 3-2 REQUIRED COLLECTION EFFICIENCIES FOR FUEL-BURNING INSTALLATIONS[a]

Unit	Size 10^6 B.t.u. per hour	REQUIRED COLLECTION EFFICIENCY (PERCENT) West Virginia	Maryland et al.	New York City, et al.
Underfeed	10	70.8	68.8	68.8
Traveling grate	50	80.2	78.6	76.0
Spreader stoker	50	92.3	91.9	90.9
	50[b]	95.0	94.6	94.0
	100	93.3	93.0	92.0
	100[b]	95.7	95.4	94.8
Cyclone	500	88.5	87.4	84.3
	5,000	90.1	92.6	89.5
	10,000	90.1	93.7	90.6
Pulverized	500	96.4	96.0	95.2
	5,000	96.9	97.7	96.8
	10,000	96.9	98.1	97.1

[a]Basis: 10 percent ash and 13,000 B.t.u. per pound.
[b]Reinjection.

Because it is common practice to compare collector performance using collection efficiency rather than total pollutant-emission rate, the collection-efficiency requirements for each of the three standards are presented in Table 3-2 for various types and sizes of equipment, based on coal with an average heat content of 13,000

Btu/lb and an ash content of 10 percent. The efficiency requirements of a sliding-scale standard increase with increasing size of the installation and also with increasing emission potential of the source.

Mass-emission rate standards also are replacing concentration standards as a meaningful method for controlling total particulate emission rates in a control region. Mass-emission-rate standards for industrial operations originated in Los Angeles County in the late 1940s. In the late 1950s a "process weight" standard was adopted by the San Francisco Bay Area Pollution Control District. This regulation, applied to all industrial process sources, limits the weight of particulate emissions per hour as a function of the total weight of raw materials introduced into a process operation (Table 3-3).

This modified regulation is now widely accepted by state and local air pollution control agencies. It is important to recognize that both the Los Angeles and Bay Area regulations were developed

TABLE 3-3 ALLOWABLE RATE OF PARTICULATE EMISSION BASED ON PROCESS WEIGHT RATE[a]

PROCESS WEIGHT RATE		*Rate of Emission,*	PROCESS WEIGHT RATE		*Rate of Emission,*
Lbs/hr	*Tons/hr*	*lbs/hr*	*Lbs/hr*	*Tons/hr*	*lbs/hr*
100	0.05	0.551	16,000	8	16.5
200	0.10	0.877	18,000	9	17.9
400	0.20	1.40	20,000	10	19.2
600	0.30	1.83	30,000	15	25.2
800	0.40	2.22	40,000	20	30.5
1,000	0.50	2.58	50,000	25	35.4
1,500	0.75	3.38	60,000	30	40.0
2,000	1.00	4.10	70,000	35	41.3
2,500	1.25	4.76	80,000	40	42.5
3,000	1.50	5.38	90,000	45	43.6
3,500	1.75	5.96	100,000	50	44.6
4,000	2.00	6.52	120,000	60	46.3
5,000	2.50	7.58	140,000	70	47.8
6,000	3.00	8.56	160,000	80	49.0
7,000	3.50	9.49	200,000	100	51.2
8,000	4.00	10.4	1,000,000	500	69.0
9,000	4.50	11.2	2,000,000	1,000	77.6
10,000	5.00	12.0	6,000,000	3,000	92.7
12,000	6.00	13.6			

[a]Data in this table can be interpolated for process weight rates up to 60,000 lb/hr by using equation: $E = 4.10\ P^{0.67}$; and can be interpolated and extrapolated for process weight rates in excess of 60,000 lb/hr by using equation: $E = 55.0\ P^{0.11} - 40$ (E = rate of emission in lb/hr; P = process weight rate in tons/hr).

for their specific areas (control regions). Also, both reflect the state-of-the-art of control technology; for Los Angeles, in the late 40s; for the Bay Area, the late 50s.

In 1967 New York State and New York City adopted mass-emission rate standards. The regulations adopted by New York City, prior to the Air Quality Act, distinguish between existing and new installations (Fig. 3-4). The New York and Pennsylvania regulations limit mass-emission rates as a function of potential emission rates and they are not limited to either particulates or industrial sources. The potential emission rate is defined as that quantity of a pollutant that theoretically would be emitted if no controls whatsoever were used. Measurement of the collection efficiency of specific source installations determines the allowable emission rate. This type regulation is relatively new and has not been tested in urban areas, but it appears promising for source categories in which the potential emission rate can be easily determined.

Pennsylvania has a variable standard. For major cities there is a potential-emission-rate standard (Fig. 3-5), but for rural areas Pennsylvania adopted property line and stack height requirements similar to the ASME Standard. Consideration is also given to pollutant toxicity, in which case lower emission limits are required.

Sulfur Oxide Emission Standards

Gaseous emission standards are controlled in very few instances and there is no single sulfur oxide emission standard that has received national acceptance. Six states have standards or regulations applicable to sulfur oxide emissions from fuel combustion sources. At the local or regional level 20 agencies have adopted standards or regulations. These generally impose limitations on the sulfur content of fuels (see subsequent section) but allow the use of stack-gas desulfurization equipment if it will achieve an equivalent degree of sulfur oxide control.

The control of industrial gaseous emissions for the most part has been limited to sulfur oxides. Regulations involve emission standards and property line concentration standards. Among the first sulfur oxide emission standards for industrial sources were those in Los Angeles County and the San Francisco Bay Area. The standard limits total sulfur oxides (sulfur dioxide, sulfur trioxide, and sulfuric acid aerosols) to 2,000 ppm (5.33 g/m^3) per unit volume of exhaust gas, calculated as sulfur dioxide. The Los Angeles sulfur oxide emission standard amounts to 50 g/100 ft^3 of gaseous fuel of sulfur compounds (calculated as hydrogen sulfide

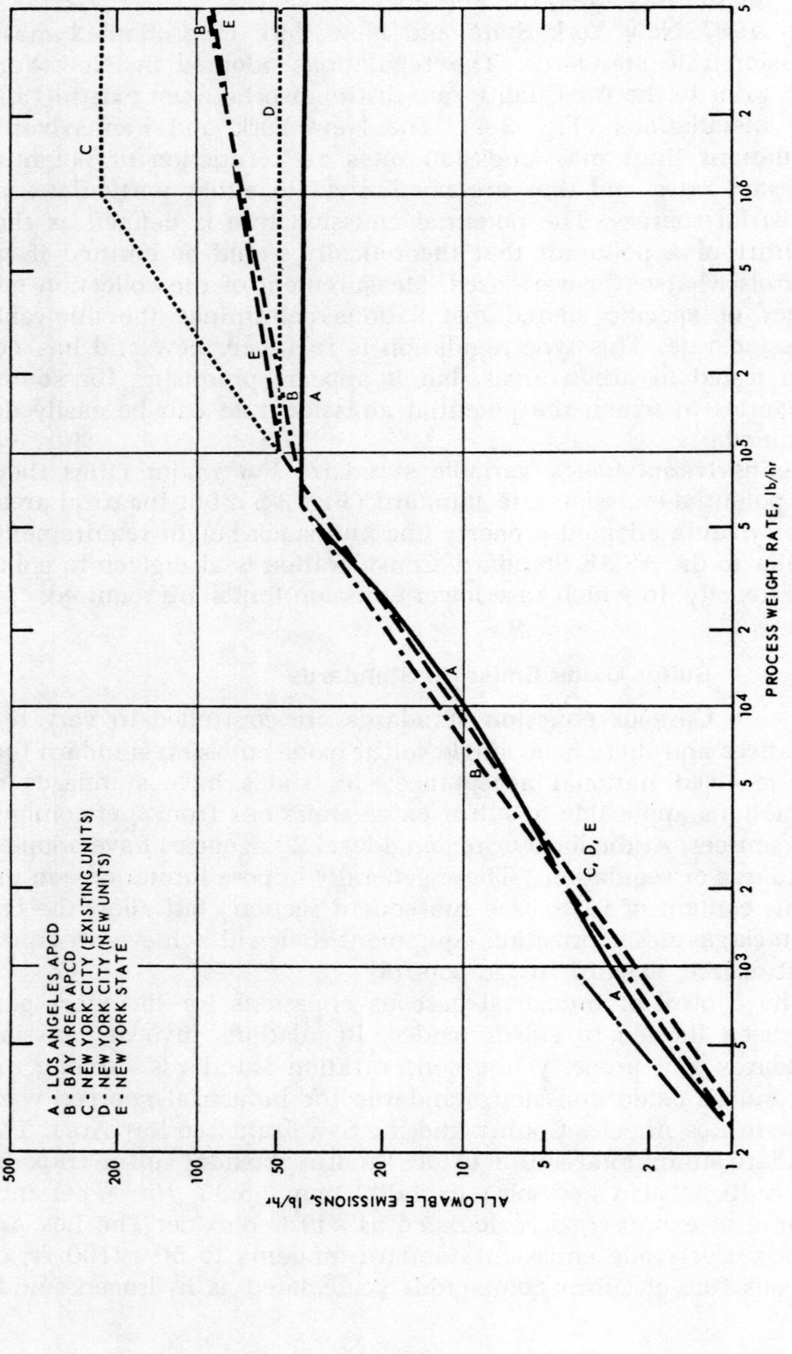

Fig. 3-4 Particulate emission regulations based on process weight rate.

at standard conditions). These areas burn very little coal, relying on natural gas and fuel oil. The Los Angeles standard is directed at petroleum refineries and by-product coke oven plants.

In 1966 St. Louis established concentration standards for sulfur dioxide and sulfur trioxide (or acid mist). Different standards were established for existing and new sources. That for existing sources is based on the Los Angeles standard. The standard for new sources was based upon reported performance of European sulfuric acid plants employing the double-contact process.

New installations are required to meet a sulfur dioxide emission standard of 500 ppm (1.33 g/m^3) and a sulfur trioxide standard of 70 μg/m^3.

For fuel-burning installations St. Louis adopted a single sulfur dioxide standard of 2.3 lb/million Btu input, for installations larger than 2,000 million Btu/hr. This standard was based upon regional air quality needs, and is the equivalent of approximately 1.4 percent sulfur coal. Smaller sources are regulated by fuel standards rather than emission standards.

Similar sulfur oxide emission standards in St. Louis have been adopted by many other control agencies. Thus, their deficiencies have been perpetuated.[36] They originated from studies on a single-source category (sulfuric acid plants) and hence have questionable applicability when applied to other major sulfur oxide sources (e.g., smelters and petroleum refineries). Also, they limit pollutant concentration rather than mass-emission rate and, thus are subject to circumvention by dilution.

A similar concentration standard was recommended by the New York–New Jersey Abatement Conference for new power plants and all nonpower generation sources, the specific standard being 0.35 lb of sulfur dioxide per million Btu input, equivalent to coal with a sulphur content of 0.2 percent and oil of about 0.3 percent. Federal regulations for U.S. government facilities in the Chicago and Philadelphia Air Quality Control Regions limit emissions to 0.65 lb of sulfur dioxide per million Btu input.

Specialized Standards

Attempts to develop generalized process weight regulations has recently led to development of a few specialized regulations: for asphalt plants (West Virginia) and ferrous foundries (New York State).

The states of Pennsylvania and New York have developed emission standards which limit the mass-emission rate as a function of the potential emission rate. The Pennsylvania standard (Fig. 3-5)

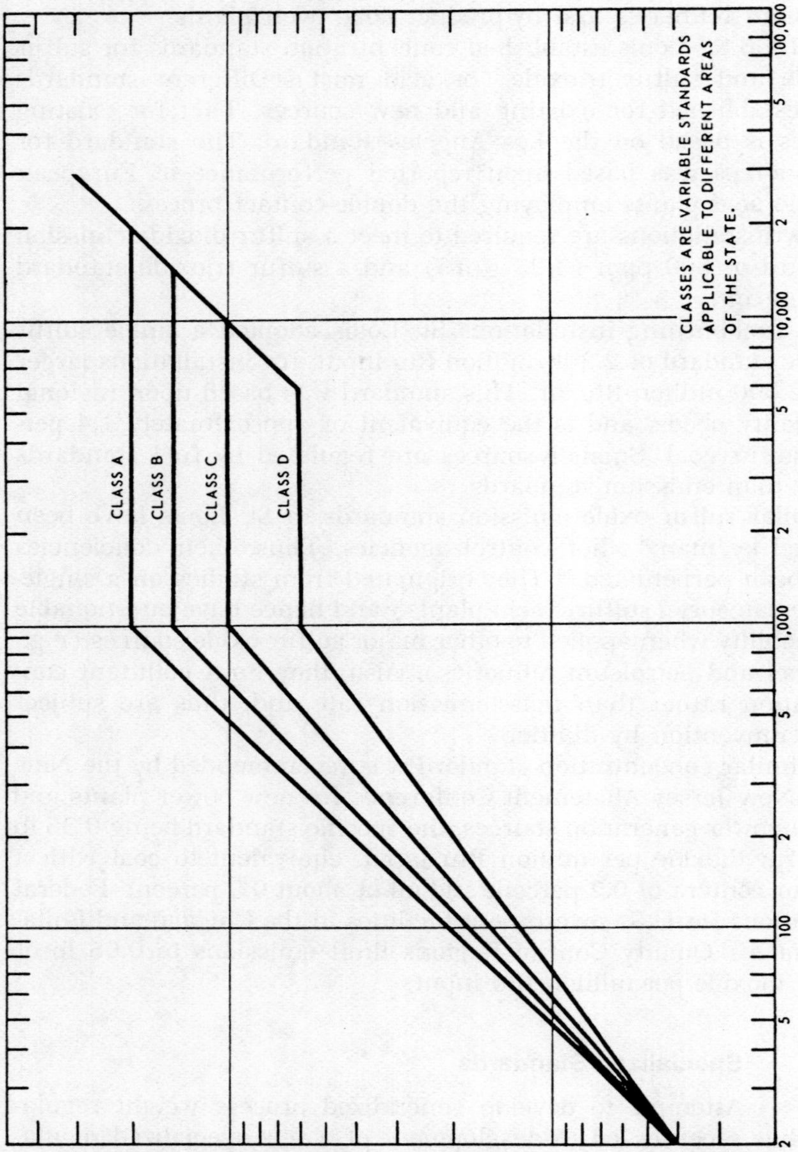

Fig. 3-5 Pennsylvania potential-emission-rate standard.

sets different standards for different area types. The regulation is readily adaptable to sources where uncontrolled emission rates are easily determined (e.g., primary smelters and sulfuric acid plants).

A specialized emission standard was established by the state of Florida to cover fluoride emissions. That state requires all phosphoric acid production plants built after 1965 to discharge less than 0.4 lb of fluoride per ton of phosphorus pentoxide design capacity. Plants existing prior to 1965 are limited to 0.6 lb of fluoride per ton of phosphorus pentoxide manufacturing capacity.

Difficulties have arisen in the application of potential emission rate standards to many industrial operations discharging particulate matter. For many operations uncontrolled rates bear no direct relationship to the amount of feed material. In these instances the potential emission rate must be determined by sampling effluent gases uncontrolled and controlled.

Since potential emissions for these sources depend upon the amount of material entrained in the exhaust gases, an alternative approach for these operations would be to assign potential emission rates based upon preestablished emission factors. Ideally, emission factors represent average measured emission rates from a number of similar installations. Therefore, such factors can serve as an equitable interim substitute for determining individual source potential emission rates until better methods are developed.

Although there will be a continued need for generalized regulations, in areas containing a significant number of similar industry types, there is a need for tailored regulations that reflect attainable emission rates for particular sources. Tailored regulations also may be based on process weight or potential emission rate.

Material Standards

Among the alternatives to emission control is the regulation of the quality of fuel or raw materials entering a process. An advantage of fuel or process material standards is that it can be enforced without time-consuming source sampling. When large numbers of sources are involved, this may well be the most practical means of enforcement. To date, this approach has centered on fuels and solvents.

Opponents will argue that fuels regulation is not the appropriate function of air pollution control agencies and that a fuel standard may not be related to available control techniques for reducing emissions. The Air Quality Act of 1967, however, provided a regulatory mechanism which required consideration of low-sulfur fuels in critical areas where necessary to provide for the protection of public health. Most fuel and process material standards allow for the use of any fuel or material if equivalent or lower emission rates are achieved by other methods, however.

The first fuel standard was adopted by St. Louis in 1937 and required the washing of mined coals which contain more than 2.0 percent sulfur or 12 percent ash. The most widely employed fuel standards relate to sulfur content and are summarized[46] in Table 3-4.

In 1958, Los Angeles County limited the sulfur content of liquid and gaseous fuels. Since 1964, several cities and states, and the federal government (federal facilities in New York, Chicago, and Philadelphia) have adopted regulations governing the sulfur content of fuels. In the District of Columbia, for example, a limitation of 1 percent took effect in July 1969. In some places, progressively more stringent limitations will take effect on specified future dates; in New Jersey, for example, a 1 percent limitation on sulfur content of bituminous coal took effect in May 1968, and a limitation of 0.2 percent in October 1971.

Regulations of maximum allowable sulfur content usually carry an alternative provision whereby any fuel may be used if flue-gas desulfurization can be shown to result in an equivalent or lower rate of sulfur oxide emission, as measured in pounds of sulfur oxides per million Btu. The "emission standard" is obtained by direct conversion from sulfur content of fuel and is based on air quality considerations.

Regulations also have been established for photochemically reactive solvents in such areas as Los Angeles and San Francisco. Such regulations even restrict the use of these in architectural coatings or their disposal by evaporation to the atmosphere.

For industrial processes that use solvents, Los Angeles County has separate emission standards for two classes of operations; those which heat solvents, and those which use specific photochemically reactive solvents. Any process that heats solvents either must limit the total discharge of organic materials to 15 lb/day or treat the exhaust gases to remove 85 percent of all organic materials discharged. Any process using specific photochemical solvents must either discharge 40 lb or less per day of organic materials or treat the effluent to remove 85 percent of the organic materials discharged. The San Francisco BAAPCB has a similar regulation.

Ambient Air Quality Standards

As of May 15, 1970, a number of states have adopted or proposed ambient air quality standards[46] for areas other than air quality control regions, designated pursuant to the Federal Air

Quality Act of 1967. These ambient air quality standards are based upon what are considered the most meaningful measurements of air pollution levels.

Ambient air quality standards for particulates[46] are given in Table 3-5 but particulate standards for settleable particulate matter (dust fall) are not shown since they are neither a meaningful index of air pollution nor widely employed. Ambient air quality standards for sulfur oxides are given in Table 3-6. Both particulate and sulfur oxide emission standards also have been officially adopted for federally established air quality control regions.

Ambient air quality standards for carbon monoxide, hydrocarbons, oxidents, and other pollutants exist in several states[46] and are presented in Table 3-7.

Receptor Standards

In contrast to ambient air quality standards, receptor standards are geared to objectionable ground level concentrations. Application is usually to the identification of nuisances such as odors, and in a few instances to pollutants causing health effects, such as fluoride compounds.

Odors. Odors constitute the most perplexing and often the most objectionable air pollution problem. They are caused by a variety of substances, many of which are detectable at trace concentrations below 1 ppb. There are many cases in which odorous substances cannot be detected by normal chemical analysis, but are detectable by the sense of smell. The human nose is, of necessity, the present standard for determining odor intensity in the ambient air and in source effluents.

The problem of odors arises in connection with the waste products of many industrial operations. Odors from all these sources generate a high level of public concern. This concern with the olfactory effects of industrial wastes is of ancient origin. Theophrastus, a pupil of Aristotle, wrote a treatise on stones about 271 B.C. in which he called attention to the objectionable odors from the combustion of coal. The odors and soot from coal smoke have chased at least two English monarchs from London. The first to go was Queen Eleanor in 1257; William III moved out about 400 years later in 1690. Queen Elizabeth I could not abide coal smoke, and Edward I became so unhappy over the smoke from London furnaces that he threatened severe penalties for any who might substitute coal for wood. The ancient notion that diseases came from "bad airs" suggests that odors and air pollution in general have long been conspicuous problems.

TABLE 3-4 STATE AND LOCAL SULFUR OXIDES CONTROL REGULATIONS APPLICABLE TO FUEL COMBUSTION SOURCES

		COAL		OIL	
State	*Area*	*Maximum Sulfur Content of Fuel*	*Effective Date*	*Maximum Sulfur Content of Fuel*	*Effective Date*
California	Monterey/Santa Cruz Co.	0.5%	7-1-69	Same as coal	
	Los Angeles Co.	0.5%	5-1-59	Same as coal	
Connecticut	Statewide	1.0%	9-1-71	Same as coal	
District of Columbia		1.0%	7-1-69	Same as coal	
Florida	Jacksonville	2.0%	10-1-68	Same as coal	
		1.5%	1-1-69	Same as coal	
		1.0%	1-1-70	Same as coal	
	Manatee Co.			2.0%	12-3-68
Illinois	Chicago domestic and commercial: existing	2.0%	7-1-70	Same as coal	
		1.25%	7-1-71	Same as coal	
		1.0%	9-1-72	Same as coal	
	New	1.0%	7-1-70	Same as coal	
	Industrial	2.0%	7-1-70	Same as coal	
		1.5%	11-1-70	Same as coal	
		1.0%	9-1-72	Same as coal	
	Utilities	1.8%	5-16-70	Same as coal	
		1.0%	1-1-72	Same as coal	

Maryland	Western area	1.0% [a]	7-1-70	1.0%	7-1-70
	Central area	1.0% [b]	7-1-70	1.0%	7-1-70
	Baltimore area	1.0%	7-1-70	1.0%	7-1-70
	Washington area	1.0%	7-1-69	1.0%	7-1-69
	Southern area	1.0% [c]	7-1-70	1.0%	7-1-70
	Eastern shore	1.0% [c]	7-10-70	1.0%	7-1-70
Massachusetts	Boston area				
	A. Inner cities and towns	0.55 lb S/10^6 Btu	10-1-70	Same as coal	
	B. Outer cities and towns	1.22 lb S/10^6 Btu	10-1-70	Same as coal	
	C. All places during adverse meteorological conditions	0.55 lb S/10^6 Btu	10-1-70	No. 2 0.17 lb/10^6 Btu 0.55 lb S/10^6 Btu	10-1-70 10-1-70
Minnesota	Minneapolis-St. Paul	2.3 lb S/10^6 Btu [d]	4-1-72	Same as coal	
	Alert levels [f]	2.0% [e]	4-1-70	Same as coal	
Missouri	St. Louis area	2.0% [e] 2.3 lb SO_2/10^6 Btu [d]	7-4-67 7-4-67	Same as coal Same as coal	
Montana	Statewide	2.0% 1.5% 1.0%	7-1-70 7-1-71 7-1-72	Same as coal Same as coal Same as coal	
Nevada	Reno, Sparks and Washoe Co.			1.2% 1.0%	9-18-69 7-1-70
New Hampshire	Statewide			0.5% (No. 2 oil)	10-1-70
	A. Existing sources	2.8 lb S/10^6 Btu [f] 2.0 lb S/10^6 Btu [g]	10-1-70	0.4% (No. 2 oil) 1.5% (No. 4 & 5 oil) 1.25% (No. 4 & 5 oil) 1.0% (No. 4 & 5 oil)	10-1-72 10-1-70 10-1-71 10-1-72

TABLE 3-4 (Continued)

State	Area	COAL Maximum Sulfur Content of Fuel	COAL Effective Date	OIL Maximum Sulfur Content of Fuel	OIL Effective Date
	B. New sources	1.5 lb $S/10^6$ Btu[f]	10-1-70	2.0% (No. 6 oil)	10-1-70
		1.0 lb $S/10^6$ Btu[g]		1.5% (No. 6 oil)	10-1-71
				1.25% (No. 6 oil)	10-1-72
				1.0% (No. 6 oil)	10-1-70
New Jersey	Statewide	1.0 (Bituminous)	5-6-68	0.3 (No. 2 oil)	5-1-68
		0.2 (Bituminous)	10-1-71	0.3 (No. 2 oil)	10-1-70
				0.2 (No. 2 oil)	10-1-71
		0.7 (Anthracite)	5-6-68	0.7 (No. 4 oil)	5-1-68
		0.2 (Anthracite)	10-1-71	0.4 (No. 4 oil)	10-1-70
				0.3 (No. 4 oil)	10-1-71
				1.0 (No. 5 & 6 oil)	5-1-68
				0.5 (No. 5 & 6 oil)	10-1-70
				0.3 (No. 5 & 6 oil)	10-1-71
New York	N.Y. City area				
	A. Utilities	1.0%	10-1-69	Same as coal	
	B. Other	1.0%	10-1-69	Same as coal	
		0.2 lb $S/10^6$ Btu	10-1-71	Same as coal	
Ohio	Cincinnati	1.25%	1-22-69	Same as coal	
	Cleveland				
	A. Existing sources	2.0%	10-15-69	Same as coal	
	B. New sources	1.0%	10-15-69	Same as coal	
	Steubenville	2.0%	10-22-68		
Pennsylvania	Allegheny Co.	2.5 lb $SO_2/10^6$ Btu	12-17-69	Same as coal	
	Philadelphia	Bituminous: 2.0	5-1-70	0.3 (No. 2 oil)	5-1-70
		1.0	7-1-71	0.3 (No. 2 oil)	10-1-72
		0.3	10-1-72	0.2 (No. 2 oil)	10-1-73
				0.7 (No. 4 oil)	5-1-70

		Anthracite: 0.7	5-1-70	0.4 (No. 4 oil)	10-1-72
		0.3	10-1-72	0.3 (No. 4 oil)	10-1-73
				1.0 (No. 5 oil)	5-1-70
				0.5 (No. 5 oil)	10-1-72
				0.3 (No. 5 oil)	10-1-73
				No. 6 oil same as No. 5	
Tennessee	Chattanooga	2.0%	3-4-69	Same as coal	
	Nashville	2.0%	3-1-70	Same as coal	
Utah	Statewide	1.5%[h]		Same as coal	
Virginia	Falls Church			1.0%	1-1-70
	Fairfax Co.			1.5%	7-1-69
	Alexandria	1.0%	7-1-69	Same as coal	
	Arlington Co.			1.0%	7-1-69

Note: Percentages refer to maximum allowable amount of sulfur in fuel, by weight, except as follows: where limitations are expressed as lb. $SO_2/10^6$ Btu (as in St. Louis area and Allegheny Co.) they refer to allowable sulfur oxide emissions, by weight, per million Btu of heat value of the fuel. Limitations applicable to all fuel combustion sources except as indicated in following footnotes.

[a]Only equipment with capacity greater than 100×10^6 Btu/hr.
[b]Only equipment with capacity greater than 500×10^6 Btu/hr.
[c]Only equipment with capacity less than 500×10^6 Btu/hr.
[d]Only equipment with capacity greater than 2×10^9 Btu/hr.
[e]Only equipment with capacity less than 2×10^9 Btu/hr.
[f]Maximum allowable sulfur content.
[g]Maximum three-month average sulfur content.
[h]Proposed.

TABLE 3-5 STATE AND LOCAL SUSPENDED PARTICULATE AMBIENT AIR QUALITY STANDARDS[46] ($\mu g/m^3$)

Political Jurisdiction	Area Type	Annual Geo. Mean (24 hr)	Annual Arith. Mean	3 mo. max.	1 mo. max.	1 wk. max.	24 HOUR max.	24 HOUR 16%	24 HOUR 10%	24 HOUR 5%	24 HOUR 1%	1 hr max.
Alabama												
Huntsville	Quarrying area						150[e]					
Arkansas		75							140			
California		60					100					
Colorado				90								
Delaware		70					200					500
Florida												
Manatee Co.	Rural	60										
	Residential	75										
	Commercial	95										
	Industrial	125										
Idaho	Residential								125			
	Commercial								150			
	Industrial								175			
	Agricultural								150			
	Recreational								100			
Kentucky		65					220				180	
Massachusetts		75					180					
Minnesota		75									200	
Montana		75										200

New Jersey		65				195			
New Mexico		60		90	110	150			
New York	Level I	45				200	70		
	Level II	55				200	85		
	Level III	65				200	100		
	Level IV	75				200	110		
Nevada, Reno—Sparks and Washoe Co's.	Residential		50						150(a)
	Res.-commercial		150						250(a)
	Comm.-industrial		150						350(a)
North Carolina		60				210			
Ohio		65				260			200
Pennsylvania		65				195			
South Carolina	Industrial	125						200	
	Residential	75						150	
Charleston Co.	Industrial	125						200	
	Residential	75						150	
Tennessee	Industrial	100(g)					200(h)		
	Commercial	75(g)					150(h)		
	Residential	60(g)					120(h)		
	Recreational	40(g)					80(h)		
Nashville, Davidson Co.	Industrial					200			
	General	50							
Utah		90							150(i)
Virginia								200(c)	200(c)
	Residential-commercial						75(f)		
	All other						150(g)		

TABLE 3-5 (Continued)

Political Jurisdiction	Area Type	Annual Geo. Mean (24 hr)	Annual Arith. Mean	3 mo. max.	1 mo. max.	1 wk. max.	24 HOUR max.	24 HOUR 16%	24 HOUR 10%	24 HOUR 5%	24 HOUR 1%	1 hr max.
Washington[d]		60						100				
West Virginia Kanawaha Valley		100									250	
Wyoming		75									200	
Texas	Residential and recreational								125			
	Commercial								150			
	Industrial								175			
	Industrial Type D								200			

[a] No 24-hour sample to exceed the specified level more than 1% of the days in any 3-mo. period.
[b] Additional standard for visibility-reducing particle states in sufficient concentration to reduce visibility to 10 miles at relative humidity of less than 70%.
[c] Not more than 1% of samples collected between April 1 and Oct. 31, nor more than 4% of the samples collected between Nov. 1 and March 31 shall exceed a concentration of 200 $\mu g/m^3$.
[d] In recognition of natural dust loading in some areas of the state exceptions are made for areas east of the Cascade Mountain crest.
[e] Above background.
[f] Plus 50 $\mu g/m^3$ background.
[g] Not to be exceeded more than 50% of the time in 30 days.
[h] Not to be exceeded more than 10% of time in 30 days.
[i] Not to be exceeded over 1 day in 100 consecutive days.

TABLE 3-6 STATE AND LOCAL SULFUR DIOXIDE AMBIENT AIR QUALITY STANDARDS[46] (ppm)

Political Jurisdiction	Area Type	Annual Geo. Mean (24 hr)	Annual Arith. Mean	1 mo. max.	24 HOUR max.	24 HOUR 10%	24 HOUR 5%	24 HOUR 1%	8 hr max.	2 hr max.	1 HOUR max.	1 HOUR 5%	1 HOUR 1%	20 min. max.	5 min. max.
Arkansas			0.02					0.10(g)					0.20(k)		0.50(m)
California					0.04(i)						0.5				
Colorado					0.1						0.5				
Delaware		0.025	0.030		0.13			0.10			0.30				
Florida															
Manatee Co.(a)	Sulfuric acid plant														
Dade Co.									0.10						
Jacksonville					0.08						0.2				0.5
Georgia					0.2						0.5				
Chatham Co.					0.2						0.5				
Fulton Co.					0.2						0.5				
Macon, Bibb Co.					0.2						0.5				
Idaho					0.42(f)										
Kentucky		0.02		0.02	0.10					0.30	0.42				
Massachusetts		0.025			0.105						0.280				
Minnesota			0.02					0.10(g)							
Montana			0.02					0.10(g)					0.25(k)		

TABLE 3-6 (Continued)

Political Jurisdiction	Area Type	Annual Geo. Mean (24 hr)	Annual Arith. Mean	1 mo. max.	24 HOUR max.	24 HOUR 10%	24 HOUR 5%	24 HOUR 1%	8 hr max.	2 hr max.	1 HOUR max.	1 HOUR 5%	1 HOUR 1%	20 min. max.	5 min. max.
Nevada—Reno, Sparks, and Washoe Co's.			0.02					0.03[g]					0.08[h]		
New Jersey		0.017	0.02		0.10			0.08			0.25				
New Mexico		0.03						0.14[c]							
New York	I, II, III, IV				0.14			0.10			0.50		0.25		
	I, II		0.02												
	III, IV		0.02												
North Carolina			0.02							0.10	0.30				
Ohio			0.015					0.10			0.30				
Pennsylvania		0.02			0.10						0.25				
South Carolina	Industrial				0.17						0.50				
	Residential				0.10						0.30				
Charleston Co.	Industrial				0.17						0.50				
	Residential				0.10						0.30				
Tennessee	Industrial							0.20[c]					0.5[d]		
	Commercial							0.15[c]					0.4[d]		
	Residential							0.10[c]					0.3[d]		
	Recreational							0.10[c]					0.3[d]		
Nashville, Davidson Co's.			0.02					0.10[c]							

Chattanooga, Hamilton Co's.				0.10				0.50	
Memphis, Shelby Co.			0.03(e)			0.10			0.25
Texas	Land use A, B, and C			0.2					
	Land use D			0.3					
Virginia		0.03			0.10				
Washington			0.02		0.10			0.4	0.25(n)
Wyoming			0.02				0.10(l)		

[a]0.10 ppm above background, not to be exceeded at any time.
[b]Not to be exceeded in a 20-min. period of any hour.
[c]Not to be exceeded 24 hr out of 100 consecutive days.
[d]Not to be exceeded 1 hr out of 100 consecutive hours.
[e]Also included is a 1-hr geometric mean standard of 0.07 ppm.
[f]Computed from total cumulative daily exposure limits for 24-hr duration.
[g]24-hr average, not to be exceeded over 1% of the days in any 3-mo. period.
[h]1-hr average, not to be exceeded for more than 1-hr in any 5 consecutive days (equivalent to 0.82 percent).
[i]Applicable in areas where the particulate matter standard is exceeded.
[j]Concentration during any 30-min. period of time, with no more than one such 30-min. period during any 12 hr.
[k]Not to be exceeded for more than 1 hr in any 4 consecutive days.
[l]24-hr average not to be exceeded over 1 day in any 3-mo. period.
[m]Not to be exceeded for more than 5 min. in any 8-hr period (equivalent to 1 percent).
[n]Not to be exceeded more than twice in any consecutive 7 days.

TABLE 3-7 STATE AND LOCAL AMBIENT AIR QUALITY STANDARDS[46]

Pollutant	Political Jurisdiction	Area Type	AMBIENT AIR QUALITY STANDARD Time	Percent	Concentration
Arsenic, Copper, and Zinc	New Mexico	—	30 day	max.	10 μg/m³
Beryllium	Montana	—	1 mo.	max.	0.01 μg/m³
	New Mexico	—	1 mo.	max.	0.01 μg/m³
	New York State	Levels I-V	1 mo.	max.	0.01 μg/m³
	Pennsylvania	—	1 mo.	max.	0.01 μg/m³
Carbon monoxide	Arizona	—	1 wk.	max.	5.2 ppm
			8 hr	max.	6 ppm
			90 min.	max.	33 ppm
	California	—	8 hr	max.	20 ppm
	Kentucky	—	8 hr	max.	8 ppm
			1 hr	max.	30 ppm
	Nevada (Reno, Sparks, and Washoe Co's.)	—	Annual	Mean (A)	2.5 ppm
			1 hr	1.0	5 ppm[a]
			24 hr	0.83	10 ppm[b]
	New York State	Levels I-V	8 hr	max.	30 ppm
				15.	15 ppm
			1 hr	1.0	60 ppm
	Ohio	—	8 hr	max.	9 ppm
	Pennsylvania	—	24 hr	max.	25 ppm
Fluorides[c] (as F)	Florida (Manatee Co.)	—	—	max.	1.0 ppb[d]

	Kentucky	—	6 mo.	max.	40 ppm
			2 mo.	max.	60 ppm
			1 mo.	max.	80 ppm
	Montana	—	[e]	max.	35 ppm
	New York State	Levels I-V	6 mo.	max.	40 ppm
			2 mo.	max.	60 ppm
			1 mo.	max.	80 ppm
	Wyoming	—	[e]	max.	35 ppm
Hydrogen fluoride	Kentucky	—	1 mo.	max.	1 ppb
			1 wk.	max.	2 ppb
			24 hr	max.	3.5 ppb
			12 hr	max.	4.5 ppb
	Montana	—	24 hr	max.	1 ppb[g]
	Pennsylvania	—	24 hr	max.	5 $\mu g/m^3$[f]
	Tennessee	—	1 mo.	max.	1.5 ppb
			1 wk.	max.	2.0 ppb
			24 hr	max.	3.5 ppb
			12 hr	max.	4.5 ppb
	Wyoming	—	24 hr	max.	1 ppb[g]
Gaseous (as F)	Montana	—	1 mo.	max.	0.3 $\mu g/m^3$
	New York State	Levels I-V	1 mo.	max.	1.0 ppb
			1 wk.	max.	2.0 ppb
			24 hr	max.	3.5 ppb
			12 hr	max.	4.5 ppb
	Wyoming	—	1 mo.	max.	0.3 $\mu g/m^3$

TABLE 3-7 (Continued)

Pollutant	Political Jurisdiction	Area Type	Time	Percent	Concentration
			AMBIENT AIR QUALITY STANDARD		
Hydrogen sulfide	California	—	1 hr	max.	0.03 ppm
	Kentucky	—	1 hr	max.	0.01 ppm
	Minnesota	—	30 min.	0.83	0.03 ppm[h]
				0.011	0.05 ppm[h]
	Missouri				
	(St. Louis)	—	30 min.	0.027	0.03 ppm[i]
				0.011	0.05 ppm[j]
	(Kansas City)	—	30 min.	0.027	0.03 ppm[i]
				0.011	0.05 ppm[j]
	Montana	—	30 min.	0.83	0.03 ppm[h]
				0.011	0.05 ppm[j]
	New Mexico	—	1 hr	max.	0.003 ppm
	New York State	Levels I-V	1 hr	max.	0.10 ppm
	Pennsylvania	—	24 hr	max.	0.005 ppm
			1 hr	max.	0.10 ppm
	Texas	Land Use A&B	30 min.	max.	0.08 ppm
		Land Use C&D	30 min.	max.	0.12 ppm
	Wyoming	—	30 min.	0.83	0.03 ppm
				0.011	0.05 ppm
Hydrocarbons (non-methane)	New York	Levels I-IV	3 hr	max.	0.18 $\mu g/m^3$
	Ohio	—	3 hr	max.	0.19 $\mu g/m^3$
Lead	Montana	—	1 mo.	max.	5 $\mu g/m^3$
	New Mexico	—	1 mo.	max.	10 $\mu g/m^3$
	Pennsylvania	—	1 mo.	max.	5 $\mu g/m^3$

Nitrogen oxides ($NO + NO_2$)	California	—	1 hr	max.	0.25 ppm(k)
	Colorado	—	1 hr	max.	0.1 ppm
	Nevada (Reno, Sparks, and Washoe Co's.)	—	Annual	Mean (A)	0.15 ppm
			24 hr	1.0	0.20(l)
			1 hr	0.83	0.30(m)
	Wyoming	—	1 hr	1.0	0.15(n)
Oxidants (total)	California	—	1 hr	0.32	0.1 ppm(o)
	Colorado	—	1 hr	max.	0.1 ppm
	Kentucky	—	24 hr	max.	0.02 (p)
			1 hr	max.	0.05 (p)
	Minnesota	—	1 hr	max.	0.15 ppm
	Missouri	St. Louis	1 hr	max.	0.15 ppm
		Kansas City	1 hr	max.	0.15 ppm
	Nevada (Reno, Sparks, and Washoe Co's.)	—	Annual	Mean (A)	0.03 ppm
			24 hr	1.0	0.05 (a)
			1 hr	0.83	0.15(b)
	New York State	Levels I-II	24 hr	max.	0.05(g)
			4 hr	max.	0.10(g)
			1 hr	max.	0.15(g)
		Level III	24 hr	max.	0.05(g)
			1 hr	max.	0.15(g)
		Levels IV-V	24 hr	max.	0.10(g)
			1 hr	max.	0.15(g)
	Ohio	—	Annual	Mean (A)	0.02 ppm
			4 hr	max.	0.04 ppm
			1 hr	max.	0.06 ppm

TABLE 3-7 (Continued)

Pollutant	Political Jurisdiction	Area Type	AMBIENT AIR QUALITY STANDARD Time	Percent	Concentration
Oxidants (cont.) total	Pennsylvania	—	1 hr	max.	0.05 ppm
	Wyoming	—	1 hr	max.	0.15 ppm
Sulfuric acid mist	Arkansas	—	12 hr	3.3	15 $\mu g/m^3$ [r]
			1 hr	max.	40 $\mu g/m^3$ [r]
	Idaho	—	24 hr	1.0	12 $\mu g/m^3$ [a]
			1 hr	1.0	30 $\mu g/m^3$ [s]
	Minnesota	—	Annual	Mean (A)	4 $\mu g/m^3$
			24 hr	1.0	12 $\mu g/m^3$
			1 hr	1.0	30 $\mu g/m^3$
	Montana	—	Annual	Mean (A)	4 $\mu g/m^3$
			24 hr	1.0	12 $\mu g/m^3$
			1 hr	1.0	30 $\mu g/m^3$
	New York State	Levels I-V	24 hr	max.	0.10 mg/m^3
	Texas	—	24 hr	max.	20 $\mu g/m^3$
			1 hr	4.2	80 $\mu g/m^3$ [t]
	Wyoming	—	Annual	Mean (A)	4 $\mu g/m^3$
			24 hr	1.0	12 $\mu g/m^3$
			1 hr	1.0	30 $\mu g/m^3$
Sulfur (reduced)	New Mexico	—	1 hr	max.	0.003 ppm
Sulfates (suspended)	Minnesota	—	Annual	Mean (A)	4 $\mu g/m^3$
			24 hr	1.0	12 $\mu g/m^3$

Missouri	Kansas City	Annual	Mean (A)	4 $\mu g/m^3$
	St. Louis	Annual	Mean (A)	4 $\mu g/m^3$
		24 hr	1.0	12 $\mu g/m^3$
		1 hr	1.0	30 $\mu g/m^3$
Montana	—	Annual	Mean (A)	4 $\mu g/m^3$
		24 hr	1.0	12 $\mu g/m^3$
Pennsylvania	—	1 mo.	max.	12 $\mu g/m^{3(u)}$
		24 hr	max.	30 $\mu g/m^{3(u)}$
Wyoming	—	Annual	Mean (A)	4 $\mu g/m^3$
		24 hr	1.0	12 $\mu g/m^3$

Note: Mean (A) = Arithmetic mean.

[a] A 24-hr average not to be exceeded over 1 percent of the days in any 3-mo. period.
[b] Not to be exceeded for more than 1 hr in any 5 consecutive days.
[c] Total fluorides on dry weight basis in and on forage for consumption by grazing ruminants.
[d] Concentration above background.
[e] Average time not specified.
[f] Total soluable fluorides as HF.
[g] Total fluorides as HF.
[h] 30-min. hydrogen sulfide concentration not to be exceeded over 2 times in 5 consecutive days.
[i] 30-min. hydrogen sulfide concentration not to be exceeded more than twice in any 5 consecutive months.
[j] 30-min. hydrogen sulfide concentration not to be exceeded over twice a year.
[k] Nitrogen dioxide standard.
[l] Not to be exceeded over 1 percent of the days in any 3-mo. period.
[m] Not to be exceeded for more than 1 hr in any 5 consecutive days.
[n] Maximum allowable 1-hr value, not to be exceeded for 1 percent of the time during any 3-mo. period.
[o] 1-hr concentration, not to be equaled or exceeded 3 days consecutively or 7 days in any 90-day period (corrected for NO_2).
[p] Expressed as ozone.
[q] Including ozone, photochemical aerosols, and other oxidant contaminants not listed separately.
[r] 12-hr concentration not to be exceeded more than twice in any 30-day period.
[s] 24-hr average concentration not to be exceeded on more than 1 day in any 3 mo.
[t] During any 1-hr period of time with no more than 1 such 1-hr period during any 24 hr.
[u] Suspended sulfates expressed as H_2SO_4.

Regardless of the area they cover, odors can damage health and deprive people of the use and enjoyment of their property. Recently, thirty-one homeowners brought suit against a paper company to recover for damages caused by odors emitted from the company's kraft pulp mill in Elkton, Md. The variety and extent of injury, especially that to health, are well illustrated by the testimony received by the court. The testimony was convincing enough that the court found for the plaintiffs in the amount of $18,703. Twenty years earlier the Supreme Court of Wisconsin upheld an award of $4,000 to a homeowner who had suffered frequently from offensive odors released by a municipal sewage disposal plant in Hartland, Wisc. An award of $1,000 for injuries caused by rendering plant odors was affirmed by the North Carolina Supreme Court in 1939. In a 1960 case a Pennsylvania court held that $5,700 was appropriate compensation for an odor nuisance created by air pollution from mine refuse dumps. A North Dakota man recovered $3,500 for harms suffered when the city of Bismarck erected a city dump 800 feet from his home. This judgment was later affirmed by the North Dakota Supreme Court.

Action against odor problems begins with measurement. Perhaps the first odor study to measure intensity as well as quality resulted from the abatement conference called in connection with interstate air pollution in the Selbyville, Del.-Bishop, Md. area. A team of observers using the scentometer device established interstate transport of odorous pollutants and found that the intensity of these odors at times reached 30 odor units per cubic foot. At such strength the emissions from the rendering plant in Bishop, Md., constituted a serious nuisance. Subsequent to this survey, the Justice Department and the state of Maryland each brought suit against the rendering plant to enforce abatement of the odors.

Current odor control regulations consist of:

1. Nuisance-type restrictions based on ambient air detection of odors.
2. Process restrictions for certain known odor-producing sources.
3. Control equipment requirements, for specific source operations.

Odor regulations are directed at measurement of odors in the ambient air. After this is done, there remains the problem of tracing the odor to its source and then specifying adequate control techniques. Control officials need a tool by which odors can be evaluated and abated before nuisance conditions develop. St. Louis adopted a regulation that allows a panel of observers to evaluate odor intensity of ambient air samples when such samples are diluted with specified quantities of odor-free air. If odors can be detected after the specified dilution has occurred, the odors are deemed objectionable. Odor-control regulations, in the form of process restrictions and control equipment specifications, have been applied to certain known odor-producing operations. Los Angeles, St. Louis, and many other agencies require that effluents from animal-matter reduction be incinerated at a temperature of 1,200°F for at least 0.3 sec. These are minimum design standards for an afterburner. Other process restrictions and control requirements seek simply to prevent unnecessary discharge of odors.

The LAAPCD has developed a quantative odor-measurement technique, based on American Society for Testing Materials Method D 1391-57. Odor concentration is expressed in odor units per standard cubic foot of flue gas. An odor unit is the quantity of odorous substances that, when completely dispersed in 1 cu ft of odor-free air, produces a threshold odor response by 50 percent of an odor panel. Although Los Angeles has not developed emission standards based on odor units per minute, they have applied the sampling procedure administratively in evaluating performance of odor-control devices and in abating nuisances.

FEDERAL-STATE STANDARDS

The federal role under the Clean Air Act is that of setting national objectives and providing the necessary scientific, technical, and financial support to launch effective state and local air quality enforcement programs. The act also provides the necessary enforcement authority for the federal government to intervene where jurisdictional impasses or breakdowns in local authority do not provide for the protection of public health or welfare.

The ultimate objective, to provide for the protection of public health in its broadest concept, was cogently described by Dr. William H. Stewart, the Surgeon General of the United States:[47]

> Thanks to many advances in protecting people against disease, we are able in the health professions to think about the positive face of health—the quality of individual living. The healthy man or woman is not merely free of specific disability and safe from specific hazard. Being healthy is not just being unsick. Good health implies to me the full and enthusiastic use by the individual of his powers of self-fulfillment.

The Air Quality Act of 1967 provided a systematic approach for achieving this goal through the development of ambient air quality standards and implementation plans to achieve these standards (Fig. 3-6). Under the provisions of the act, the Secretary of Health, Education and Welfare was to complete three actions: (1) desig-

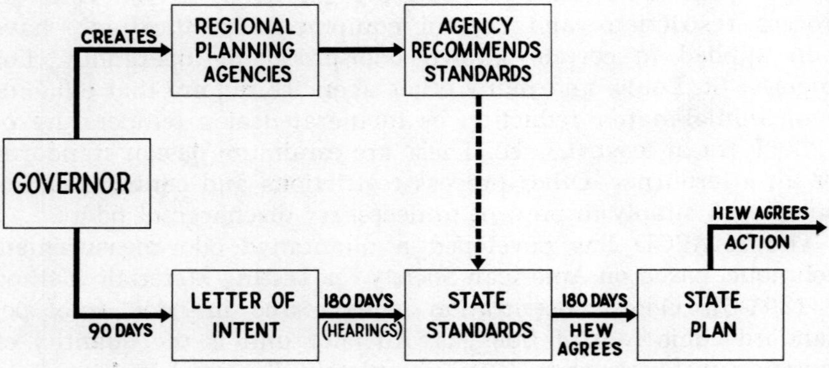

Fig. 3-6 Standard setting procedure pursuant to air quality act of 1967.

nate the air quality control regions; (2) develop and issue air quality criteria; and (3) develop and issue information on recommended control technologies. Effective December 3, 1970, this authority was transferred to the administrator of the newly constituted Environmental Protection Agency.

Upon completion of these three actions the states were to begin developing air quality standards and plans for implementing these standards in designated air quality regions. This procedure was modified in December 1970 with enactments of the Clean Air Amendments[24] which provided procedures for expediting (1) the designation of air quality control regions; (2) the issuance of air quality criteria and recommended control techniques, and (3) the establishment of ambient air quality standards. These modifications are discussed subsequently. In addition to providing for the promulgation of national primary and secondary ambient air quality standards the 1970 amendments provided for the promulgation of new stationary source performance standards and emission standards for hazardous materials.

Air Quality Control Regions

The Air Quality Act of 1967 provided the first step in the development of a regional approach to air pollution control. Recognition of the regional nature of air pollution was preserved in the 1970 amendments which required that all standards and implementation plans be developed on a statewide basis. Significantly, existing air quality control regions were preserved, and a mechanism was provided for the designation of additional air quality control regions deemed necessary or appropriate for the attainment and maintenance of ambient air quality standards.

Under the Air Quality Act, before a region was designated, an engineering evaluation was conducted to determine the nature and extent of the problem. This included the establishment of an emission inventory, collection of meteorological data, and collection of ambient air quality data. Wherever applicable an additional step was taken involving application of a meteorological diffusion model which estimates air quality levels within a proposed region. Preliminary boundaries were also announced in the *Federal Register* along with the time and location for the consultation hearing with affected state and local authorities.

Under the procedures in the Air Quality Act of 1967 some 108 Air Quality Control Regions (AQCRs) were designated and an additional 63 proposed as of January 1971. Following the accelerated procedures in the Clean Air Amendments of 1970 a total of 235 Air Quality Control Regions were designated (Table 3-8), 90 of which are interstate.

Interstate Compacts. During 1968, the Senate Committee on Public Works held hearings on congressional conditions of consent to interstate air pollution control compacts. It was the committee's opinion[48] that cooperative action by the states for the purpose of preventing and controlling air pollution and securing the enhancement of the quality of ambient air is to be encouraged to the fullest extent possible. It was further indicated that a compact must:

- provide broad authority on the part of the compact commission to undertake activities designed to prevent and control air pollution and should at a minimum include authority to (a) develop and implement ambient air quality standards, (b) provide monitoring of ambient air levels and (c) provide for reassessment of standards to reflect new and improving technology;
- provide effective enforcement against all sources of pollution within any air quality control region commonly included in the geographical boundaries of the participating states;
- provide an effective enforcement mechanism; and
- provide a prohibition against any party state or any political subdivision of such state in areas covered by a commission's jurisdiction establishing standards less restrictive than those set by the commission.

Legislation was introduced in the 92nd Congress (1971) granting preconsent of the Congress to interstate environmental compacts, permitting signatories to enter into supplementary agree-

TABLE 3-8 STATUS OF AIR QUALITY CONTROL REGIONS

State	Air Quality Control Region	Designated	Other Cities or States Included
Alabama	Alabama and Tombigbee Rivers	Mar. 31, 1971	
	Birmingham	Apr. 1, 1971	Tuscaloosa
	Columbus, Ga. (I)	Mar. 31, 1971	Phoenix City, Ala.
	East	Mar. 31, 1971	
	Florence (I)	Mar. 31, 1971	Corinth, Miss. and Tenn.
	Mobile (I)	Mar. 31, 1971	Pensacola-Panama City, Fla., & Southern Miss.
	Southeast	Mar. 31, 1971	
	Tennessee River Valley (I)	Mar. 31, 1971	Cumberland Mountains (Tennessee)
Alaska	Anchorage	Aug. 12, 1970	Cook Inlet
	Northern	Mar. 26, 1971	
	South Central	Mar. 26, 1971	
	Southeastern	Mar. 26, 1971	
Arizona	Four Corners (I)	Feb. 9, 1971	Colo., N. M., and Utah
	Las Vegas, Nev. (I)	Mar. 31, 1971	Clark-Mojave and Kingman, Ariz.
	Phoenix-Tucson	Apr. 24, 1970	
	Southern Border (I)	Mar. 31, 1971	New Mexico
Arkansas	Central	Mar. 9, 1971	Little Rock-Pine Bluff-Hot Springs
	Fort Smith (I)	Mar. 9, 1971	Muskogee, Okla.
	Joplin, Mo. (I)	Dec. 8, 1970	N.E. Okla., S.E. Kans., and Fayetteville, Ark.
	Memphis, Tenn. (I)	Apr. 29, 1970	Mississippi
	Monroe–El Dorado (I)	Dec. 8, 1970	Louisiana
	Northeast	Mar. 9, 1971	Blytheville
	Northwest	Mar. 9, 1971	
	Shreveport-Texarkana-Tyler (I)	Dec. 9, 1970	N.E. Texas, S.W. Ark., Okla., and N.W. La.
California	Great Basin Valley	Mar. 31, 1971	
	Los Angeles	Jan. 29, 1969	

	North Central Coast	Mar. 31, 1971	Santa Cruz-Salinas-Monterey
	North Coast	Mar. 31, 1971	Eureka
	Northeast Plateau	Mar. 31, 1971	
	Sacramento Valley	Mar. 31, 1971	Sacramento-Redding
	San Diego	Mar. 31, 1971	
	San Francisco	May 1, 1969	
	San Joaquin Valley	Mar. 31, 1971	Bakersfield-Fresno-Stockton
	South Central Coast	Mar. 31, 1971	San Luis Obispo-Santa Maria
	Southeast Desert	Mar. 31, 1971	El Centro-Palm Springs
Colorado	Comanche	Mar. 19, 1971	Northeastern
	Denver	Mar. 19, 1971	
	Four Corners (I)	Feb. 9, 1971	Ariz., N. M., and Utah
	Grand Mesa	Mar. 19, 1971	Grand Junction
	Pawnee	Mar. 19, 1971	Southeastern
	San Isabel	Mar. 19, 1971	Pueblo-Colorado Springs
	San Luis	Mar. 19, 1971	Alamosa
	Yampa	Mar. 19, 1971	Northwestern
Connecticut	Eastern	Mar. 31, 1971	New London-Willimantic
	Hartford-New Haven-Springfield (I)		Waterbury and Meriden, Conn. and Mass.
	New York, N.Y. (I)	Feb. 13, 1971	New Jersey
	Northwestern	Mar. 31, 1971	Torrington
Delaware	Philadelphia, Pa. (I)	Dec. 17, 1968	Camden, N. J., and Wilmington, Del.
	Southern	Mar. 19, 1971	
D.C.	National Capital (I)	Oct. 1, 1968	Maryland and Virginia
Florida	Central	Dec. 11, 1970	Orlando
	Jacksonville (I)	Mar. 31, 1971	Brunswick, Ga., and Tallahassee, Fla.
	Miami	Dec. 11, 1970	West Palm Beach (Southeast Florida)

TABLE 3-8 (Continued)

State	Air Quality Control Region	Designated	Other Cities or States Included
	Mobile, Ala. (I)	Mar. 31, 1971	Pensacola-Panama City, Fla., & Southern Miss.
	Southwest	Dec. 11, 1970	Sarasota-Ft. Myers
	West Central	Dec. 11, 1970	Tampa-St. Petersburg
Georgia	Atlanta, Ga.	Mar. 31, 1971	
	Augusta, Ga. (I)	Mar. 31, 1971	Aiken, S. C.
	Central	Mar. 31, 1971	
	Chattanooga, Tenn. (I)	Mar. 31, 1971	
	Columbus, Ga. (I)	Mar. 31, 1971	Phoenix City, Ala.
	Jacksonville, Fla. (I)	Mar. 31, 1971	Brunswick, Ga., and Tallahassee, Fla.
	Northeast	Mar. 31, 1971	
	Savannah, Ga. (I)	Mar. 31, 1971	Beaufort, S. C.
	Southwest	Mar. 31, 1971	
Hawaii	Entire State	Aug. 13, 1970	
Idaho	Boise	Nov. 3, 1970	
	Eastern	Mar. 31, 1971	
	Eastern Washington-N. Idaho (I)	Feb. 9, 1971	Lewiston-Clarkston & Spokane-Coeur d'Alene
Illinois	Burlington-Keokuk, Iowa (I)	Mar. 31, 1971	Missouri
	Chicago (I)	Mar. 31, 1971	Indiana
	Cincinnati, Ohio (I)	June 2, 1969	Kentucky
	East Central	Mar. 31, 1971	
	Dubuque, Iowa (I)	Dec. 10, 1970	Wisconsin
	North Central	Mar. 31, 1971	
	Paducah, Ky. (I)	Mar. 31, 1971	Cairo, Ill.
	Quad Cities (I)	Dec. 8, 1970	Rock Island-Moline, Ill. and Davenport, Iowa
	Rockford (I)	Aug. 27, 1970	Janesville-Beloit, Wisc.

	St. Louis, Mo. (I)	Mar. 31, 1971	Illinois
	Southeast	Mar. 31, 1971	
	West Central	Mar. 31, 1971	
Indiana	Chicago, Ill. (I)	Mar. 31, 1971	
	Cincinnati, Ohio (I)	Mar. 31, 1971	Kentucky
	East Central	Mar. 26, 1971	
	Evansville (I)	Mar. 31, 1971	Owensboro-Henderson, Ky.
	Indianapolis	Sept. 18, 1969	
	Louisville, Ky. (I)	Dec. 6, 1969	
	Northeast	Mar. 26, 1971	
	South Bend (I)	Aug. 29, 1970	Benton Harbor, Mich.
	Southern	Mar. 26, 1971	
	Wabash Valley	Mar. 26, 1971	
Iowa	Burlington-Keokuk (I)	Mar. 31, 1971	Illinois
	Dubuque (I)	Dec. 10, 1970	Illinois and Wisconsin
	North Central	Mar. 26, 1971	
	Northeast	Mar. 26, 1971	
	Northwest	Mar. 26, 1971	
	Omaha, Nebr. (I)	June 11, 1970	Council Bluffs
	Quad Cities (I)	Dec. 8, 1970	Rock Island-Moline, Ill., and Davenport, Iowa
	Sioux City, Iowa (I)	Nov. 13, 1970	Nebraska and S.D.
	Sioux Falls, S.D. (I)	Mar. 31, 1971	South Dakota
	South Central	Mar. 26, 1971	
	Southeast	Mar. 26, 1971	
	Southwest	Mar. 26, 1971	
Kansas	Joplin, Mo. (I)	Dec. 8, 1969	N.E. Okla., S.E. Kans., and Fayetteville, Ark.
	Kansas City, Kan. (I)	July 19, 1969	Missouri
	North Central	Mar. 26, 1971	
	Northeast	Mar. 26, 1971	

TABLE 3-8 (Continued)

State	Air Quality Control Region	Designated	Other Cities or States Included
	Northwest	Mar. 26, 1971	
	South Central	Mar. 26, 1971	
	Southeast	Mar. 26, 1971	
	Southwest	Mar. 26, 1971	
Kentucky	Cincinnati, Ohio (I)	Mar. 31, 1971	Illinois
	Appalachian	Mar. 31, 1971	
	Bluegrass	Mar. 31, 1971	
	Evansville, Ind. (I)	Mar. 31, 1971	Ownesboro-Henderson, Ky.
	Huntington, W. Va. (I)	Mar. 31, 1971	Ashland, Ky., and Portsmouth-Ironton, Ohio
	Louisville (I)	Dec. 6, 1969	Indiana
	North Central	Mar. 31, 1971	
	Paducah (I)	Mar. 31, 1971	Cairo, Ill.
	South Central	Mar. 31, 1971	
Louisiana	New Orleans (I)	Feb. 9, 1971	Beaumont-Port Arthur, Tex., and Baton Rouge
	Monroe–El Dorado (I)	Dec. 8, 1970	Arkansas
	Shreveport-Texarkana-Tyler (I)	Dec. 9, 1970	N.E. Texas, S.W. Ark., Okla., and N.W. La.
Maine	Androscoggan Valley (I)	Mar. 19, 1971	New Hampshire
	Aroostook	Mar. 19, 1971	Presque Isle-Houlton
	Down East	Mar. 19, 1971	Bangor-Machias
	Northwest	Mar. 19, 1971	
	Portland	Sept. 22, 1970	
Maryland	Baltimore	Aug. 16, 1969	
	Central	Mar. 31, 1971	Hagerstown-Frederick
	Cumberland (I)	Mar. 31, 1971	Keyser, W. Va.
	Eastern Shore	Mar. 31, 1971	

	National Capital (I)	Oct. 1, 1968	District of Columbia and Virginia
	Southern	Mar. 31, 1971	
Massachusetts	Boston	Mar. 9, 1971	
	Berkshire	Mar. 9, 1971	Pittsfield, Mass.
	Central	Mar. 9, 1971	Worcester-Fitchburg-Leominster
	Hartford-New Haven-Springfield, Conn. (I)	Mar. 9, 1971	Waterbury and Meriden
	Merrimack Valley (I)	Nov. 6, 1970	Manchester, N. H., & Lawrence-Lowell, Mass.
	Providence, R.I. (I)	Mar. 9, 1971	Fall River, Mass., and Pawtucket, R.I.
Michigan	Central	Mar. 31, 1971	
	Detroit-Port Huron	Dec. 17, 1969	
	Lake Michigan (I)	Mar. 31, 1971	Menomine-Escanaba, Mich. and Marinette, Wisc.
	South Central	Mar. 31, 1971	
	South Bend, Ind. (I)	Aug. 29, 1970	Benton Harbor, Mich.
	Toledo, Ohio (I)	Apr. 17, 1970	
	Upper	Mar. 31, 1971	
Minnesota	Duluth (I)	Mar. 31, 1971	Superior, Wisc. (N.W. Wisc.)
	Central	Mar. 31, 1971	
	Fargo, N. D. (I)	Oct. 31, 1970	Moorhead, Minn.
	La Crosse, Wisc. (I)	Mar. 31, 1971	Winona-Rochester, Minn.
	Minneapolis-St. Paul	Aug. 16, 1969	
	Northwest	Mar. 31, 1971	
	Sioux Falls, S. D. (I)	Nov. 13, 1970	Minnesota
	Southwest	Mar. 31, 1971	
Mississippi	Delta	Mar. 31, 1971	Greensville-Clarksdale
	Florence, Ala. (I)	Mar. 31, 1971	Corinth, Miss., N.E. Miss., and Tennessee

TABLE 3-8 (Continued)

State	*Air Quality Control Region*	*Designated*	*Other Cities or States Included*
	Memphis, Tenn. (I)	Apr. 29, 1970	Arkansas
	Mobile, Ala. (I)	Mar. 31, 1971	Pensacola-Panama City, Fla., & Southern Miss.
Missouri	Joplin (I)	Dec. 8, 1970	N.E. Okla., S.E. Kans., and Fayetteville, Ark.
	Kansas City (I)	July 19, 1969	Kansas
	Northern	Mar. 31, 1971	Columbia-Jefferson City-Hannibal
	St. Louis (I)	Mar. 31, 1971	Illinois
	Southeast	Mar. 31, 1971	Popular Bluff-Cape Guardeau
	Southwest	Mar. 31, 1971	Springfield
Montana	Billings	Mar. 31, 1971	
	Great Falls	Mar. 19, 1971	
	Helena	Mar. 19, 1971	Helena-Butte
	Miles City	Mar. 19, 1971	E. Montana
	Missoula	Mar. 19, 1971	
Nebraska	Lincoln-Beatrice-Fairbury	Mar. 31, 1971	
	Omaha (I)	June 11, 1970	Council Bluffs and Nebraska
	Sioux City, Iowa (I)	Nov. 13, 1970	Iowa and S.D.
Nevada	Las Vegas (I)	Mar. 31, 1971	Clark-Mojave and Kingman, Ariz.
	Northwestern	Jan. 7, 1971	Reno-Carson City
New Hampshire	Androscoggin Valley (I)	Mar. 19, 1971	Maine
	Merrimack Valley (I)	Aug. 6, 1970	Manchester, N.H., & Lawrence-Lowell, Mass.
New Jersey	Allentown-Bethlehem-Easton, Pa. (I)	——————	Phillipsburg, N. J. (N.E. Penn. and Upper Delaware Valley)
	New York, N.Y. (I)	Feb. 13, 1971	Connecticut
	Philadelphia, Pa. (I)	Dec. 17, 1968	Wilmington, Del., and Camden, N. J.

New Mexico	Alburquerque	Mar. 31, 1971	Mid-Rio Grande
	El Paso, Tex. (I)	Mar. 31, 1971	Las Cruces and Alamogordo, N. M.
	Four Corners (I)	Feb. 9, 1971	Arizona, Colorado, and Utah
	Northeastern Plains	Mar. 31, 1971	
	Pecos–Permian Basin	Mar. 31, 1971	
	Southern Border (I)	Mar. 31, 1971	Arizona and New Mexico
	Southwestern Mountains–Augustine Plains	Mar. 31, 1971	
	Upper Rio Grande Valley	Mar. 31, 1971	
New York	Buffalo	May 1, 1969	(Niagara Falls)
	Central	Feb. 13, 1971	Syracuse-Utica-Watertown
	Champlain Valley (I)	Feb. 13, 1971	Burlington, Vt., and Plattsburg, N. Y.
	Genesee-Finger Lakes	Feb. 13, 1971	Rochester
	Hudson Valley	Feb. 13, 1971	Albany-Poughkeepsie
	New York City (I)	Feb. 13, 1971	N. J. and Conn.
	Southern Tier East	Feb. 13, 1971	Binghamton
	Southern Tier West	Feb. 13, 1971	Jamestown-Elmira
North Carolina	Charlotte (I)	Mar. 26, 1971	South Carolina
	Eastern Mountain	Mar. 26, 1971	Morgantown-Lenoir
	Eastern Piedmont	Mar. 26, 1971	Raleigh-Durham-Wilson-Rocky Mount
	Greenville-Spartanburg	Dec. 9, 1970	
	Northern Coastal Plain	Mar. 26, 1971	Greenville-Elizabeth City
	Northern Piedmont	Mar. 26, 1971	Greensboro-High Point-Winston Salem
	Sandhills	Mar. 26, 1971	Fayetteville
	Southern Coastal Plain	Mar. 26, 1971	Wilmington-New Bern
	Western Mountain	Mar. 26, 1971	Asheville
North Dakota	Fargo (I)	Oct. 31, 1970	Moorhead, Minn.
Ohio	Cincinnati (I)	Mar. 31, 1971	Kentucky and Indiana
	Cleveland	May 23, 1969	

TABLE 3-8 (Continued)

State	Air Quality Control Region	Designated	Other Cities or States Included
Ohio	Columbus	Mar. 31, 1971	
	Dayton	Dec. 17, 1969	Springfield
	Huntington, W. Va. (I)	Mar. 31, 1971	Ashland, Ky., and Portsmouth-Ironton, Ohio
	Mansfield-Marion	Mar. 31, 1971	
	Northwest	Mar. 31, 1971	
	Parkersburg, W. Va. (I)	Oct. 15, 1970	Marietta, Ohio
	Sandusky	Mar. 31, 1971	
	Steubenville/Weirton (I)	Dec. 6, 1969	Wheeling, W. Va.
	Toledo (I)	Apr. 17, 1970	
	Wilmington-Chillicothe-Logan	Mar. 31, 1971	
	Youngstown (I)	Dec. 11, 1970	Erie, Pa. (N.W. Penn.)
	Zanesville-Cambridge	Mar. 31, 1971	
Oklahoma	Eastern	————	Ardmore-Ada
	Fort Smith, Ark. (I)	Mar. 9, 1971	Muskogee, Okla.
	Joplin, Mo. (I)	Dec. 8, 1970	N.E. Okla., S.E. Kan., and Fayetteville, Ark.
	North Central	Mar. 31, 1971	Enid-Ponca City-Stillwater
	Northeastern	Mar. 31, 1971	Tulsa, Okla.
	Northwestern	Mar. 31, 1971	
	Oklahoma City	Mar. 31, 1971	Central Oklahoma
	Shreveport-Texarkana-Tyler (I)	Dec. 9, 1970	N.E. Texas, S.W. Ark., Okla., and N.W. La.
	Southeastern	Mar. 31, 1971	Lawton
	Southwestern	Mar. 31, 1971	
Oregon	Portland (I)	June 11, 1970	Washington and Salem-Eugene
	Central	Mar. 26, 1971	
	Eastern	Mar. 26, 1971	
	Northwest	Mar. 26, 1971	

Pennsylvania	Allentown-Bethlehem-Easton (I)	Dec. 11, 1970	Phillipsburg, N. J. (N.E. Pa. and Upper Delaware Valley)
	Central	Dec. 9, 1970	Altoona-Johnstown-Williamsport
	Philadelphia (I)	Dec. 17, 1968	Wilmington, Del., and New Jersey
	Pittsburgh	Dec. 11, 1970	(S.W. Penn.)
	South Central	Dec. 9, 1970	Harrisburg-Lancaster-York
	Youngstown, Ohio (I)	Dec. 11, 1970	Erie, Pa. (N.W. Penn.)
	Southwest	Mar. 26, 1971	
Rhode Island	Providence (I)	Mar. 9, 1971	Fall River, Mass., and Pawtucket, R. I.
South Carolina	Augusta, Ga. (I)	Mar. 31, 1971	Aiken, S. C.
	Camden-Sumter	Dec. 9, 1970	
	Charleston	Dec. 9, 1970	
	Charlotte, N.C. (I)	Mar. 26, 1971	South Carolina
	Columbia	Dec. 9, 1970	
	Florence	Dec. 9, 1970	
	Georgetown	Dec. 9, 1970	
	Greenwood	Dec. 9, 1970	
	Greenville-Spartanburg	Dec. 9, 1970	
	Savannah, Ga. (I)	Mar. 31, 1971	Beauford, S. C.
South Dakota	Rapid City	Mar. 26, 1971	
	Sioux City, Iowa (I)	Nov. 13, 1970	Iowa and Nebr.
	Sioux Falls, S.D. (I)	Mar. 31, 1971	Iowa
Tennessee	Bristol, Va. (I)	Mar. 31, 1971	Johnson City and Knoxville, Tenn. (E. Tenn. and S.W. Va.)
	Chattanooga (I)	Mar. 31, 1971	Georgia
	Florence, Ala. (I)	Mar. 31, 1971	Corinth, Miss., and Tenn.
	Memphis (I)	Apr. 29, 1970	Arkansas and Mississippi
	Middle	Mar. 31, 1971	Nashville

TABLE 3-8 (Continued)

State	Air Quality Control Region	Designated	Other Cities or States Included
Tennessee	Tennessee River Valley (Ala.)	Mar. 31, 1971	Cumberland Mountains
	Western	Mar. 31, 1971	Jackson
Texas	Abilene-Wichita Falls	Feb. 9, 1971	
	Amarillo-Lubbock	Feb. 9, 1971	
	Austin-Waco	Feb. 9, 1971	
	Brownsville-Laredo	Feb. 9, 1971	
	Corpus Christi-Victoria	Feb. 9, 1971	
	Dallas-Ft. Worth	Feb. 9, 1971	
	El Paso (I)	Mar. 31, 1971	Las Cruces and Alamogardo, N. M.
	Houston-Galveston	Feb. 9, 1971	Texas City
	Midland-Odessa-San Angelo	Feb. 9, 1971	
	New Orleans, La. (I)	Feb. 9, 1971	Beaumont-Port Arthur, Tex., and Baton Rouge
	San Antonio	Feb. 9, 1971	
	Shreveport-Texarkana-Tyler (I)	Dec. 9, 1970	N.E. Texas, S.W. Ark., Okla., and N.W. La.
Utah	Four Corners	Feb. 9, 1971	Ariz., Colo., and N. M.
	Wasatch Front	Aug. 12, 1970	Salt Lake City
Vermont	Champlain Valley (I)	Feb. 13, 1971	Burlington, Vt., and Plattsburg, N. Y.
Virginia	Bristol (I)	Dec. 31, 1970	Johnson City-Knoxville, Tenn. (E. Tenn. and S.W. Va.)
	Central	Mar. 19, 1971	Lynchburg-Danville
	Hampton Roads	Mar. 19, 1971	
	National Capital (I)	Oct. 1, 1968	Maryland and D.C.
	Norfolk	———	
	Northeastern	Mar. 19, 1971	Charlottesville-Fredericksburg
	State Capital	Mar. 19, 1971	Richmond-Petersburg
	Valley	Mar. 19, 1971	Roanoke-Shenandoah Valley

Washington	Eastern Washington-N. Idaho (I)	Feb. 9, 1971	Lewiston-Clarkston-Spokane-Coeur d'Alene
	Northern	Mar. 31, 1971	
	Olympic-Northwest	Mar. 31, 1971	
	Portland, Ore. (I)	Mar. 31, 1971	Salem-Eugene
	Puget Sound	Oct. 25, 1969	Seattle-Everett-Tacoma
	South Central	Mar. 31, 1971	
West Virginia	Allegheny	Mar. 31, 1971	
	Central West	Mar. 31, 1971	
	Cumberland, Md. (I)	Mar. 31, 1971	Keyser, W. Va.
	Eastern Panhandle	Mar. 31, 1971	
	Huntington (I)	Mar. 31, 1971	Ashland, Ky., and Portsmouth-Ironton, Ohio
	Kanawha Valley	Mar. 31, 1971	
	North Central	Mar. 31, 1971	
	Southwest	Mar. 31, 1971	
	Parkersburg (I)	Oct. 15, 1970	Marietta, Ohio
	Steubenville-Wierton, Ohio (I)	Dec. 6, 1969	Wheeling, W. Va.
Wisconsin	Dubuque, Iowa (I)	Dec. 10, 1970	Illinois
	Duluth, Minn. (I)	Mar. 31, 1971	Superior, Wisc. (N.W. Wisc.)
	La Crosse (I)	Mar. 31, 1971	Winona-Rochester, Minn.
	Lake Michigan (I)	Mar. 31, 1971	Menominee-Escanaba, Mich. and Marinette, Wisc.
	North Central	Mar. 31, 1971	Wausau
	Rockford, Ill. (I)	Aug. 27, 1970	Janesville-Beloit, Wisc.
	Southeastern	Mar. 31, 1971	Milwaukee
	Southern	Mar. 31, 1971	Madison
Wyoming	Casper	Mar. 26, 1971	
	Cheyenne	Oct. 31, 1970	
Guam			
Puerto Rico	Entire Commonwealth	Aug. 11, 1970	
Virgin Islands	Entire Territory	June 11, 1970	

(I) Interstate Compact.

ments on interstate environmental pollution problems. The legislation was intended to comply specifically with the federal requirements of congressional consent to interstate compacts in the fields of air pollution, water pollution, and solid wastes. The proposed legislation, the Interstate Environment Compact Act of 1972, also would permit supplementary agreements in the fields of land use, costal zone management, energy production and transmission, and related activities.

Air Quality Criteria

The second step to be taken by the secretary (now the administrator of EPA) under the Air Quality Act of 1967 was the issuance of air quality criteria to aid the states in developing ambient air quality standards. The Clean Air Act of 1963, provided for the development of "air quality criteria" which accurately reflect current knowledge on the quantitative relationships between air pollution exposures and health and welfare effects. This approach was reaffirmed in the Air Quality Act of 1967 and the Clean Air Amendment of 1970. This legislation also provided for the issuance for each contaminant "recommended control techniques." Both "air quality criteria" and "recommended control techniques" documents are to be used by the states in the establishment of regional ambient air quality standards.

As envisioned, air quality criteria are to consist of the latest available knowledge of medical and scientific observations concerning known or knowable health and welfare effects of air pollution. They are intended to delineate, on the basis of scientific and medical evidence, the effects of individual contaminants, combinations of contaminants, or categories of contaminants on the constantly changing, somewhat indeterminate, environment of man. Thus, economic and technological considerations of control are not relevant in the establishment of ambient air quality criteria. Congress intended that economic and technical considerations should relate to implementation and time schedules.

Clearly formulated air quality criteria provide a quantitative basis for air quality standards, but criteria and standards are not synonymous. Their relationships was made clear in testimony on the Air Quality Act of 1967 when Dr. John T. Middleton, director of the then National Center for Air Pollution Control stated:[49]

> Air quality criteria are an expression of the scientific knowledge of the relationship between various concentrations of pollutants in the air and their adverse effects on man, animals, vegetation, materials, visibility and so on.

> Air quality criteria can and should be used in developing air quality standards. Criteria and standards are not synonymous. Air quality criteria are descriptive; that is, they describe the effects that can be expected to occur whenever and wherever the ambient air level of a pollutant reaches or exceeds a specific figure for a specific time period.

Dr. Middleton also said:

> Air quality standards are prescriptive; they are prescribe pollutant levels that cannot legally be exceeded during a specific time in a specific geographic area.

As expressions of scientific knowledge, air quality criteria are first a compilation of the experience of experts to guide those less experienced. Second, such criteria are crystallizations of past experiences to facilitate future action. As such, these criteria indicate quantitatively the relationship between various levels of exposure to a given contaminant or combination of contaminants of inorganic, organic, bacteriological, or radioactive nature, whether man-made or natural, and the short-term or long-term effects on health or welfare.

It must be remembered, however, that cause-effect relations involving health are studied by empirical tools and empirical methods. There are no sharp lines defining cause and effect. The suggestion that "threshold effects" exist is deceiving. Many effects are not observed because study techniques are either insensitive or the effects unsuspected. As knowledge improves on the cause-effect relation, new and more refined techniques will be required to better delineate dose-response relationships. The techniques used 10 years ago are not good enough today and there is no reason to assume that analytical methods and techniques used today will suffice tomorrow. Both new and more positive research methods combined with new and constantly broadening scientific and social concepts are needed.

Ultimately, air quality criteria must include indirect social costs. For example, it is generally assumed that air pollution tends to depress property values. A recently completed analysis of property values in relation to air pollution was performed in Washington, D.C., Kansas City, and St. Louis. In each community, a comparison was made of the selling prices of homes in neighborhoods with different levels of sulfur oxides and particulate pollution. Even after allowance was made for other relevant factors, such as size of homes, proximity to schools, and character of the neighborhood, it was found that homes in the areas of higher air pollution levels generally sold for $300 to $500 less.

The issuance of criteria and the accompanying reports on air pollution control techniques is the signal for state governments to begin the air quality standard-setting process in the air quality control regions. On February 11, 1969, NAPCA published the first such air quality criteria documents. These deal with two of the most common air pollutants—sulfur oxides and particulate matter.

In early 1970, air quality criteria for carbon monoxide, photochemical oxidants, and hydrocarbons were issued. Those for nitrogen oxides were published in 1971. Prior to enactment of the Clean Air Amendments of 1970, a tentative schedule for the issuance of criteria for odors (including toxological and corrosion aspects of hydrogen sulphide), lead, fluorides, polynuclear organics, asbestos, hydrogen chlorides, beryllium, chlorine, gas, arsenic, nickel, vanadium, barium, boron, chromium (including chromic acid), mercury, selenium, pesticides, and radioactive substances was set up. The effect of the 1970 amendments may well be that some of these pollutants will be covered by national emission standards for hazardous pollutants or the performance standards for new stationary sources rather than the criteria approach.

As knowledge increases on the effects of chronic long-term exposures to atmospheric contaminants, or environmental contaminants, the concept of "threshold effects" will be supplanted with the concept of minimizing the risks associated with exposure to environmental contaminants. The cause-and-effect relationships involving health are studied through use of empirical tools and empirical methods. As new and more defined research techniques evolve, it will be possible to delineate dose-response relationships and their associated risks.

In applying these criteria, it must be recognized by all parties involved that, in dealing with environmental health problems, responsible public policy cannot wait upon a perfect knowledge of cause and effect. Where the available evidence indicates a health or welfare effect from environmental exposures, government must move to minimize these exposures. This involves interposing barriers between the public and the causative agents producing stress. Environmental quality requires minimization of man's exposure to stress and, of necessity, action is based upon best current evidence of causation available. This philosophy was appropriately expressed by Sir Austin B. Hill when he wrote:[50]

> All scientific work is incomplete—whether it be observational or experimental. All scientific work is liable to be upset or modified by advancing knowledge. That does not confer upon us a freedom to ignore the knowledge we already have, or to postpone the action that it appears to demand at a given time.

Historically, general health and safety practices are characterized by empirical decisions, by an eternally persistent reappraisal of public health standards against available knowledge of causation by consistently giving the public the benefit of the doubt, and by ever striving for improved environmental quality.

Initial public health standards have invariably preceded full scientific understanding and acceptance. Predicated upon observed associations between human exposures and disease, initial criteria have been subsequently reinterpreted and refined, and new standards established in light of new knowledge and experience.

Particulates. Atmospheric particulate concentrations represent an index of a diverse class of substances often referred to as aerosols. These materials, both solid and liquid, can be classified by size, although this is not the current practice. The more abundant atmospheric particulate contaminants can be placed in four categories: (1) ash particles, primarily the noncombustible components of fuels, containing oxides, sulphates, chlorides, carbonates, nitrates, silicates, and many trace metals, (2) carbon, (3) sulphuric acid, and (4) tar particles containing hydrogen, organic acids, bases, phenols, etc. Typical annual average particulate levels in nonurban areas range from 10 to 60 $\mu g/m.^3$

There is a tendency to consider all particulates together; however, this results in measured atmospheric levels of particulates that are not a direct measure of human exposure. Three factors must be taken into consideration. First, particulate size rather than the quantity of particulates is the governing factor in affecting human exposure. Due to the filtration mechanism of the respiratory tract particles of greater than 5 to 10μ in diameter do not reach the lung.

Second, the contaminants considered as particulates vary widely in their toxicity. Many are known to be directly toxic to man (e.g., fluorides, beryllium, lead, asbestos). For this reason separate air quality criteria are under development for lead, fluorides, asbestos, beryllium, arsenic, cadmium, copper, manganese, nickel, vanadium, zinc, barium, boron, chromium, selenium and pesticides.

Third, additive and synergistic effects are often produced when particulates occur along with gaseous contaminants.

The air quality criteria for particulates[51] treats particulate matter as a whole in considering the effects associated with atmospheric exposure. The particulate criteria document reviews and summarizes results of approximately 350 studies of the effects of particulate air pollution.

The Air Quality Criteria for Particulate Matter indicates health effects were observed at 80 $\mu g/m^3$ as increased incidence of death for persons over 50 years of age. Additionally, at 100 $\mu g/m^3$ children residing in such an area are likely to experience increased incidence of respiratory disease.

At annual geometric mean particulate levels of 70 $\mu g/m^3$, in the presence of other contaminants, there is an increased public awareness and concern for air pollution. At a particulate level of 60 $\mu g/m^3$, in the presence of sulfur oxides and moisture, there is accelerated corrosion of steel and zinc.

Health effects are observed when the average 24-hr concentration of particulates reaches 300 $\mu g/m^3$ for 3 or 4 days. Under these conditions increased hospital admissions and increased job absenteeism are observed, particularly among older persons.

Visibility may be reduced to as low as 5 miles with a concentration of 150 $\mu g/m^3$.

The Criteria are careful to point out that criteria indicate cause-effect relationships. "It is reasonable and prudent to conclude that when promulgating ambient air quality standards, consideration should be given to requirements for margins of safety which would take into account long-term effects on health, vegetation, and materials occurring below these levels."

Sulfur Oxides. Atmospheric sulfur dioxide concentrations represent an index of exposure to a wide spectrum of sulfur oxide contaminants arising mainly from the combustion of fuels. Although the index of exposure is the measurment of sulfur dioxide, the effects cited in the *criteria* result from a combined exposure to sulfur dioxide, sulfur trioxide, and the corresponding acids and salts (sulfites and sulfates).

The 1963 Clean Air Act provided for the development of air quality criteria on "air pollution agents" indicating their likely adverse effects if present in the air in varying quantities. The first criteria for oxides of sulfur was issued on March 23, 1967,[52] while congressional hearings were in progress on the Air Quality Act of 1967. Immediately, a public debate ensued as to the accuracy of the sulfur oxides criteria, spurred in large part over a concern by industry that the ambient air concentrations at which effects were cited could not be achieved with existing control technology.

During this same period the administration undertook to expand its research program to develop control methods for sulfur oxide emissions[53] and requested additional funds from the Congress for this purpose.

The Congress also recognized this need and provided for an expanded research program—Section 104 of the Air Quality Act of 1967. The act directs that special emphasis be given to research and development into new and improved methods for the prevention and control of air pollution resulting from the combustion of fuels. In reporting the Air Quality Act the Senate Committee on Public Works stated:[54]

> The oxides of sulfur controversy is indicative of the need more precisely to define the relationship between pollution and health and welfare. Because the committee is concerned with both long and short-term hazards as well as the need for valid scientific data to substantiate the correlation between pollution and health and welfare, the Secretary is urged to move forward with diligence and perseverance in the area of scientific analysis as well as research into ways feasibly and effectively to control potentially dangerous emissions.

Because of the debate on the validity of the sulfur oxides criteria and in recognition of its critical role in the development of air quality standards, the Congress also directed that the original air quality criteria be reevaluated, and if necessary revised and reissued. Authority to establish a technical advisory committee for this purpose was provided. Subsequently, the revised air quality criteria for sulfur oxides was reissued.[55]

The *Air Quality Criteria for Sulfur Oxides* indicate that when sulfur oxide levels, measured as sulfur dioxide, reach an annual arithmetic average of 115 $\mu g/m^3$, increased mortality from bronchitis and lung cancer may occur.

When sulfur dioxide levels reach 300 $\mu g/m^3$ for 3 to 4 days, increased hospital admissions and increased job absenteeism from work are observed, particularly among older persons.

At a level of 286 $\mu g/m^3$ visibility may be reduced to as low as 5 miles.

At levels in excess of 145 $\mu g/m^3$ for short-term exposures (e.g., 4 hours) sulfur oxides may react synergistically with either ozone or nitrogen oxides to produce moderate to severe injury to sensitive plants. At an annual mean concentration 85 $\mu g/m^3$ chronic plant injury and excessive leaf drop may occur.

At a sulfur dioxide level of 345 $\mu g/m^3$, in the presence of high particulate levels, the corrosion rate of steel may be increased by 50 percent.

Carbon Monoxide. The most common and widely distributed of atmospheric contaminants is carbon monoxide. In urban areas the primary source is the incomplete combustion of carbonaceous materials used as fuels for vehicles, space heating, industrial processing, and the burning of refuse. Background levels in the lower atmosphere range from 0.01 to 0.2 $\mu g/m^3$ (0.01 to 0.2 ppm).[56]

The Carbon Monoxide Air Quality Criteria indicates:[57]

> An exposure of 8 or more hours to a carbon monoxide concentration of 12 to 17 mg/m^3 (10–15 ppm) will produce a blood carboxyhemoglobin level of 2.0 to 2.5 percent in nonsmokers. This level of blood carboxyhemoglobin has been associated with adverse health effects as manifested by impaired time interval discrimination. Evidence also indicates that an exposure of 8 or more hours to a CO concentration of 35 mg/m^3 (30 ppm) will produce blood carboxyhemoglobin levels of about 5 percent in nonsmokers. Adverse health effects as manifested by impaired performance on certain other psychomotor tests have been associated with this blood carboxyhemoglobin level, and above this level there is evidence of physiologic stress in patients with heart disease.
>
> There is some epidemiological evidence that suggests an association between increased fatality rates in hospitalized myocardial infarction patients and exposure to weekly average CO concentrations of the order of 9 to 16 mg/m^3 (8–14 ppm).
>
> Evidence from other studies of the effects of CO does not currently demonstrate an association between existing ambient levels of CO and adverse effects on vegetation, materials, or other aspects of human welfare.
>
> It is reasonable and prudent to conclude that, when promulgating air quality standards, consideration should be given to requirements for margins of safety that would take into account possible effects on health that might occur below the lowest of the above levels.

It is significant to note that cigarette smokers generally have a median carboxyhemoglobin level of 5 percent, in excess of the level for adverse health effects as identified in the *criteria*. Whereas nonsmokers usually have about 0.5 percent carboxyhemoglobin.

The need for further research was noted in the criteria document on (1) physiology of carbon monoxide in the body, (2) effects on enzyme systems and tissue oxygenation, (3) effects on human behavior and performance, and (4) relationship between CO exposure and cardiovascular disease.

The CO air quality criteria also provides data on currently experienced ambient air levels. It is of interest that for those cities presented ambient air levels exceed the air quality criteria levels at which health effects have been observed.

Hydrocarbons. The class of atmospheric contaminants known as hydrocarbons includes all the compounds composed of hydrogen and carbon, except for the oxides of carbon, the carbides, and the carbonates. All hydrocarbons are organic compounds.

Estimates of background levels of methane in nonurban air are 0.7–1.0 $\mu g/m^3$ (1.0–1.5 ppm). For other hydrocarbons the estimate is less than 0.1 ppm each. Atmospheric sources of hydrocarbon emissions are primarily from the inefficient combustion of volatile fuels and from their use as process chemicals.

The air quality criteria for hydrocarbons is concerned primarily with gas-phase hydrocarbons and certain of their oxidation products, particularly aldehydes, that are associated with photochemical air pollution. Criteria for particulate hydrocarbons, specifically polynuclear hydrocarbons, were scheduled for publication in 1971.

The Air Quality Criteria for Hydrocarbons states:[58,59]

> Studies conducted thus far of the effects of ambient air concentrations of gaseous hydrocarbons have not demonstrated direct adverse effects from this class of pollution on human health. However, it has been demonstrated that ambient levels of photochemical oxidant, which do have adverse effects on health, are a direct function of gaseous hydrocarbon concentrations; and when promulgating air quality standards for hydrocarbons, their contribution to the formation of oxidant should be taken into account.
>
> Analysis of 3 years of data collected in three American cities shows that on those several days a year when meteorological conditions were most conducive to the formation of photochemical oxidant, non-methane hydrocarbon concentrations of 200 $\mu g/m^3$ (0.3 ppm C) for the 3-hr period from 6.00 to 9.00 A.M. might produce an average 1-hr photochemical oxidant concentration of up to 200 $\mu g/m^3$ (0.10 ppm) 2–4 hr later. The hydrocarbon measurements were confined to 200 $\mu g/m^3$ (0.3 ppm C), or above, because of instrumentation limitations. However, if the functional relationship between the hydrocarbon and photochemical oxidant measurements were extended to include the lowest levels at which photochemical oxidant has been observed to adversely affect human health, the corresponding hydrocarbon concentration would be approximately *130 $\mu g/m^3$ (0.2 ppm C).*

Photochemical Oxidants. Photochemical oxidants result from a complex series of atmospheric reactions when reactive organic substances, including hydrocarbons and nitrogen oxides, accumulate in the atmosphere and are exposed to sunlight. An array of new compounds are formed, among which are found oxidants, ozone, and peroxyacyl nitrates. Nitrogen dioxide is also an oxidant; however, nitrogen oxides were the subject of an air quality criteria document issued in Jan. 1971.

The Air Quality Criteria Document for Photochemical Oxidants states[60,61] that

> Under the conditions prevailing in the areas where studies were conducted, adverse health effects, as shown by impairment of performance of student athletes, occurred *when the hourly average oxidant concentrations exceeded 130 μg/m³ (0.07 ppm).* An increased frequency of asthma attacks in a small proportion of subjects with this disease was shown on days when oxidant concentration exceeded peak values of *490 μg/m³ (0.25 ppm),* a level that would be associated with an hourly average concentration *as low as 300 μg/m³ (0.15 ppm).* Adverse health effects, as manifested by eye irritation, were reported by subjects in several studies when photochemical oxidant concentrations reached instantaneous levels of about 200 μg/m³ (0.10 ppm). Adverse affects on sensitive vegetation were observed from exposure to photochemical oxidant concentrations of about 100 μg/m³ (0.05 ppm) for 4 hr. Adverse effects on materials from exposure to photochemical oxidants have not been precisely quantified, but have been observed at the levels presently occurring in many urban atmospheres. It is reasonable and prudent to conclude that, when promulgating ambient air quality standards, consideration should be given to requirements for margins of safety that would take into account possible effects on health, vegetation, and materials that might occur below the lowest of the above levels.

The criteria document recommended further research on (1) identification of the substances which cause eye irritation, (2) effects on patients with chronic respiratory disease, (3) systemic effects of ozone (headache, fatigue, impaired oxygen transport, inability to concentrate, etc.), (4) air-pollution-associated congenital malformations, stillbirths, hospital admissions for miscarriage, and alterations in the sex ratio of new borns, and (5) effects of air pollution on children—to name a few areas.

Nitrogen Oxides. Usually nitrogen oxides originate from high-temperature combustion. Their presence in the atmosphere is associated with a variety of respiratory diseases and is essential to the production of photochemical oxidants.

The *Air Quality Criteria for Nitrogen Oxides*[62] indicate that adverse health effects, as evidenced by a greater incidence of acute bronchitis among infants and school children, have been observed when nitrogen dioxide levels over a 24-hr period varied from 118 to 156 μg/m³ (0.063 to 0.083 ppm) over a 6-month period.

Adverse effects also have been observed on vegetation as leaf abscission and decreased yield of navel oranges during fumigation studies when the nitrogen dioxide level was 470 μg/m³ (0.25 ppm) during an 8-month period. Nitrate compounds have been identified with corrosion and failure of electrical components.

National Ambient Air Quality Standards

The Air Quality Act of 1967 provided for a procedure where the initial selection of ambient air quality standards became a state and local decision. The federal responsibility was limited to insuring that the standard was adequate to protect public health. Using federally developed air quality criteria, state and local governments were to decide how clean an environment they desired.

The Clean Air Amendments of 1970 provided for the federal promulgation of national primary and secondary ambient air quality standards as a means of expediting the standard setting procedure. Within 90 days of the issuance of ambient air quality criteria the EPA administrator is required to promulgate national primary and secondary ambient air quality standards. The primary standard is to provide for the protection of public health, while the secondary standard is for the protection of public welfare. (This procedure was to be initiated for existing criteria by January 3, 1971.)

The primary significance in this approach is the distinction between public health and public welfare effects with the protection of public health being the minimum federal requirement. Under the 1970 Amendments there is an additional requirement that the primary or public health standard be implemented within 3 years. This is discussed subsequently in detail.

Considerable insight into how this distinction transpired can be gained from reviewing implementation of the Air Quality Act of 1967. Under this act, state-proposed ambient air quality criteria standards were submitted to the Secretary of Health, Education and Welfare (now the EPA administrator for federal approval).

In approving the standards the secretary was to review the standards for

1. consistency with the Clean Air Act;
2. consistency with the air quality criteria for the contaminants in question;
3. public health and welfare implications of the air quality standards intended for adoption; and
4. comparison with existing air quality.

Of significance was the emphasis placed upon consideration for the protection of public health when establishing ambient air quality standards. The Senate Public Works Committee[18] was explicit on this point when it stated:

> The Committee feels that under any circumstances protection of health should be considered a minimum requirement, and wherever possible standards should be established which enhance the quality of the environment.

The report also stated:

> Consideration of technology and economic feasibility, while important in helping to develop alternative plans and schedules for achieving goals of air quality, should not be used to mitigate against protection of the public health and welfare.

Therefore, it must be concluded that, to be consistent with the purposes enunciated in the Air Quality Act, air quality standards must at a minimum protect public health and this must be achieved regardless of cost. The effect of the Clean Air Amendments of 1970 was to affirm this public policy with provision for a primary or health standard with specific time schedules for their achievement and a secondary or welfare standard for which the time schedule for compliance is determined by state and local governments.

On January 25, 1971, the EPA administrator proposed[63] national primary and secondary ambient air quality standards for sulfur oxides, particulates, carbon monoxide, photochemical oxidants, hydrocarbons, and nitrogen oxides. After revision final standards were promulgated in April 1971[64] which are discussed in the subsequent subsection on federal-state ambient air quality standards.

Federal-State Ambient Air Quality Standards

Existing federal statute requires the issuance of both "air quality criteria" and "recommended control techniques" to assist the states in the development of ambient air quality standards and plans of implementation. It must be recognized that the states are provided with an opportunity to establish more stringent standards or timetables than those advocated by federal policy.

The complex procedure provided in the Air Quality Act of 1967 is presented graphically in Fig. 3-6. This procedure was expedited in the Clean Air Amendments of 1970, eliminating the requirements for a "letter of intent" from governors and separate filing of proposed ambient air quality standards and plans of implementation, maintenance, and enforcement of the national primary ambient air quality standard within 9 months after the promulgation of the standard. The administrator then has 4 months to approve the plan in whole or in part.

The same time schedule is provided for submission of an implementation plan for the national secondary ambient air quality standard unless the administrator determines additional time up to 18 months is justified.

A procedure is provided for the administrator to develop an implementation plan for an area where

- a state fails to submit an implementation plan;
- the submitted plan does not meet the requirements of the Clean Air Act, as amended; or
- the state fails to revise a plan within 60 days of notification by the administrator.

Upon approval by the administrator, the standards and implementation plans become joint federal-state standards and plans for each air quality control region.

In addition to meeting the federal requirement for the protection of public health, they reflect any additional quality of air that a community wants to breathe and for which it is willing to pay.

The public hearing is the mechanism employed for determining the social, economic, and political desires of a community regarding ambient air quality standards. The Air Quality Act required a public hearing prior to the adoption of air quality standards for an air quality control region. The Clean Air Amendments of 1970 provided for a single hearing on both the standards and the plan for their implementation. A hearing is required even though the air quality standards under consideration may have been the subject of prior hearings under state or local law.

Following the public hearing, the proposed standards and a copy of the hearing transcript are submitted to the administrator for his approval.

To aid in the establishment of air quality standards NAPCA published *Guidelines*[65] in 1969. The *Guidelines* recommended that proposed standards be accompanied by data on current levels of contaminants for use in reviewing proposed standards.

The *Guidelines* also recommended that air quality standards be expressed in a manner that facilitates determination of their consistency with air quality criteria and national ambient air quality standards. Standards were to be expressed in terms of maximum concentrations for any given average time (e.g., a 24-hour maximum concentration).

The *Guidelines* also recommended that ambient air quality standards be formulated in terms of maximum concentrations for any given average time. Both a long-term standard directed at chronic exposure effects and a short-term standard (e.g., 3 to 4 days) directed at acute effects were recommended.

The secretary (now administrator) had to approve state standards and plans after considering a number of factors. He had to determine that standards were "consistent with the air quality criteria." A determination had to be made that the plan of implementation was "consistent with the purposes of the act," insofar as it assures achieving such standards of air quality within a reasonable time, "before a plan of implementation becomes effective." The three key phrases are "consistent with the criteria," "consistent with the purposes of the act," and "within a reasonable time."[66]

Air quality criteria describe the health and welfare effects of air contaminants. Criteria issued to date describe the adverse effects of air pollution following short-term, relatively brief exposures. To be *consistent with the criteria* ambient air quality standards must consider the effects of both short-term and long-term exposures.

A major purpose of the Air Quality Act was to "protect and enhance the quality of the nation's air resource so as to promote the public health and welfare and the productive capacity of its population." A question then arises on what is meant by *consistent with the purposes of the act.*

In its report[18] on the Air Quality Act of 1967, the Senate Committee on Public Works stated, with respect to the purpose of the statute:

> Its objective is the enhancement of air quality and the reduction of harmful emissions consistent with maximum utilization of an expanding capacity to deal with them effectively.

In turn the report[67] of the House Committee on Interstate and Foreign Commerce stated:

> The important objective of the bill is to insure that air pollution problems will, in the future, be controlled in a systematic way.

From these two statements, it is possible to conclude that one major purpose of the act was to achieve a "systematic reduction of harmful emissions." The primary mechanism for achieving this was to proceed from ambient air quality standards to a plan which contains emission standards and a time schedule for their implementation.

Critical to this procedure was the setting of air quality standards. In establishing an air quality standard, a state must define the quality of air that its residents are prepared to breathe. In effect it describes a desired maximum level of pollutants in the ambient air.

While providing for the protection of public health an additional purpose was "to protect and *enhance* the quality of the nation's air resource." Air quality standards which, even if fully implemented would permit continuing deterioration of air quality in any major portion of an air quality control region or would fail to enhance air quality would clearly be in conflict with the expressed purpose of the law.

The third key term was "reasonable time." An expressed intent of the Air Quality Act of 1967 was attainment of air quality standards within a reasonable time. Exactly what may be considered a "reasonable time" for attainment of an air quality standard depends on a number of factors including the availability of applicable control techniques and particularly, the nature and seriousness of the adverse effects of the pollutants involved. Under the Air Quality Act, where a threat to public health is involved, air quality standards should be attained within the shortest possible time. The Clean Air Amendments of 1970 provide a 3-year time schedule.

Every implementation plan includes a timetable for reaching compliance with the projected requirements for the prevention, abatement, and control of air pollution. This timetable provides for meaningful increments of progress over relatively short intervals, such as 1-year or 2-year periods, during the total timespan covered by the plan.

The air quality standards and plan of implementation must provide for this systematic reduction in emissions to receive federal approval. There is provision made for recognition that this time schedule is flexible, however.

Two basic tenets of the federal policy warrant additional discussion. First is the basic premise that prevention and control of air pollution at its source is the primary responsibility of state and local governments. Second, the avowed purpose to "encourage and assist the development of regional air pollution programs." Yet, there is the requirement that standards for an air quality control region be adopted by state action, and the various restrictions included in an implementation plan must likewise be enforceable by state action.

This does not mean that the states must assume exclusive responsibility for enforcement; they may rely on the capability of local and regional agencies. The states must, however, have the legal authority to conduct enforcement activities. To comply with the Clean Air Act, as amended, this authority must be broad enough to permit a state to enforce the orderly application of control techniques in accordance with the timetable set forth in the implementation plan and, when necessary, to curtail pollution emissions on an emergency basis to prevent the occurrence of short-episodes of high pollutant concentrations.

This suggests that the state program must be built around the existing strengths of the various political jurisdictions in an air quality control region, with the state serving in a coordinative and supplemental role.

The basic objective must be for unified programs to assure consistency in the application of the "systematic plan for emission reduction" and to achieve the level of air quality that citizens demand. This unified program must provide equitably for control of all the sources of air pollution emissions.

Particulate—Air Quality Standards. The air quality criteria and the recommended control techniques for particulates were issued in February 1969. The governors affected immediately moved to establish both long-term and short-term ambient air quality standards.

The *Guidelines* recommend that the long-term air quality standard for particulates be expressed as an annual geometric mean that cannot be exceeded over a 1-year period. The *Guidelines* also recommended that the short-term particulate standard be expressed in terms of a 24-hr maximum average concentration.

Proposed and approved ambient air quality standards for particulates, as of January 1971 when the Clean Air Amendments of 1970 took effect, are shown in Table 3-9. A considerable variation is noted among the various air quality control regions in long-term particulate standards—between annual geometric means of 45 to 80 $\mu g/m^3$.

The *Air Quality Criteria for Particulate Matter* indicates an increased incidence of death among persons over 50 years of age at this level.

The *Criteria* document is careful to point out that it defines levels at which health and welfare effects occur. The document concludes that it is reasonable and prudent to conclude that, when promulgating ambient air quality standards, consideration should be given to margins of safety which take into account long-term effects on health and materials occurring below the criteria levels.

The short-term particulate standards were to be expressed in terms of maximum 24-hr concentrations. Some governors established standards for even shorter time intervals (e.g., 30-min or 5-min). The maximum 24-hr particulate standards range from 100 to 286 $\mu g/m^3$.

The *Criteria* document indicates that when the average 24-hr concentration of particulates reaches 300 $\mu g/m^3$ for 3 or 4 days, health effects occur. Under these conditions there are increased hospital admissions and job absenteeism, particularly among older persons.

The first two air quality control regions to have their particulate and sulfur oxide standards approved were Philadelphia and Chicago. The Philadelphia region includes the states of Pennsylvania, New Jersey, and Delaware who agreed to move toward a 1975 average annual particulate ambient air quality standard of 65 to 70 $\mu g/m^3$. This represents a 45–55 percent reduction in particulate levels over the estimated 1970 exposure levels in the region. They agreed on a maximum 24-hr particulate ambient air quality standard of 195 to 200 $\mu g/m^3$.

The Chicago region, including the states of Illinois and Indiana, established a 1972 ambient air quality goal for particulates of 75 $\mu g/m^3$. This represents a 48–55 percent reduction in particulate levels for the region.

Subsequently, the EPA administrator promulgated national ambient air quality standards for particulates.[64] Consistent with the *Guidelines*[65] both long-term and short-term standards were proposed, in addition to reference monitoring and analytical procedures. The *long-term* primary ambient air quality standard for particulates was expressed as an annual geometric mean of 75 $\mu g/m^3$. The *short-term* primary ambient air quality standard was expressed as a maximum 24-hr concentration of 260 $\mu g/m^3$ not to be exceeded more than once per year.

The *long-term* secondary ambient air quality standard for particulates is an annual geometric mean concentration of 60 $\mu g/m^3$. The *Criteria* document indicates that at an annual geometric mean particulate matter level of 70 $\mu g/m^3$, in the presence of other contaminants, there is an increased public awareness and concern for air pollution. At a particulate level of 60 $\mu g/m^3$, in the presence of sulfur oxides and moisture there is accelerated corrosion of steel and zinc. The *short-term* particulate secondary ambient air quality standard is a maximum 24-hr concentration of 150 $\mu g/m^3$ not to be exceeded more than once per year. The *Criteria* indicate that visibility may be reduced to as low as 5 miles with a concentration of 150 $\mu g/m^3$.

Sulfur Oxides—Air Quality Standards. The air quality criteria for sulfur oxides and the recommended control techniques[68] were issued by the secretary in February 1969. As for particulates, the *Guidelines* recommend both short-term and long-term ambient air quality standards. It is suggested that the long-term sulfur oxide standard be expressed in terms of an annual arithmetic average.

TABLE 3-9 FEDERAL-STATE PARTICULATE AMBIENT AIR QUALITY STANDARDS ($\mu g/m^3$)

Air Quality Control Region	State	Stds. Due	Attainment Date (Notes)	Annual Geo. Mean (24 hr)	Annual Arith. Mean	24 HOUR				1-hr max.
						max.	16%	10%	1.%	
Atlanta	Georgia	1/25/71		70.		200.				
Baltimore	Maryland	5/13/70	1973		75.	160.				
			1980		65.	140.				
Birmingham	Alabama	12/2/70	1975	75.		260.			193.	
			1978	65.		212.			166.	
Boston	Massachusetts	1/7/70	1975(g)	75.		180.				
Buffalo	New York	1/26/70	1974(b)	45.		250.	70.			
			(c)	55.		250.	85.			
			(d)	65.		250.	100.			
			(e)	75.		250.	110.			
Champlain Valley	New York	3/8/71	(b)	45.		250.	70.			
			(c)	55.		250.	85.			
			(d)	65.		250.	100.			
			(e)	75.		250.	110.			
	Vermont	Do								
Chattanooga	Tennessee	1/5/71		80.						
	Georgia	Do		70.						
Chicago	Illinois	11/10/69	1972	75.		260.			200.	
	Indiana	Do	1972	75.		260.			200.	
Cincinnati	Ohio	1/27/70	1973(f)	65.		260.			200.	
	Indiana	Do	1975	75.		260.			200.	
	Kentucky	Do		65.		220.			180.	
Cleveland	Ohio	2/17/70	1975(f)	65.		260.			200.	
Dayton	Ohio	9/14/70	1973	65.					200.	

Dallas-Ft. Worth	Texas	10/19/70	[h]	60.				125.
			[i]	65.				150.
			[j]	70.				175.
Denver	Colorado	5/7/70	1973		70.	200.		
			1976		55.	180.		
			1980		45.	150.		
Evansville-Owensborough	Kentucky							
	Indiana							
Hartford-Springfield	Connecticut	6/30/70		75.		210.	105.	
	Massachusetts	Do	[a]	70.		180.		
Houston-Galveston	Texas	10/19/70	[h]	60.				125.
			[i]	65.				150.
			[j]	70.				175.
			[k]	75.				200.
Huntington-Ashland	West Virginia							
	Ohio							
	Kentucky							
Indianapolis	Indiana	6/15/70		75.		220.		180.
	Missouri	Do						
Kansas City	Kansas	4/15/70	[g]	60.				150.
	Missouri	Do	[f]	60.				150.
Los Angeles	California	11/10/69			60.		100.	
Louisville	Kentucky	9/2/70						
	Indiana	Do		65.		220.		180.
Memphis	Tennessee	1/25/71		80.		240.		
	Arkansas	Do						
	Mississippi	Do						
Miami	Florida	3/8/71		50.		180.		
Milwaukee	Wisconsin	6/15/70		75.		260.		200.
Minneapolis-St. Paul	Minnesota	5/13/70		75.				200.

TABLE 3-9 (Continued)

Air Quality Control Region	State	Stds. Due	Attainment Date (Notes)	Annual Geo. Mean (24 hr)	Annual Arith. Mean	24 HOUR max.	24 HOUR 16%	24 HOUR 10%	24 HOUR 1.%	1-hr max.
Mobile-Gulfport-Pensacola	Alabama	5/10/71								
	Florida	Do		50.		180.				
	Mississippi	Do								
National Capital	D.C.	11/10/69	1973		75.				160.	
			1977		65.				140.	
	Maryland	Do	1973		75.	160.				
			1977		65.	140.				
	Virginia	Do	[a]	60.				100.		
New York	New York	11/10/69	1973[b]	45.		250.	70.			
			[c]	55.		250.	85.			
			[d]	65.		250.	100.			
			[e]	75.		250.	110.			
	New Jersey	Do		65.		195.				
	Connecticut	Do		70.		210.	105.			
Oklahoma City	Oklahoma	3/8/71			75.				180.	
Omaha-Council Bluffs	Nebraska	3/8/71		60.		195.			150.	
	Iowa	Do								
Paducah	Kentucky									
	Illinois									
Philadelphia	Pennsylvania	11/10/69		65.		195.				
	Delaware	Do	1975	70.		200.				500.
	New Jersey	Do		65.		195.				
Phoenix-Tucson	Arizona	1/19/71		70.		100.				
Pittsburgh	Pennsylvania	1/26/70		65.		195.				

Providence- Paw- tucket	Rhode Island	9/2/70	1973	60.	168.		
			1975	50.	130.		
	Massachusetts	Do	(g)	75.	180.		
Puget Sound	Washington	7/22/70		60.		100.	
St. Louis	Missouri	1/6/70	(f)	75.			200.
	Illinois	Do	1972	75.	260.		200.
San Antonio	Texas	10/19/70	(h)	60.			125.
			(i)	65.			150.
			(j)	70.			175.
			(k)	75.			200.
San Francisco	California	1/26/70		60.	100.		
South Bend-Benton Harbor	Indiana						
	Michigan						
Steubenville	Ohio	9/2/70	1973	65.	260.		200.
	West Virginia	Do		65.	260.		200.

[a]24-hr concentration not to be exceeded more than 10% of the time in any 30-day period.
[b]Standards applicable to Land Use, New York State Level I Category.
[c]Standards applicable to Land Use, New York State Level II Category.
[d]Standards applicable to Land Use, New York State Level III Category.
[e]Standards applicable to Land Use, New York State Level IV Category.
[f]24-hr concentration not to be exceeded over 1 day in any 3-month period.
[g]The annual average value shall be the arithmetic average of four consecutive seasonal geometric means.
[h]Applicable to land use type A.
[i]Applicable to land use type B.
[j]Applicable to land use type C.
[k]Applicable to land use type D.

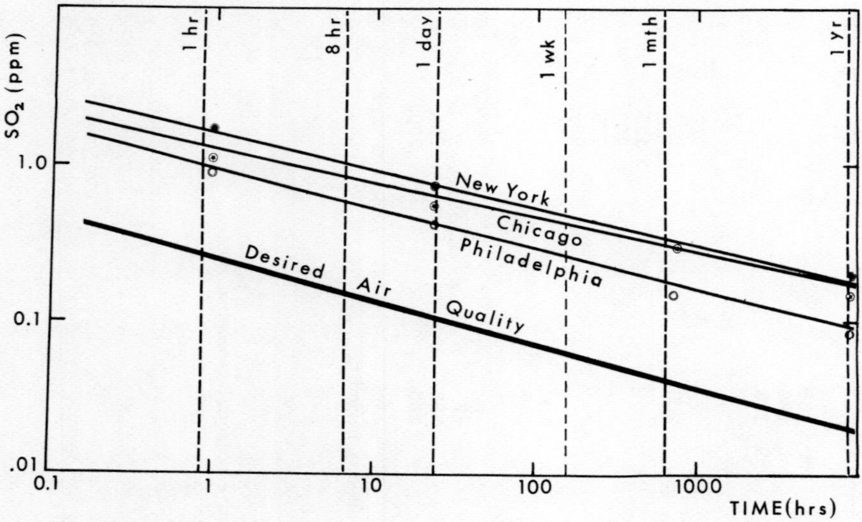

Fig. 3-7 Average maximum sulfur dioxide concentrations by various averaging times in New York, Philadelphia and Chicago compared with desirable air quality.

This standard is to be accompanied by a short-term standard or standards formulated in such a manner as to assure that, at a minimum, the 3-to-4 day health effect level is not exceeded. This requires the establishment of a 24-hr standard. Short-term standards are also advised for shorter average times to serve for "alert" monitoring and protection against short-term effects. The relationship between short-term sulfur oxide levels and long-term levels in the ambient air is graphically presented in Fig. 3-7.

It is important to recognize that the sulfur oxide standard is expressed in terms of sulfur dioxide, an index of the atmospheric level of this category of contaminants. The effects observed and reported in the *Criteria* document were observed when atmospheric sulfur dioxide levels reached the specified level of sulfur dioxide. Obviously, the cumulative level of all sulfur oxides is a higher value. Total sulfur oxides cannot be measured routinely with existing capabilities. The effect occurs, however, as a result of the category of contaminants when the index—sulfur dioxide—reaches the cited value.

The *Guidelines* recommend that the long-term sulfur oxide standard be expressed as an annual arithmetic average sulfur dioxide concentration. The proposed and adopted ambient air quality standards for sulfur oxides, as of January 1971 when the Clean Air Amendments of 1970 took effect, appear in Table 3-10.

The annual arithmetic averages vary from 43 to 114 $\mu g/m^3$. At a level of 43 $\mu g/m^3$ it is conceivable that the maximum 24-hr levels would range from 142 to 332 $\mu g/m^3$.

At a level of 114 $\mu g/m^3$, however, it is conceivable that the 24-hr level could reach 376–890 $\mu g/m^3$ based upon observed frequency distributions (Fig. 3-7). The *Criteria* indicate that when sulfur dioxide levels reach 300 $\mu g/m^3$ for 3 or 4 days there are increased hospital admissions and job absenteeism, particularly among older persons. Recognizing this fact, maximum 24-hr standards have been established, along with emergency action plans.

The *Guidelines* recommend a short-term sulfur oxide standard formulated so as to assure that, at a minimum, the 3-4-day health effect level is not exceeded. This standard is to be expressed as a 24-hr sulfur dioxide concentration.

The *Guidelines* also recommend other short-term standards for shorter time periods or averaging times for "alert" purposes, estimating compliance with standards, and affording protection against short-term effects.

In establishing maximum 24-hr sulfur oxide standards, consideration has been given to short-term, high-atmospheric levels of sulfur oxides. The maximum 24-hr standards for sulfur oxides range from 486 to 114 $\mu g/m^3$. Many regions have established 1-hr and 5-min "alert" standards, also. The test will be whether the governors will act when the crunch occurs. As for particulates, the actions in Philadelphia and Chicago to control sulfur oxides are indicative of a national trend to reduce ambient air quality levels of sulfur oxides.

Subsequent to enactment of the Clean Air Amendments of 1970, the EPA administrator promulgated national ambient air quality standards for sulfur oxides along with reference monitoring or analytical procedures.[64] Consistent with the *Guidelines* both long-term and short-term standards were promulgated.

The *long-term* primary ambient air quality standard for sulfur oxides is 80 $\mu g/m^3$ (0.03 ppm) annual arithmetic mean. The *Criteria* document indicates that when sulfur dioxide levels reach an annual arithmetic average of 115 $\mu g/m^3$, increased mortality from bronchitis and lung cancer may occur.

The *short-term* primary ambient air quality standard for sulfur oxides is a maximum 24-hr concentration of 365 $\mu g/m^3$ (0.14 ppm) not to be exceeded more than once per year. With a maximum restriction of 365 $\mu g/m^3$ (24 hr), the 4-day average could vary from an estimated 225 $\mu g/m^3$ to 185 $\mu g/m^3$. These values must be compared against the health effect level cited.

The *long-term* secondary ambient air quality standard for sulfur

TABLE 3-10 FEDERAL-STATE SULFUR DIOXIDE AMBIENT AIR QUALITY STA

Air Quality Control Region	State	Stds. Due	Attainment Date (notes)	Annual Geo. Mean (24-hr)	Annual Arith. Mean	1-mo. max.	3-day max.
Atlanta	Ga.	1/25/71			43.		
Baltimore	Md.	5/13/70	1973(j)		86.		
			1980(j)		43.		
Birmingham	Ala.	12/2/70	1975		29.		
Boston	Mass.	1/7/70	1975(h)	73.			
Buffalo	N.Y.	1/26/70	(e)		57.		
			(f)		86.		
Champlain Valley	N.Y.	3/8/71	(e)		57.		
	Vt.	Do	(f)		86.		
Chattanooga	Tenn.	1/5/71			43.		
	Ga.	Do			43.		
Chicago	Ill.	11/10/69	1972	43.			
	Ind.	Do	1972		43.		
Cincinnati	Ohio	1/27/70	1973(i)		43.		
	Ind.	Do	1975		43.		
	Ky.	Do		57.		143.	
Cleveland	Ohio	2/17/70	1973(i)		43.		
Dayton	Ohio	9/14/70	1973		43.		
Dallas-Fort Worth	Tex.	10/19/70	(k)		57.		
			(l)		72.		
Denver	Colo.	5/7/70	1973(d)		60.		
			1976(d)		25.		
			1980		10.		
Hartford-Springfield	Conn.	6/30/70			57.		
	Mass.	Do	(h)		73.		
Evansville-Owensborough	Ky.						
	Ind.						
Houston-Galveston	Tex.	10/19/70	(k)		57.		
			(l)		72.		
Huntington-Ashland	W. Va.						
	Ky.						
	Ohio						
Indianapolis	Ind.	6/15/70		43.			
	Mo.	Do					
Kansas City	Kans.	4/15/70	(i)	43.			
	Mo.	Do	1969(i)	43.			

NDARDS ($\mu g/m^3$)

24 HOUR max.	24 HOUR 1%	4-hr max.	2-hr max.	1 HOUR max.	1 HOUR 1.%	1 HOUR 0.1%	30-min 4.2%	5-min 1.%	5-min 0.1%
286.				858.					
286.					572.			1430.	
143.					286.			715.	
	86.				143.				
300.				800.					
400.	286.			1430.	715.				
400.	286.			1430.	715.				
400.	286.			1430.	715.				
400.	286.			1430.	715.				
286.				858.					
286.				858.					
486.	286.			1201.					
286.				1201.					
	286.			858.					
286.				1201.					
286.			858.	1201.					
	286.			858.					
	286.			858.					
	286.						1144.		
	572.						1430.		
300.						800.			
150.						300.			
55.									
428.	286.			1425.					
300.				800.					
	286.						1144.		
	572.						1430.		
286.				1201.					
	215.			1001.					
	215.			1001.					

TABLE 3-10 (Continued)

Air Quality Control Region	State	Stds. Due	Attainment Date (notes)	Annual Geo. Mean (24-hr)	Annual Arith. Mean	1-mo. max.	3-day max.
Los Angeles	Calif.	11/10/69					
Louisville	Ky.	9/2/70		43.			
	Ind.	Do			43.		
Memphis	Tenn.	1/25/71					
	Ark.	Do					
	Miss.	Do					
Miami	Fla.	3/8/71			9.		
Milwaukee	Wisc.	6/15/70		72.			
Minneapolis-St. Paul	Minn.	5/13/70	[i]		57.		
Mobile-Gulfport-Pensacola	Ala.	5/10/71					
	Fla.	Do			29.		
	Miss.	Do					
National Capital	D.C.	11/10/69	1973		75.		
			1975		57.		
	Md.	Do	1973[b]		86.		
			1975[b]		43.		
	Va.	Do	[a]	57.			
New York	N.Y.	11/10/69	[e]		57.		
			[f]		86.		
	N.J.	Do	[c]	49.	57.		
	Conn.	Do		57.			
Oklahoma City	Okla.	8/8/71			57.		
Omaha-Council Bluffs	Nebr.	8/8/71	1973		43.		
			1976		29.		
	Iowa	Do					
Paducah	Ky.						
	Ill.						
Philadelphia	Penn.	11/10/69		57.			
	Del.	Do	1975	72.	86.		
	N.J.	Do	[c]	49.	57.		
Phoenix-Tucson	Ariz.	1/19/71			50.		120.
Pittsburgh	Penn.	1/26/70		57.			
Providence-Pawtucket	R.I.	9/10/70	1973	72.			
			1975	57.			
	Mass.	Do	[h]	73.			
Puget Sound	Wash.	7/22/70			57.		

24 HOUR		4-hr	2-hr	1 HOUR			30-min	5-min	
max.	*1%*	*max.*	*max.*	*max.*	*1.%*	*0.1%*	*4.2%*	*1.%*	*0.1%*
114.									
286.				1201.					
286.				858.					
29.		57.							
	286.			1201.					
	286.			715.					
57.		143.		286.					
285.				860.					
225.				715.					
286.						572.			1430.
143.						286.			715.
	286.								
400.	286.			1430.	715.				
400.	286.			1430.	715.				
286.	229.			715.		572.			
428.	286.			1425.					
	285.								
150.				400.					
129.				321.					
286.				715.					
373.	287.			860.					
286.	229.			715.		572.			
250.				850.					
286.				715.					
358.				800.					
286.				686.					
300.				858.					
286.				1144.	715.				

TABLE 3-10 (Continued)

Air Quality Control Region	State	Stds. Due	Attainment Date (notes)	Annual Geo. Mean (24-hr)	Annual Arith. Mean	1-mo. max.	3-day max.
St. Louis	Mo.	1/6/70	([g])		57.		
	Ill.	Do	1972	43.			
San Antonio	Tex.	10/19/70	([k])		57.		
			([l])		72.		
San Francisco	Calif.	1/26/70					
South Bend-Benton Harbor	Ind.						
	Mich.						
Steubenville	Ohio	9/2/70	1973		43.		
	W. Va.	Do		57.			

[a]24-hr concentration not to be exceeded more than 10% of the time in any 30-day period, or not more than 1 percent of the time in 1 yr.
[b]1-hr concentration not to be exceeded more than once per calendar month; 5-min. concentration not to be exceeded more than twice per calendar week.
[c]1-hr concentration may attain but not exceed 572 $\mu g/m^3$ no more than 9 times during any consecutive 12-mon. period; 24-hr concentration may attain but not exceed 229 $\mu g/m^3$ no more than 4 times during any consecutive 12-mon. period.
[d]The 1-hr maximum arithmetic mean for any 24-hr period must not be exceeded more than once in any 1-mon. period.
[e]Standards applicable to land use, New York State Levels I and II.
[f]Standards applicable to land use, New York State Levels III and IV.

oxides is 60 $\mu g/m^3$ (0.02 ppm) annual arithmetic average. The *Criteria* document indicates that chronic plant injury and excessive leaf drop may occur at an annual arithmetic average sulfur dioxide level of 85 $\mu g/m^3$.

Two *short-term* secondary ambient air quality standards were established for sulfur oxides. A maximum 24-hr concentration of 260 $\mu g/m^3$ (0.1 ppm) was established, not to be exceeded more than once a year. The *Criteria* document indicates that at a sulfur oxide level of 286 $\mu g/m^3$ visibility may be reduced to as low as 5 miles. A second standard was adopted to minimize the impact from large sources during brief periods of adverse weather conditions. This second standard for sulfur oxides is a maximum 3-hr concentration of 1300 $\mu g/m^3$ (0.5 ppm) not to be exceeded more than once a year.

Carbon Monoxide. The air quality criteria and recommended control techniques for carbon monoxide were issued in March 1970 and the affected governors moved to establish ambient air quality standards.

24 HOUR		4-hr	2-hr	1 HOUR			30-min	5-min	
max.	*1%*	*max.*	*max.*	*max.*	*1.%*	*0.1%*	*4.2%*	*1.%*	*0.1%*
	286.			1030.					
486.	286.			1201.					
	286.						1144.		
	572.						1430.		
114.									
	286.			858.					
	286.			858.					

[g]24-hr concentration not to be exceeded over 1 day in any 3-mon. period.
[h]The annual average value shall be the arithmetic average of 4 consecutive seasonal geometric means.
[i]Not to be exceeded on more than one 24-hr period in any 3 consecutive calendar months.
[j]1-hr concentration not to be exceeded more than 8 times per calendar month; 5-min concentration not to be exceeded more than 20 times per calendar week.
[k]Standards applicable to land use types A, B, & D; 30-min. concentration with no more than one such period during any 12 hrs.
[l]Standards applicable to land use type C; 30-min. concentration with no more than one such period during any 12 hrs.

Ambient air levels of carbon monoxide exhibit diurnal, weekly, and seasonal variations. Both diurnal and weekly variations correlate with community traffic patterns, which is expected, since the internal combustion engine as the principal source accounts for 58 percent of total CO emissions. Forest fires account for an estimated 17 percent of CO emissions; industrial processes 11 percent; solid waste disposal 8 percent; and fuel combustion 2 percent.

Seasonal variables appear most dependent on meteorological factors. Macrometeorological factors can lead to air stagnation phenomena, which produce elevated CO levels in communities. Shorter term micrometeorological factors, such as mechanical turbulence caused by automobiles and air currents around buildings, are important in determining CO levels at street-side.

Consideration of ambient air quality levels alone is not sufficient in determing air quality standards for carbon monoxide. Analyses indicate CO concentrations inside automobiles in downtown traffic can approach three times those in the central urban area and five times those in residential areas. Occupants of vehicles traveling along expressways and arterial routes are exposed to carbon monoxide at levels somewhere between those in the central city and in downtown traffic.

In underground garages, in tunnels, and in buildings constructed over highways CO concentrations exceeding 100 mg/m^3 (87 ppm) have been measured. Background levels of carbon monoxide are estimated to range from 0.029 to 1.15 mg/m^3 (0.025–1.00 ppm).

This wide variation in exposure levels between downtown traffic, urban street-side, and residential exposure levels must be considered in developing ambient air quality standards for carbon monoxide. The averaging time of most importance is 8 hr. Human exposure studies involving continuous exposure to a constant CO level indicate blood carboxyhemoglobin levels approach equilibrium after about 8 hr of exposure. Approximately 80 percent of the equilibrium value is approached after 4 hr of exposure. This is accelerated, however, through physical exercise.

Following the procedures outlined in the *Guidelines* document, the governors proposed or adopted carbon monoxide ambient air quality standards (Table 3-11) prior to enactment of the Clean Air Amendments of 1970. Subsequently the EPA administrator promulgated identical primary and secondary ambient air quality standards for carbon monoxide.[64] The *long-term* ambient air quality standard is a maximum 8-hr average concentration of 10 mg/m^3 (9 ppm), not to be exceeded more than once a year. The *Criteria* document indicates that continuous exposure to carbon monoxide at 12 to 17 mg/m^3 will produce blood carboxyhemoglobin levels (2 percent) comparable to those associated with impairment in time-interval discrimination.

The *short-term* ambient air quality standard for carbon monoxide is a maximum 1 hr concentration of 40 mg/m^3 (35 ppm), not to be exceeded more than once per year. The *Criteria* document does not indicate the occurrence of effects for a 1-hr exposure period.

Hydrocarbons and Photochemical Oxidants. As mentioned earlier, *the Air Quality Criteria for Hydrocarbons* focuses on gas-phase hydrocarbons and certain of their oxidation products—particularly aldehydes—that are associated with the manifestation of photochemical air pollution. Future criteria will address particulate hydrocarbons, more specifically polynuclear hydrocarbons.

Concern for hydrocarbons in the *Criteria* document is as to their role as precursors of other compounds formed in the atmospheric photochemical system. Primary concern is not for the direct effects of hydrocarbons themselves.

Ambient air quality levels of hydrocarbons are observed to reach maximum 1 hr values of 8 to 17 ppm (as carbon). A large fraction, at least half, is probably methane. Other predominant classes of hydrocarbons include alkanes, aromatics, olefins, acetylenes, and alicyclics.

The diurnal variation of hydrocarbon levels in the ambient air resembles that of carbon monoxide. There is a pronounced maximum which appears between 6 and 8 A.M.

The elusive problem is one of relating emission levels of hydrocarbons to ambient air quality levels and then to the secondary oxidant products. The model currently used is empirical, consisting of comparing 6 to 9 A.M. average hydrocarbon levels against maximum 1-hr oxidant levels attained later in the day. This approach has validity only because of the dominating influence of macrometeorological variables on photochemical smog reactions.

When a large number of days are considered, a useful, direct relationship is observed if early morning hydrocarbon levels are plotted against maximum daily oxidant levels. The upper limit appears to vary between metropolitan areas only to the extent that meteorological differences exist between these areas.

This upper limit is given in the *Criteria* document. In order to keep the ambient air oxidant level below a daily maximum 1-hr average concentration of 200 $\mu g/m^3$ (0.1 ppm C) it is necessary to keep the 6 to 9 A.M. average non-methane hydrocarbon level below 200 $\mu g/m^3$ (0.3 ppm C). It is anticipated this maximum oxidant concentration will be exceeded about 1 percent of the time.

The *Criteria* document for oxidants provides a summary of ambient air concentration of oxidants for 12 monitoring stations over 4 years. Approximately 75 percent of the photochemical reactions are so minimal as to result in 1-hr average ozone concentrations approaching zero. This is due to the dependency of photochemical reactions on both the intensity and duration of sunlight, as well as temperature. Therefore, yearly average standards for oxidants are not adequate when applied to photochemical oxidants. For the 12 stations cited in the *Criteria* document, daily maximum 1-hr average oxidant concentrations equaled or exceeded 290 $\mu g/m^3$ (0.15 ppm) up to 41 percent of the time, ranging from 250 to 1,140 $\mu g/m^3$ (0.13 to 0.58 ppm). Shorter-term peaks were observed as high as 1,310 $\mu g/m^3$ (0.67 ppm).

Keeping the relationship between hydrocarbons and photochemical oxidants in mind, and being aware of their diurnal variability, the governors proposed or adopted in accordance with the procedures of the Air Quality Act ambient air quality standards for

TABLE 3-11 FEDERAL-STATE CARBON MONOXIDE AMBIENT AIR QUALITY STANDARDS (mg/m³)

Air Quality Control Region	States	Stds. Due	Attainment Date (notes)	Annual Geo. Mean (24-hr)	Annual Arith. Mean	7-day	24-hr max.	8 HOUR max.	8 HOUR 0.37%	90-min max.	1 HOUR max.	1 HOUR 0.14%
Atlanta	Georgia	1/25/71						11.5			50.	
Baltimore	Maryland											
Birmingham	Alabama	12/14/70						10.4			20.7	
Boston	Massachusetts	12/14/70						11.5				
Buffalo	New York	12/14/70						9.2				
Champlain Valley	New York	3/8/71						9.2				
	Vermont	Do										
Chattanooga	Tennessee	1/5/71						12.			50.	
	Georgia	Do						11.5			50.	
Chicago	Illinois	12/14/70		6.9				11.5			34.5	
	Indiana	Do						11.5			58.	
Cincinnati	Ohio	12/14/70										
	Indiana	Do						11.5			58.	
	Kentucky	Do						9.2			34.5	
Cleveland	Ohio											
Dallas-Fort Worth	Texas	12/14/70						10.				
Denver	Colorado	12/14/70			2.3				11.5			23.8
Evansville-Owensborough	Indiana	6/14/71						11.5			58.	
	Kentucky	Do						9.2			34.5	
Hartford-Sprinfield	Connecticut	12/14/70						11.5				
	Massachusetts	Do						11.5				
Houston	Texas	11/14/70						10.				

Huntington-Ashland	W. Virginia	6/14/71				
	Ohio	Do				
	Kentucky	Do			9.2	34.5
Houston	Texas	11/14/71			10.	
Indianapolis	Indiana					
	Missouri					
Kansas City	Kansas	12/14/70				
	Missouri	Do			10.4	
Los Angeles	California					
Louisville	Kentucky	12/14/70			9.2	34.5
	Indiana	Do			11.5	58.
Memphis	Tennessee	1/25/71				
	Arkansas	Do				
	Mississippi	Do				
Miami	Florida	3/8/71			9.2	13.8
Milwaukee	Wisconsin					
Minneapolis-St. Paul	Minnesota					
Mobile-Gulfport-Pensacola	Alabama	5/10/70			9.2	13.8
	Florida	Do				
	Mississippi	Do				
National Capital	D.C.					
	Maryland					
	Virginia					
New York	New York	12/14/70			9.2	
	New Jersey	Do				
	Connecticut	Do			11.5	
Oklahoma City	Oklahoma	3/8/71			34.5	
Omaha-Council Bluffs	Nebraska	3/8/71	4.6	10.4		19.6
	Iowa	Do				

TABLE 3-11 (Continued)

Air Quality Control Region	States	Stds. Due	Attainment Date (notes)	Annual Geo. Mean (24-hr)	Annual Arith. Mean	7-day	24-hr max.	8 HOUR max.	8 HOUR 0.37%	90-min max.	1 HOUR max.	1 HOUR 0.14%
Paducah	Kentucky	6/21/71						9.2			34.5	
	Illinois	Do										
Philadelphia	Pennsylvania	11/14/70						11.5			69.	
	Delaware	Do										
	New Jersey	Do										
Phoenix	Arizona	1/19/71				6.		7.		38.		
Pittsburgh	Pennsylvania											
Providence-Pawtucket	Rhode Island	12/14/70						9.2				
	Massachusetts	Do						11.5				
Puget Sound	Washington											
St. Louis	Missouri	12/14/70		6.9				11.5			34.5	
	Illinois	Do										
San Antonio	Texas	12/14/70						10.				
San Francisco	California											
South Bend-Benton Harbor	Indiana	5/26/71						11.5			58.	
	Michigan	Do										
Steubenville	Ohio											
	W. Virginia											

hydrocarbons (Table 3-12) and photochemical oxidants. Subsequent to enactment of the Clean Air Amendments of 1970 national ambient air quality standards were established.[64]

The photochemical oxidants primary and secondary national ambient air quality standards are identical: A maximum 1-hr concentration of 160 $\mu g/m^3$ (0.08 ppm) not to be exceeded more than once per year. The limit proposed[63] in January 1971 was 125 $\mu g/m^3$ (0.06 ppm). Evidence was produced, however, that such levels are reached naturally in some areas. Questions were also raised regarding the data contained in the *Criteria* document.

The *Criteria*, however, indicate an increased frequency of asthma attacks in a small proportion of subjects with this disease at levels associated with hourly average concentrations ranging from 100 to 120 $\mu g/m^3$. Adverse health effects, manifested by eye irritation, were reported at levels associated with average hourly concentrations ranging from 60 to 100 $\mu g/m^3$.

A single national ambient air quality standard was promulgated for hydrocarbons: A maximum 3-hr concentration (6 to 9 A.M.) of 160 $\mu g/m^3$ (0.24 ppm), not to be exceeded more than once per year.

Concern for hydrocarbons centers on their role as a precursor of other compounds formed in the atmosphere particularly in photochemical smog reactions. A diurnal variation is observed in atmospheric levels with a pronounced maximum between 6 and 8 A.M. The proposed standard[63] in January 1971 was 125 $\mu g/m^3$ (0.19 ppm). Adjustment of the final standard was necessary to reflect revision of the photochemical standard and the interrelationship between atmospheric levels of photochemical oxidants and hydrocarbons.

Nitrogen Oxides. Prior to enactment of the Clean Air Amendments of 1970, state ambient air quality standards were not proposed. National ambient air quality standards were proposed, however.[63] A single primary and secondary ambient air quality standard of 100 $\mu g/m^3$ (0.05 ppm) annual arithmetic mean was promulgated.[64] A proposed 24-hr nitrogen oxide standard of 250 $\mu g/m^3$ not to be exceeded more than once a year was not finalized. The annual average standard alone was judged adequate for the protection of public health. An analysis of yearly average nitrogen dioxide levels from major metropolitan areas[62] indicates that the health effect related value of 113 $\mu g/m^3$ (0.06 ppm) was exceeded in 10 percent of the cities with populations less than 50,000; in 54 percent of the cities with populations between 50,000 and 500,000, and in 85 percent of the cities with populations over 500,000.

TABLE 3-12 FEDERAL-STATE HYDROCARBON AMBIENT AIR QUALITY STANDARDS ($\mu g/m^3$)

Air Quality Control Regions	State	Stds. Due	Attainment Date (notes)	Annual Geo. Mean	Annual Arith. Mean	24-hr max.	8-hr 0.37%	3-hr max.	1 HOUR max.	1 HOUR 1%	1 HOUR 0.14%
Atlanta	Georgia										
Baltimore	Maryland										
Birmingham	Alabama	12/14/70						98.			
Boston	Massachusetts	12/14/70						118.			
Buffalo	New York	12/14/70						118.			
Champlain Valley	New York	3/8/70						118.			
	Vermont	Do									
Chattanooga	Tennessee	1/5/71	[a]					100.			
	Georgia	Do	[a]					100.			
Chicago	Illinois	12/14/70	[b]	655.						1310.	
	Indiana	Do	[b]						131.		
Cincinnati	Ohio	12/14/70									
	Indiana	Do	[b]					131.			
	Kentucky	Do	[a]					98.			
Cleveland	Ohio										
Dayton	Ohio										
Dallas-Ft. Worth	Texas	12/14/70						130.			
Denver	Colorado	12/14/70	[b]		1310.		3275.				
Evansville-Owensborough	Kentucky	6/14/71	[a]					98.			
	Illinois	Do	[b]					131.			
Hartford-Springfield	Connecticut	12/14/70						52.			
	Massachusetts	Do						118.			
Houston	Texas	12/14/70						130.			

Huntington-Ashland	W. Virginia	6/14/71					
	Kentucky	Do	[a]			98.	
	Ohio						
Indianapolis	Indiana						
	Missouri						
Kansas City	Kansas	12/14/70					
	Missouri	Do	[b]			260.	
Los Angeles	California						
Louisville	Kentucky	12/14/70	[a]			98.	
	Indiana	Do	[b]			131.	
Memphis	Tennessee	1/25/71	[a]			100.	
	Arkansas	Do					
	Mississippi	Do					
Miami	Florida						
Milwaukee	Wisconsin						
Minneapolis-St. Paul	Minnesota						
Mobile-Gulfport-Pensacola	Alabama						
	Florida						
	Mississippi						
National Capital	D.C.		1973				
			1977				
	Maryland		1973				
			1977				
	Virginia		1970[a]				
New York	New York	12/14/70				118.	
	New Jersey	Do					
	Connecticut	Do				52.	
Oklahoma	Oklahoma						
Omaha-Council Bluffs	Nebraska	3/8/71	[b]	655.	983.		1310.
	Iowa	Do					

TABLE 3-12 (Continued)

Air Quality Control Regions	State	Stds. Due	Attainment Date (notes)	Annual Geo. Mean	Annual Arith. Mean	24-hr max.	8-hr 0.37%	3-hr max.	1 HOUR max.	1 HOUR 1%	1 HOUR 0.14%
Paducah	Kentucky	6/21/70	[a]					98.			
	Illinois	Do									
Philadelphia	Pennsylvania	12/14/70						131.			
	Delaware	Do									
	New Jersey	Do									
Phoenix-Tucson	Arizona	1/19/71						80.			
Pittsburgh	Pennsylvania										
Providence-Pawtucket	Rhode Island	12/14/70	1973					118.			
	Massachusetts	Do						118.			
Puget Sound	Washington										
St. Louis	Missouri	12/14/70									
	Illinois	Do	[b]	655.						1310.	
San Antonio	Texas	12/14/70						130.			
South Bend-Benton Harbor	Indiana										
	Michigan										
San Francisco	California										
Steubenville	Ohio										
	W. Virginia										

Note: Standards are expressed for non-methane hydrocarbons unless otherwise noted; figures are for 25°C and 760 mm Hg (ppm × 655 $\mu g/m^3$ as CH_4).

[a] 3-hour (morning average).

[b] Standards are for total hydrocarbons.

TABLE 3-13 FEDERAL-STATE PHOTOCHEMICAL OXIDANT AMBIENT AIR QUALITY STANDARDS ($\mu g/m^3$)

Air Quality Control Region	States	Stds. Due	Attainment (notes) Date	Annual Geo. Mean (24-hr)	Annual Arith. Mean	24-hr max.	8-hr 0.37%	4-hr max.	1 HOUR max.	1 HOUR 1%	1 HOUR 0.14%	1 HOUR Peak
Atlanta	Georgia	1/25/71							60.			
Baltimore	Maryland											
Birmingham	Alabama	12/14/70							39.			
Boston	Massachusetts	12/14/70							118.			
Buffalo	New York	12/14/70							118.			
Champlain Valley	New York	3/8/71							118.			
	Vermont	Do										
Chattanooga	Tennessee	1/5/71							60.			
	Georgia	Do							60.			
Chicago	Illinois	12/14/70		9.8					98.			
	Indiana	Do										
Cincinnati	Ohio	12/14/70										
	Indiana	Do				59.			137.			
	Kentucky	Do				39.			98.			
Cleveland	Ohio											
Dayton	Ohio											
Dallas-Ft. Worth	Texas	12/14/70						80.	100.			200.
Denver	Colorado	12/14/70			20.		59.				98.	
Evansville-Owensborough	Kentucky	6/14/71				39.			98.			
	Indiana	Do				59.			137.			
Hartford-Springfield	Connecticut	12/14/70							49.			
	Massachusetts	Do							118.			

TABLE 3-13 (Continued)

									1 HOUR			
Air Quality Control Region	*States*	*Stds. Due*	*Attain-ment (notes) Date*	*Annual Geo. Mean (24-hr)*	*Annual Arith. Mean*	*24-hr max.*	*8-hr 0.37%*	*4-hr max.*	*max.*	*1%*	*0.14%*	*Peak*
Houston	Texas	12/14/70						80.	100.			200.
Huntington-Ashland	W. Virginia	6/14/71										
	Kentucky	Do				39.			98.			
	Ohio	Do										
Indianapolis	Indiana	12/14/70										
	Missouri	Do										
Kansas City	Kansas	12/14/70							130.			
	Missouri	Do										
Los Angeles	California											
Louisville	Kentucky	12/14/70				39.			98.			
	Indiana	Do				59.			137.			
Memphis	Tennessee	1/25/71										
	Arkansas	Do										
	Mississippi	Do										
Miami	Florida											
Milwaukee	Wisconsin											
Minneapolis-St. Paul	Minnesota											
Mobile-Gulfport-Pensacola	Alabama											
	Florida											
	Mississippi											
National Capital	D.C.											
	Maryland											
	Virginia											

New York	New York	12/14/70					118.		
	New Jersey	Do							
	Connecticut	Do					49.		
Oklahoma City	Oklahoma	3/8/71				100.	100.		
Omaha-Council Bluffs	Nebraska	3/8/71		39.	59.		98.		
	Iowa	Do							
Paducah	Kentucky	6/21/71			39.		98.		
	Illinois	Do							
Philadelphia	Pennsylvania	12/14/70							
	Delaware	Do					196.	118.	
	New Jersey	Do							
Phoenix-Tucson	Arizona	1/19/71					80.		150.
Pittsburgh	Pennsylvania								
Providence-Pawtucket	Rhode Island	12/14/70					118.		
	Massachusetts	Do					118.		
Puget Sound	Washington								
St. Louis	Missouri	12/14/70							
	Illinois	Do	9.8				98.		
San Antonio	Texas	12/14/70				80.	100.		200.

NATIONAL EMISSION STANDARDS

The Air Quality Act of 1967 directed that an investigation be conducted to determine the need for national emission standards for stationary sources. In response to this directive data were assembled on the nature and geographic distribution of major stationary sources of air pollution; the nature and magnitude of air pollution from such sources; the effects of stationary sources emissions on public health and welfare and the environment; the impact of stationary source emissions on ambient air quality and how it can be lessened through the application of emission limitations; the means of selecting and calculating appropriate emission limitations; the typical emission control regulations already in effect at state and local levels; and the prospective cost of reducing emissions from stationary sources.

The major findings were[69]

- Large stationary sources of air pollution can represent significant contributors to the deterioration of regional ambient air qualities.
- Large stationary sources of air pollution can present a serious threat to localized air quality levels although their regional effect is relatively small;
- A significant fraction of large stationary sources are located in nonurban areas.

The report presented three alternative approaches to the adoption of and enforcement of national emission standards for major stationary sources of air pollution. The first approach suggested was to apply national emission standards to such sources. These standards could be set with or without reference to resultant ambient air quality standards.

The second approach presented was essentially a modification and extension of the federal-state system in the Air Quality Act. The concept calls for state rather than federal designation of additional air quality control regions. Plans of implemenation would be adopted as under the Air Quality Act; however, the states would be expected to include in all implementation plans provision for insuring that potential new major sources of air pollution are designed and equipped to prevent and control emissions to the fullest extent possible with the technology available at the time of their construction. An additional feature would be the establishment of national emission standards for pollutants which are or may be extremely hazardous to public health. State governments would then be expected to enforce such national emission standards.

The third approach, proposed by the administration in 1970, included features of the other two. It also proposed the establishment of national ambient air quality standards. It was advocated that these standards provide for the protection of public health and guard against adverse environmental and economic effects. The states would then be expected to adopt statewide implementation plans with priority given to interstate areas where air pollution is serious. Previously established standards would remain in effect where they were more stringent than natural ones.

It must be recognized that in establishing emission limitations by source category, control of pollutants may be required where applicable air quality criteria do not exist. In such instances the Clean Air Amendments of 1970 require that source emission standards be established for existing sources in the same categories for which new source performance standards have been established. The administrator (of EPA) is required to establish a procedure under which the states can develop implementation and enforcement plans. The plans must then be submitted for federal approval, as are ambient air quality standard implementation plans.

New Source Performance Standards. The Clean Air Amendments of 1970 provide for the federal promulgation of standards of performance for new stationary sources where a category of sources may contribute significantly to air pollution which causes or contributes to the endangerment of public health. As envisioned, "standards of performance" are to reflect the maximum degree of emission control achievable when consideration is given to the processes, operation, direct emission control, and other methods available.

The concept of new source performance standards implies the application of emission limitations based on the greatest degree of emission control achieved by an industry. Inherent in the concept is consideration of technical and economic feasibility; an "available" technology must be within the economic reach of an industrial plant. In requiring the maximum degree of emission control "achievable," an industry is asked to examine and evaluate its own as well as all other possible techniques for achieving maximum control.

Of significance is the inclusion of modification to existing facilities in the definition of new sources. "Modification" is defined to include both physical and operational changes in or to a stationary source which increases emissions or results in the emission of pollutants not previously emitted.

Effective August 17, 1971, the EPA administrator proposed[70] standards of performance for new stationary sources for steam generators, portland cement plants, incinerators, nitric acid plants, and sulfuric acid plants. The proposed standards included emission limitations for one or more of the following four pollutants; particulate matter, sulfur dioxide, nitrogen oxides, and sulfuric acid mist. The proposal included requirements for performance testing, stack gas monitoring, record keeping and reporting, and EPA procedures for preconstruction review. The final standards were published[71] on December 23, 1971 for any plant which commenced modification or construction after August 17, 1971.

Incinerators. The EPA administrator established new source performance standards for solid waste incinerators with a charging rate in excess of 50 tons per day.[71] The particulate emisssion standard is a maximum 2-hr average concentration of 0.08 g/s.c.f. (0.18 g/NM^3) corrected to 12 percent carbon dioxide.

Nitric Acid Plants. The new source performance standard established for nitric acid plants[71] is a maximum 2-hr average nitrogen oxide emission of 3 lbs/ton of acid produced (1.5 kg/metric ton), expressed as nitrogen dioxide. This standard is applicable to any unit producing 30 to 70 percent nitric acid by either pressure or atmospheric pressure processes.

Portland Cement. New source performance standards were promulgated[71] by the EPA administrator for the manufacture of portland cement by either the wet or dry process for both kilns and clinker coolers. The particulate matter standards for kilns are a maximum 2-hr average emission of 0.30 lbs per ton of feed (0.15 kg/metric ton) and an opacity standard not greater than 10 percent except were the presence of uncombined water is the reason for violation. The new source particulate matter standards for clinker coolers are an opacity of 10 percent or greater and a maximum 2-hr average emission in excess of 0.10 lb/ton of feed to the kiln (0.050 kg/metric ton).

Steam Electric Power Plants. Particulate matter, sulfur oxide, and nitrogen oxide new source performance standards were promulgated[71] for fossil fuel-fired steam generating units with a heat input in excess of 250 million Btu/hr. The particulate standard is a maximum 2-hr average concentration of 0.10 lb/million Btu heat input (0.18 g/million cal). The opacity standard is 20 percent except that 40 percent opacity is permitted for not more than 2 min in any hour.

The new source performance standards for sulfur dioxide were promulgated for liquid and solid fossil fueled units. The liquid fossil fuel standard is a maximum 2-hr average sulfur dioxide emission of 0.80 lb/million Btu input (1.4 g/million cal). The solid fossil fuel standard is a maximum 2-hr average concentration of 1.2 lb/million Btu input (2.2 g/million cal). Where a combination of fossil fuels are burned simultaneously compliance with the sulfur oxide standard is determined using the following formula:

$$\frac{0.80\ Y + 1.2\ Z}{X + Y + Z}$$

The symbols X, Y, and Z represent the percentages of total heat input derived from gaseous, liquid, and solid fossil fuels, respectively.

Nitrogen oxide new source performance standards also were established for each fuel; however, lignite-fired units were exempted for lack of sufficient information. The standards are a maximum 2-hr nitrogen dioxide standard for gaseous fossil fuel-fired units of 0.20 lb/million Btu input (0.36 g/millon cal); for liquid fossil fuel-fired units, 0.30 lb/million Btu input (0.54 g/million cal); and for solid fossil fuel-fired units 0.70 lb/million Btu input (1.26 g/million cal). Where a combination of fossil fuels are burned simultaneously compliance with the nitrogen oxide standard is determined by the following formula:

$$\frac{0.20\ X + 0.30\ Y + 0.70\ Z}{X + Y + Z}$$

The symbols X, Y, and Z represent the percentages of total heat input derived from gaseous, liquid, and solid fossil fuels, respectively.

Sulfuric Acid Plants. New source performance standards for sulfuric acid production apply to facilities employing the contact process but do not apply to facilities where the purpose of the unit is the control of atmospheric emissions of sulfuric dioxide or other sulfur compounds. The sulfur dioxide standard is a maximum 2-hr average emission of 4 lbs/ton of acid produced (2 kg/metric ton). An acid mist standard also was established at a maximum 2-hr average emission of 0.15 lb/ton of acid produced (0.75 kg/metric ton).

Hazardous Air Pollutants. Provision was made also in the 1970 Amendments for national emission standards for hazardous air pollutants. Included are those materials to which existing standards are not applicable and which, in the judgment of the administrator may cause, or contribute to, serious irreversible, or incapacitating reversible, illness.

The first standards to be proposed by the EPA administrator for hazardous pollutants were for asbestos, beryllium, and mercury.[72] These standards had not been finalized at the time this text went to press.

It is intended that these standards apply to all sources and provide ample margins of safety to assure public health protection. A maximum 2-year exemption can be provided by the President only where he finds that the necessary control technology is unavailable and that operation of the source is required for national security.

While enforcement of the new source performance standards and national emission standards for hazardous materials is primarily a federal responsibility, an opportunity is provided for delegation of this authority to any state which demonstrates an enforcement capability.

PLANS OF IMPLEMENTATIONS

The Air Quality Act of 1967 provided for the development of implementation plans for each air quality control region. The plan was to represent a blueprint of the steps that must be taken to achieve established air quality standards.

The emission standards were to be included in the implementation plan derived from the ambient air quality standards. Consideration was to be given to both long-term control and short-term requirements. An integral part of the plan of implementation, emission standards prescribe legal limits that individual sources cannot exceed.

There are two basic philosophies held by regulatory or control officials when establishing plans of implementation that deserve comment here. One philosophy is that emission standards must be capable of enforcement today. These same emission requirements must be economically reasonable.

This necessitates a regulatory determination of economic feasibility, and such determination may well reflect the "backbone" of the regulatory agency. A "minimum" standard might well be established that reflects the average emission controls required in the United States.

A "good practice" standard could also be established which reflected the emission controls incorporated in the better installations in the world. Some dedicated control officials go one step further, proposing the establishment of the "best feasible" emission standard: one that reflects the highest degree of control available with current technology and incorporated in the best installations in the world. There are situations, however, where even the best feasible practice is not good enough to achieve the desired ambient air quality standard.

When the best feasible practice is insufficient, alternative control approaches quite frequently are possible (e.g., land-use planning). These alternatives are reflected in the Clean Air Amendments of 1970. It was the Senate's intent[22] that these alternatives include land-use and transportation controls.

The second regulatory or control philosophy advocates that emission standards be determined according to technical and economic feasibility alone. Frequently, when compliance is impossible, a procedure is provided for granting variances. The variance contributes nothing to the reduction of the problem and is, in effect, a license to violate the law.

Federal-State Emission Standards

The Air Quality Act of 1967 provided a systematic approach to the establishment of emission standards. Minimum emission standards for new sources and hazardous materials were provided, however, with the Clean Air Amendments of 1970. The air quality control region approach was preserved and there still remained the need to establish regional plans to implement ambient air quality standards. The Senate report on the Air Quality Act of 1967 stated:[18]

> The achievement of the established ambient air quality standards is contingent upon the application of meaningful emission controls on the various sources of air pollution, within a given air quality control region.
>
> Such emission control requirements are established for the purpose of achieving specific air quality standards. They may include such alternative courses of action as process changes, flue gas stack controls, stack height requirements, fuel use limitations, or plan location rules, but in any event should include the best available technology required to achieve the desired level of ambient air quality.
>
> Any emission control should be based on a survey of the emission inventory within the air quality control region. This survey should include the amounts and types of pollutants being emitted, an evaluation of those meteorological factors that will influence their dispersal and transport, and a consideration of the various other factors that influence air quality.

Subsequently, these emission inventories will need to be updated as new sources of pollution are established within the air quality control region. Consequently, it is desirable that the emission inventory survey include a projection of increased urbanization, industrialization, and population trends within the air quality control region.

It should be noted that when emission control requirements are set on the basis of emission inventories for a given air quality control region, those emission control requirements could conceivably vary from one air quality control region to another, even though their application may result in the same degree of ambient air quality.

Since emission control requirements are legally enforceable, limitations on the amount of pollution that a single source or category of source may discharge into the atmosphere, they must consider technology and the economics of control. Current methods of control that are technologically and economically feasible may not be as effective as required to achieve the desired ambient air quality. Therefore, as technology advances, new emission control requirements must be implemented on a continually more restrictive basis until such time as the established ambient air quality standards is achieved. Such an approach necessitates that the best available control technology be applied at the time it is developed.

Assurance that ambient air quality standards are achieved and maintained requires that implementation plans consider increased urbanization, industrialization, and population growth.

While developing implementation plans each state must hold a public hearing on their proposed plan to implement the national standards or more stringent state standards. Implementation plans are then submitted to the federal administrator for approval. The plans must provide for implementation, maintenance, and enforcement of applicable standards.

Detailed implementation plan requirements are enunciated in the Clean Air Amendments of 1970. A plan of implementation must include

- emission limitations, schedules, and time tables for compliance with such limitations;
- measures to insure attainment and maintenance of applicable standards, including land-use and transportation controls;
- provision for environmental monitoring and the compilation and analysis of data on ambient air quality;
- a procedure for review of the location of new sources subject to national performance standards, prior to construction or modification;
- requirements for monitoring and periodic reports on the nature and amounts of emission by the owners or operators of stationary sources;

- assurances that the state will provide adequate personnel, funding, and authority to carry out the implementation plan.

It is possible to envision cases where currently available control techniques are not adequate to achieve and maintain ambient air levels which protect public health; for example, in the case of carbon monoxide and nitrogen oxides. In these instances emission controls must be established on a continually more restrictive basis as technology improves, until air quality levels that provide for the protection of public health are achieved and maintained.

The federal policy requires that implementation plans be developed and implemented to achieve protection of public health or the national primary ambient air quality standard 3 years after plans are approved. Two procedures were available for extending the 3-year deadline, however.

Prospectively, at the request of a governor, the administrator may extend the 3-year period up to two additional years if a determination is made that the necessary technology or other alternatives are not available and reasonably available alternatives were considered and applied. Provision must be made for reasonable interim emission control measures, however.

A 1-year extension for any stationary source or class of sources is available, also, where the administrator determines

1. good faith efforts have been made to comply;
2. the necessary technology or alternative methods are not available;
3. any available alternative operating procedures and interim control measures were employed;
4. continued operation of such source is essential to national security or to the public health or welfare.

Thus, the administrator has the authority to extend compliance schedules up to three years. Any determination is subject to judicial review in the U.S. Court of Appeals.

An effective approach to the implementation of ambient air quality standards is the Rollback Technique, presented graphically in Fig. 3-8. Under the Rollback Technique, new or revised emission restrictions are applied on a continually more restrictive basis. This is in effect the procedure that must be provided in achieving the national secondary ambient air quality standard once the primary standard is achieved.

Consideration is given to the obsolescence rate of existing sources, the replacement rate of sources phased out by obsolescence, and the addition of new sources. Increasingly in the regulatory process officials and industry are becoming aware that possibly accelerated obsolescence or compulsory land-use plans, in

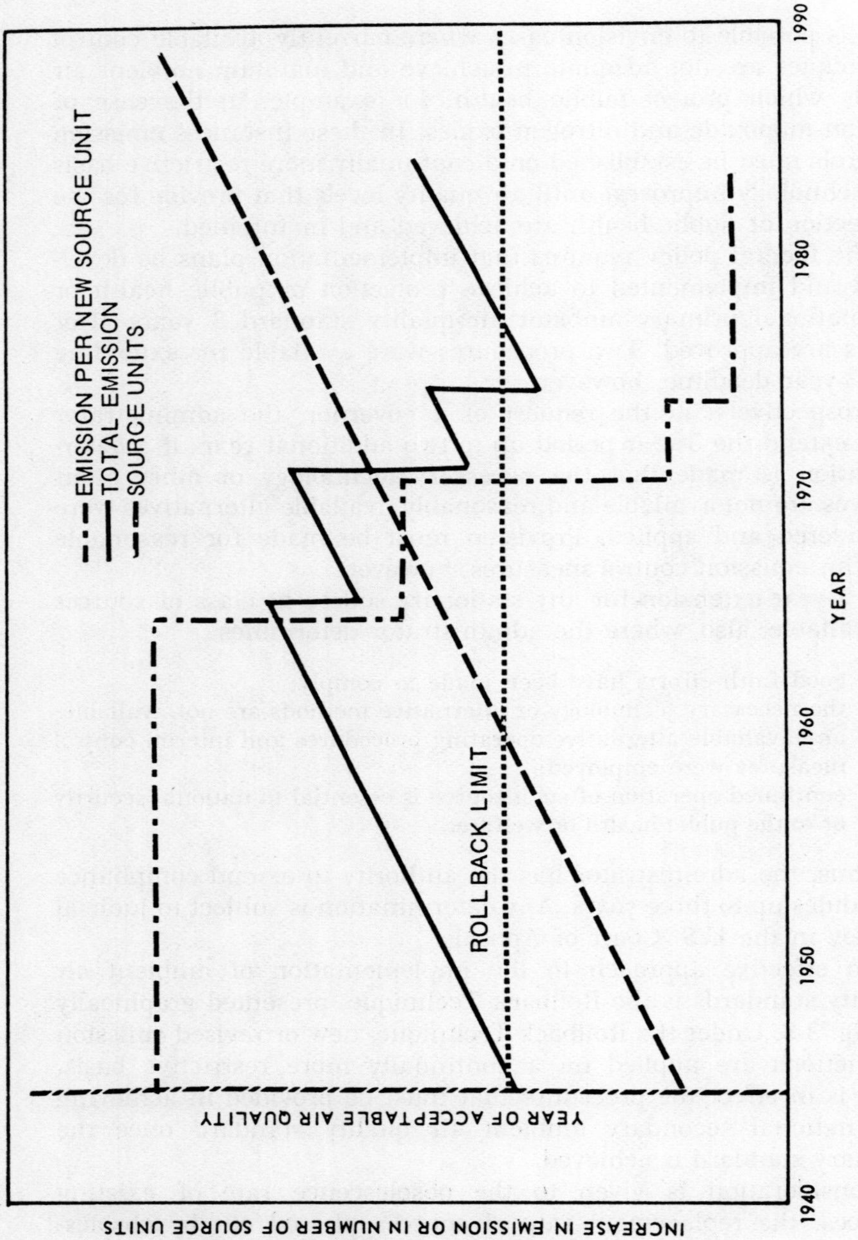

Fig. 3-8 Rollback technique.

some air quality control regions, will be the only effective methods for achieving and maintaining ambient air qualities that protect public health.

An interesting example of the Rollback Technique is the control of automotive emissions in California. In Los Angeles in the early 1940s records of automobile registration and gasoline consumption and measured emission rates from vehicles were used to estimate the total emissions by automobiles in the "before smog" era. Carbon monoxide, hydrocarbons, and nitrogen oxides were considered. From estimates of the number of vehicles operating at future dates and from the acceptable total emission limit, maximum allowable emissions were calculated.

The difficulty is in providing a time schedule that is reasonable and equitable. Too long a time schedule for implementation can prove as much a deterrent to success as one too short. The English consider planned obsolescence within 7 years reasonable in their enforcement of emission controls.

There are basically four measures[69] that can be adopted to control atmospheric emissions within a region.

(1) It can limit or prohibit the operation of specific processes within its boundaries, such as rendering of animal matter, tanning of leather, or making of steel. Such a local or regional control measure would force the proscribed process or industry into the hinterland beyond the limiting region, but as a state or national measure it presumably would deprive the state or nation of necessary products or services and thus would not be feasible. The limitation on operation may be conditional depending on location, distance from land-use zones, or time of day (e.g., it may prohibit the burning of leaves during specific hours.)

(2) It can limit or prohibit the importation of specific substances such as fuels of a specified composition, dead animals, hides for tanning, or certain types of ore. This control measure is more feasible for regional than for state or national application. The limitation can be with or without a condition that depends on time or location.

(3) It can limit or prohibit the way processes are operated and the way substances are used. It may require that all incinerators be multiple-chamber, that stacks on certain processes have a specified minimum height, or that coal not be burned on hand-fired grates. This measure is pactical for regional, state, or national application since it does not prohibit the operation of processes or the use of substances; it merely prescribes how they must be operated or used.

(4) It can limit or prohibit the emission of specific pollutants

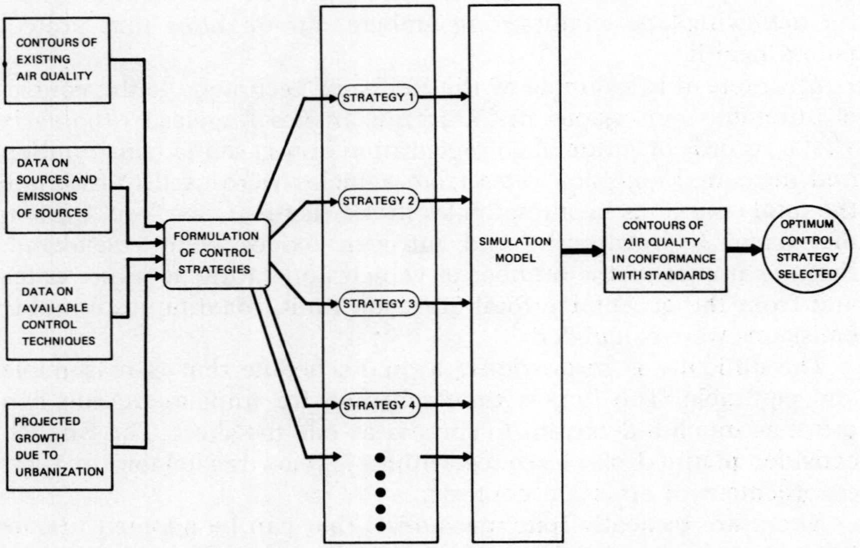

Fig. 3-9 Flow diagram for use of modeling in the development of implementation plans.

from sources in accordance with air quality requirements, in the light of source density. This measure is practical for regional, state, or national applications; practically is determined by the characteristics of the process and the raw materials, not by the demographic characteristics of the area. For example, the limitation can prohibit the use of a process in a large city but allow its use in a sparsely populated country if the process is necessary to the economy and does not endanger health or welfare. This measure stresses the need for flexibility in the structure of emission standards applied to large areas. Although a jurisdiction can, by legislative enactment, limit the emission of all pollutants and include in the enactment a definition of a pollutant, some agency subsequently must apply this definition to specific emissions to determine whether or not they fall within intent of the enactment.

These four types of measures are interdependent. Any regional control scheme may well involve all four elements. This is presented schematically in Fig. 3-9.

As a condition for federal approval of the plan of implementation the various requirements included must be enforceable by state action. At the same time this material was prepared, some states did not have sufficiently broad enough enforcement authority (Table 3-1), and implementation plans were not established for any air quality control regions. It is possible to speculate, however, on minimum emission standards that will be required and the form they will take.

Particulates. Particulate emission standards will be formed to limit the total of these on the basis of mass rate, rather than concentration.[73] Most likely, for fuel-burning equipment, regulations comparable to those of the state of Maryland (Fig. 3-1) will be required. For incinerators, standards similar to that for New Jersey for new installations will be required (Fig. 3-2).

For process sources, a process weight rate regulation modeled after that of the San Francisco Bay Air Pollution Control District is anticipated. This regulation limits the weight of particulate emissions per hour as a function of the total weight of raw materials introduced into a process operation.

Sulfur Oxides. For fuel-burning equipment it is reasonable to assume that sulfur oxide emission standards will be similar to those now in force in St. Louis and Los Angeles. For large steam electric power plants the average national sulfur oxide emission will approach 1.46 lb sulfur dioxide per million Btu input; this is equivalent to a coal with a 1.0 percent sulfur content by weight or 1.38 percent sulfur content oil. In critical areas, however, the sulfur oxide emission standard could easily approach 0.9 lb sulfur dioxide per million Btu input for installations larger than 2,000 million Btu/hr.

For industrial installations, such as petroleum refineries and by-product coke ovens, the national average sulfur dioxide emission standard will probably approach a concentration standard of 500 ppm sulfur dioxide. Regional emission standards may well approach 50 grains of sulfur compounds (calculated by hydrogen sulfide at standard conditions) per 100 cu ft of gaseous fuel.

For small multiple sources sulfur oxide regulations will continue to be based on restricting the sulfur content of fuel, because this is easily enforced by regulations on the importation, distribution, and sale of high-sulfur fuels. Emission-testing of numerous small sources is not feasible.

Carbon Monoxide. It has been estimated that carbon monoxide emission standards will have to be reduced by at least 95 percent by 1975 through the treatment of all exhaust gases.[68]

Hydrocarbons. On a national basis transportation represents the largest source category for emissions of hydrocarbons and related organic compounds, accounting for 52 percent. Other major sources include organic solvent evaporations (27 percent) industrial processes (14 percent), solid waste disposal (5 percent), and fuel combustion in stationary sources (2 percent).

Under local variations transportation sources account for from 37 to 99 percent of local emissions, and process losses account for from 1 to 63 percent. Hydrocarbon emissions, therefore, originate primarily from the inefficient combustion of fossil fuels and from their use as process raw materials.

Although emission standards for hydrocarbons have not been established, estimates are possible.[73] The control of hydrocarbons from process sources employs the same four basic principles applied to the control of automotive emissions:

1. combustion process optimization;
2. recovery by mass transfer principles;
3. restriction of evaporative losses;
4. process material and fuel substitution.

For process sources a 90-percent removal of organic material by weight is envisioned by 1975.

It is anticipated that petroleum product storage tanks, reservoirs, and containers of more than 40,000-gallon capacity and vapor pressure must be equipped with floating roofs, vapor recovering systems, or other equally efficient control equipment. In addition, submerged filling inlets will be required on all gasoline storage tanks of 250-gallon capacity, or more.

Estimated Impact

Following enactment of the Air Quality Act of 1967 a national target date of 1975 was suggested for compliance with ambient air quality standards for particulate, sulfur oxides, carbon monoxide, hydrocarbons, and photochemical oxidant standards. This objective was incorporated as public policy in the Clean Air Amendments of 1970 which provided for national primary ambient air quality standards providing for the protection of public health.

In 1969, estimates[74] indicated that this would require approximately 25 percent compliance by 1973, an additional 50 percent compliance by 1974, and 100 percent compliance by mid-1975 for the 100 urban areas selected.

When estimating the impact of federal, state, and local air pollution control programs, several categories of stationary sources were selected. The first included solid waste disposal sources which emit primarily particulates and hydrocarbons; the second, industrial processes which emit particulates, sulfur oxides, hydrocarbons, and carbon monoxide.

TABLE 3-14 NATIONAL INDUSTRIAL EMISSIONS: PARTICULATES, SULFUR OXIDES, AND OTHER POLLUTANTS

		PARTICULATES		Sulfur Oxides,	OTHER POLLUTANTS	
Industry	*Production Per Year*	*Control (percent)*	*Emissions (tons per year)*	*Emissions (tons per year)*	*Emissions (tons per year)*	*Pollutant*
Carbon black	1,242,000 tons	96	103,000		285,000	Carbon monoxide.
Cement	74,000,000 tons	90	870,000			
Coal cleaning		45	160,000			
Coal refuse	190,000,000 tons		400,000	600,000	1,200,000	Do.
Coke (by product)	66,000,000 tons	45	94,000	600,000	1,500,000	Hydrocarbons.
Cotton ginning	10,825,500 bales	50	30,000			
Ferroalloy	2,488,000 tons	20	275,000			
Grain milling and handling		50	1,122,000			
Gray iron foundry	23,800,000 tons	12	190,000		3,500,000	Carbon monoxide.
Iron and steel[a]	214,100,000 tons	65	1,700,000		75,000	Do.
Nitric acid	6,100,000 tons	70	50,000		110,000	Nitrogen oxides.
Nonferrous metallurgical:						
Aluminum reduction	3,967,000 tons	80	92,500	54,000		
Copper smelting	2,293,000 tons	50	60,000	2,950,000		
Lead smelting	944,000 tons	50	55,500	263,000		
Zinc smelting	1,373,000 tons	50	28,500	640,000		
Petroleum refining	11,080,000 bbls. per day	60	85,000	2,100,000	2,030,000	Carbon monoxide.
					1,950,000	Hydrocarbons.
Phosphate fertilizer		90	259,000		53,000	Fluorides.
Phosphoric acid	4,500,000 tons	92	5,000		1,100	Acid mists.
Pulp:						
Kraft (sulfate)	22,300,000 tons	90	630,000	84,000	2,600,000	Carbon monoxide.
Sulfuric acid	24,700,000 tons	96	31,000	590,000	23,000	Nitrogen oxides.
Municipal incineration	16,000,000 tons[b]	10	136,000			
Steam-electric power:						
Coal fired	266,400,000 tons[b]		5,500,000	14,000,000	2,975,000	Nitrogen oxides (coal and oil).
Oil fired	144,300,000 bbls.[c]	80	30,000	1,000,000		

[a]Includes: Pig iron, 87,000,000; raw steel, 127,100,000. [b]Tons of waste incinerated. [c]Quantity of fuel consumed.

TABLE 3-15 ESTIMATED IMPACT OF CLEAN AIR ACT, AS AMENDED, ON MAJOR INDUSTRY CATEGORIES IN 1975 ·

		EXPECTED EMISSION LEVEL (THOUSANDS OF TONS PER YEAR)[a]				ASSOCIATED EMISSION CONTROL LEVEL (PERCENT)[a]			
Industry Category	*Year*	*Parti-culates*	*SO_x*	*HC*	*CO*	*Parti-culates*	*SO_x*	*HC*	*CO*
Asphalt batching plants	1967	206				80.0			
	1975	82				92.0			
Cement plants	1967	525				86.0			
	1975	37				99.0			
Coal-cleaning plants	1967	27				57.5			
	1975	3				94.6			
Commercial-institutional heating plants	1967	64				0	0		
	1975	31				51.5	14.0		
Grain mills	1967	96				35.0			
	1975	1				99.0			
Grain elevators	1967	1,100				35.0			
	1975	17				99.0			
Gray iron foundries	1967	94			1,250	12.0			18.0
	1975	11			76	90.0			95.0
Iron and steel plants	1967	1,060				52.5			
	1975	101				95.5			
Industrial boilers	1967	1,540				62.0	0		
	1975	50				98.8	63.0		
Kraft (sulfate) pulp plants	1967	109				81.0			
	1975	21				96.3			
Lime plants	1967	251				60.0			
	1975	25				96.0			

Primary nonferrous metallurgical plants	1967	12/10	144			95/85.0	49.0		
	1975	12/3	4			95/96.0	98.8		
Secondary nonferrous metallurgical plants	1967	45				48.5			
	1975	4				95.0			
Petroleum refining plants	1967	48	293	587	1,080	67.0	37.0	48.5	47.0
	1975	26	737	79	102	82.0	50.0	93.0	95.0
Petroleum products storage plants	1967			636				53.5	
	1975			343				75.0	
Phosphate fertilizer plants	1967	11				97.0			
	1975	4				99.0			
Residential heating plants	1967	163	1,043			0.0	0.0		
	1975	57	541			65.2	48.1		
Rubber (tires and inner tubes) plants	1967								
	1975								
Solid waste disposal	1967	844		160	1,950	53.5		63.0	63.0
	1975	167		43	262	91.0		95.0	90.0
Steam-electric power plants	1967	2,190	6,860			86.0	0		
	1975	198	2,500			98.7	63.6		
Sulfuric acid plants	1967	40	480			45.7	0.0		
	1975	24	56			67.0	86.0		
Varnish plants	1967			5				18.0	
	1975			1				90.0	

[a]For levels of particulates, sulfur oxides (SO_x), hydrocarbons (HC), carbon monoxide (CO). Blanks in the table indicate the emission levels meet the applicable regulations without additional control.

Four types of fuel combustion sources were considered: steam-electric power plants, industrial boilers, commercial-institutional heating plants, and residential heating plants. These four account for an estimated 32 percent of the particulates and 74 percent of the sulfur oxides emitted in the nation by all sources (Table 3-14).

Industrial processes, in turn, included the following: kraft (sulfate) pulp, iron and steel, gray iron foundry, sulfuric acid, petroleum refining, asphalt batching, cement, primary nonferrous metallurgical smelting, phosphate fertilizer, lime, coal, cleaning, petroleum products storage, grain milling and handling, varnish, rubber (tires and inner tubes), and secondary nonferrous metallurgical recovery.

The 16 industries or industrial groups mentioned contribute approximately 21 percent particulates, 22 percent sulfur oxide, 7 percent hydrocarbons, and 7 percent carbon monoxide emissions nationally. These figures, however, do not account for industrial growth.

In estimating emission requirements, regulations already in effect in various places in the United States were considered. Using the target date of 1975, the estimated percentage reductions[74] are provided in Table 3-17.

In the following discussion, estimates are presented of current emissions and anticipated percentage reductions required by 1975 (Table 3-15). No effort is made to present regional differences. These figures do not reflect requirements arising from the Clean Air Amendments of 1970, which provide for new source performance standards and emission standards for hazardous materials.

Asphalt. The production of asphalt, a road surface paving mixture, results in the emission of a fine particulate dust. Major emissions occur when the stone aggregate is crushed, dried, and heated.

Almost all producers have primary collection devices which yield a particulate emission control level of 80 percent; however, this control level will not meet anticipated process weight rate regulations. Overall emission control will most likely have to increase to an average of 92 percent. This will require the installation of secondary collection devices, such as electrostatic precipitators.

Carbon-black. Carbon black, an ultrafine soot, is produced by two methods: (1) thermal decomposition, or (2) burning gaseous or liquid hydrocarbons in a limited air supply. In 1967, there were 31 plants, the majority of which—87 percent—were outside the first 57 air quality control regions (AQCR's).

The carbon black industry is expected to undergo a 35 percent growth by 1979.[75] This industry most likely will be subjected to national emission standards; however, thermal and furnace processes for carbon black production are amenable to emission control with present-day technology. No satisfactory technique exists for emission control of the channel process, but it is believed that such plants will be phased out in the near future. This phaseout will result in a 97-percent reduction of present emissions. Estimates of the impact of emission standards on the carbon black industry are not available.

Cement. Portland cement, which accounts for 98 percent of the cement in the United States, is produced by both dry and wet processes. In 1968, there were 178 plants, but only 28 percent are in the first 57 AQCR's. These are currently operating at 77 percent capacity and a 50-percent growth rate is expected in the next 10 years.[76] Four major operations are involved: (1) quarrying and crushing, (2) grinding and blending, (3) clinker production, and (4) finish grinding and packaging. Particulate emissions occur for each operation, but the major source is from the kilns, which operate at 2700°F and produce an intermediate product which is later ground into a fine powder. Dry-process kilns emit about 21 percent more dust than wet-process ones.

Currently it is believed that all kilns have at least a primary dust collector with an overall collection efficiency of 86 percent. By 1975, process weight regulation for particulate control will probably require about 99 percent overall collection efficiency. This will require the installation of secondary collectors on all kilns.

Coal Cleaning. The demand for coal cleaning—the removal of noncombustible materials from coal—has increased steadily and by 1965 amounted to 65 percent of the coal produced. The industry is expected to exhibit a 100-percent growth by 1979, however, due in large part to air pollution regulations. There are 667 locations in the United States cleaning coal; however, only 5.9 percent are in the first 57 AQCR's and only 18 percent in SMSA's.

The cleaning operation is accomplished by washing the coal with air or water. Particulate dust is the major atmospheric contaminant and arises from three major sources: (1) flash driers, (2) fluidized-bed dryers, and (3) pneumatic cleaners. The best information currently available indicates that the present overall control of particulates in the industry is around 45 percent; however, within metropolitan areas the overall efficiency is around 57.5 percent. About 87 percent of all thermal driers (flash and fluidized-bed) are controlled at a level of 80 percent efficiency and about

16 percent of the pneumatic cleaners are controlled at an 80-percent level. To upgrade the control of particulates to meet the estimated process weight rate regulation would increase the industry's overall collection efficiency to 95 percent.

Commercial-Institutional Heating Plants. It is estimated that there are more than 999,000 heating plants that provide heat and steam for hotels, retail stores, schools, hospitals, and such. About 600,000 of these are within 100 metropolitan areas. The average capacity of these plants is less than that of an industrial boiler.

Currently, emissions from these boilers are under little or no control. Because relatively simple-to-operate heating equipment is used in commercial and institutional establishments, the emissions can be reduced only by changing to fuels with low ash and sulfur content—namely, natural gas and oil. By this method particulates and sulfur oxides emissions can be reduced by 52 and 14 percent, respectively.

Cotton Ginning. There are approximately 4,400 cotton gins in the United States which produce from 200 to 20,000 bales per year. It is reasonable to assume that the majority will not be confined to AQCR's. There is very little information available on current air pollution control practices. Emission controls are available, but due to low capitalization of many ginning companies, the cost of control in general has been prohibitive.

Grain Handling and Milling. Particulate wastes occur in the handling, storage, and cleaning of grains. Dusts are generated also in the milling operations which process grain into flour, livestock feeds, cereals, corn syrup, and various bread and pastry mixes.

For the grain-handling facilities the wastes amount to an estimated 0.3 percent of the bulk material handled. Of the 11,000 grain elevators in the United States only 15 percent are within SMSA's.

In the milling of grain the primary air contaminant is particulates; however, other contaminants occur in the processing of linseed and soybean oil, or the malting of barley. In the milling of other than livestock feeds the resulting dusts are controlled because of their value. Approximately 65 percent of the 618 flour mills in this country are outside AQCR's.

Control of dust emitted from milling and handling can be accomplished with air-cleaning devices, such as baghouses, to give 99 percent or better control as compared to the present average collection efficiciency of 35 percent. Such controls will bring the levels of pollution into compliance with the process weight rate regulation for particulate control.

Gray Iron Foundry. Atmospheric emissions from foundries result from the melting of metals to produce castings. Three types of furnaces are used: cupolas, electric arc, and electric induction. Cupolas are used to produce 93 percent of the gray iron casting. In the cupola furnace coke is burnt in direct contact with the metal. The most abundant emissions are particulate dust and carbon monoxide. Good control of cupolas is possible with present-day technology.

Electric furnaces, which are generally used to make special alloys, do not require fuel and as a result are cleaner. Emission control from electric arc furnaces is presently difficult, however, due to overhead hooding limitations. This is another industry which is exemplified by many small operations, approximately 1,700, and the cost of emission controls may be prohibitive. These foundries are generally small and competitive businesses; about half of them are captive (owned and controlled by other businesses).

Gas-cleaning equipment, such as wet scrubbers, in combination with afterburners can reduce cupola emission levels to achieve compliance with the process weight standard for particulates and carbon monoxide. These assumed regulations will require the industry to increase its present average particulate removal efficiency of 12 percent to 90 percent. For carbon monoxide the present 18 percent control will have to be increased to 95 percent.

Industrial Boilers. In the United States there are an estimated 307,000 industrial boilers that supply steam for material processing, space heating, and, in some large industrial complexes, electric power generation. Approximately 219,000 are within the 100 major metropolitan areas. Though the boilers involved generally are smaller than those in steam-electric power plants, they cause a significant amount of pollution in highly industrialized areas.

In 1967, boilers in industrial plants emitted 3.0 million tons of particulates and about 5.3 million tons of sulfur oxides. Average control was estimated at 20 percent for particulates and there was little control of sulfur oxide emissions. As for steam-electric power plants it is expected that the equivalent of the Maryland particulate emission control regulation will be required by 1975. This can be met by using various types of gas-cleaning equipment. Such control devices have little effect on sulfur oxides, however. To achieve sulfur oxide emission standards, industrial boilers shift from high-sulfur to low-sulfur oil to oil which does not exceed 1.38 percent sulfur content. Such a change in fuel will reduce both particulates and sulfur oxides to comply with the applicable regulations.

Iron and Steel. There are four major processes which contribute to air pollution in the iron and steel industry: coking, sintering, melting, and scarfing. Nearly all the coke used is produced in by-product ovens. There are 65 plants, which are primarily within AQCR's regions. Although new plant design features hold promise for emission control, current technology is not satisfactory.

Major types of furnaces are blast, open hearth, basic oxygen, and electric. The resultant emissions are particulates, carbon monoxide, small amounts of nitrogen oxides, and other common combustion gases. Fluorides can be emitted both as gaseous and particulate matter. There are 224 blast furnaces producing pig iron; all but 6 are within SMSA's. Data was not available on these installations or scarfing operations. The steel industry is expected to undergo a 45-percent growth rate by 1979. Currently there are 138 steel plants of which 86 percent are within SMSA's. Good controls appear practicable with current technology.

Particulates are now controlled to an overall efficiency of about 52 percent. Process weight emission standards by 1975 will require an estimated 95 percent control. This will require the application of gas-cleaning equipment such as electrostatic precipitators and wet scrubbers.

Kraft (Sulfate) Pulp Industry. Among the various methods employed to manufacture pulp from wood and other materials for making paper and related products those of air pollution concern are the chemical methods. Sulfite and sulfate (kraft) pulp plants account for 75 percent of the industry's pulp output. The pulp industry is expected to experience an 80 percent growth by 1979. Sulfite plants are not included in the figure.

The sulfate (kraft) process is carried out at 113 plants, mostly in the southern states. Less than 10 percent of both the kraft and sulfite plants are within the first 57 AQCRs. Sulfate plants emit significant amounts of particulates from four devices (1) recovery furnaces, (2) smelt dissolving tanks, (3) time recovery kilns, and (4) bark boilers. Although recovery furnaces also emit sulfur oxides, the best estimates indicate emissions are less than the assumed regulation limit of 500 ppm.

The necessary reductions in particulate emissions to meet process weight regulations can be accomplished with gas-cleaning equipment. The application of these devices will allow the 1967 national average collection efficiency of 81 percent to be increased to 96 percent by 1975 (Table 3-15). Further research is needed, however, on both better control methods and new less-odorous pulping processes.

Lime Production. Lime is produced in three forms from quarried lime-bearing stone: quicklime, hydrated lime, and burned dolomite. Lime products are used in construction, the manufacture of cement, and in the manufacture of products such as steel. The principal lime-manufacturing step is a high-temperature kiln operation that converts limestone to lime. The major atmospheric emissions from this operation are particulates, primarily lime dust, carbonates of calcium and magnesium, calcium sulfate, and fly ash.

Kilns cause the most severe air pollution problem. Only minor amounts are emitted during limestone quarrying, transportation, preparation, lime-handling, hydrating, and other operations which are not discussed here.

Rotary kilns accounts for 80 percent of production and emit approximately 10 times as much particulates as vertical kilns. Although rotary kilns are generally controlled to a level of about 80 percent, they need additional controls to meet the projected regulations.

In general, vertical kilns are uncontrolled. Particulate emissions will need to be reduced 10 to 50 percent of present levels, depending on kiln capacity, to comply with projected process weight regulation. Ten percent of present emissions would be comparable to an overall collection efficiency of a 96-percent level. Gas-cleaning equipment is assumed to provide the additional emission control needed for both kiln types.

Nitric Acid. Nitric acid production is expected to increase by 100 percent by 1979, from the 1967 production of 6 million tons. The major portion of the output is used in the production of ammonium nitrate fertilizer.

There are 72 nitric acid plants with annual production capacities from 25,000 to 425,000 tons. The majority of these plants, 57 percent, are outside SMSA's, while 87 percent are outside the first 57 AQCR's.

Nitrogen oxide emissions are not currently covered by the Air Quality Act of 1967. It is reasonable to assume available emission control techniques are not adaquate to meet the needs of this industry. However, better emission control techniques are being developed rapidly and emissions are expected to decrease markedly.

Nonferrous Metallurgy—Primary. The primary nonferrous metallurgical industry produces aluminum, copper, lead, and zinc from raw or refined ores. The annual production of aluminum, 3.5

million tons in 1968, is greater than for any other nonferrous metal. Seventy percent of the existing 24 primary aluminum reduction plants in the United States are outside SMSA's and 83 percent are outside the first 57 AQCR's. In addition, six companies are constructing plants with a combined capacity of 240,000 tons by 1970, 532,000 tons by 1971, and 657,000 tons by 1972.

Aluminum production is based on the chemical reduction of alumina (refined aluminum ore). The reduction takes place in electrolytic cells (pots) and large amounts of particulates, mostly fluorides, are emitted. Sulfur oxide emissions are considered low. Emissions during aluminum refining are unknown, but are thought to be small in comparison with those from reduction.

To meet the particulate process weight rate regulation, aluminum producers will have to upgrade the present average particulate collection efficiency of 85 percent to an average efficiency of 96 percent. This will probably require the use of wet-scrubbing equipment on the effluent gases from aluminum reduction pots.

During the smelting of copper, lead, and zinc from ores and the subsequent refining of these metals into higher purity metals, both particulate and sulfur oxides are emitted, along with metallic fumes of small particle size, including either elemental or compounds of copper, zinc, lead, arsenic, antimony, and other trace elements.

Even though the potential for particulate emissions is significant, particulates are currently controlled at a level of 95-percent efficiency, which meets the requirements of the process weight rate standard. Sulfur oxides, however, are of concern.

The current average sulfur dioxide removal efficiencies are 49 percent for copper, lead, and zinc plants. This can feasibly be increased to 98.8 percent with the installation of gas-cleaning equipment and acid recovery plants, which will reduce sulfur dioxide emissions to a level of 1,000 to 1,800 ppm. However, this will not fully comply with the anticipated requirement of a 500 ppm maximum sulfur oxide content.

Of 19 primary copper smelters whose output in 1967 was 1,133,000 tons are outside the first 57 AQCR's, while 79 percent are outside SMSA's.

Of the 21 primary lead and zinc smelters (or roasters) 38 percent are within SMSA's and only 19 percent are within the first 57 AQCR's. Capacity in 1967 amounted to 1,294,200 tons of zinc slab and 401,151 tons of primary and some secondary lead.

Nonferrous Metallurgy—Secondary. The secondary nonferrous industry is concerned primarily with the recovery of aluminum, copper, lead, and zinc from scrap materials. The recovered metal is often refined and sold to other firms that use the metal for further processing.

There are 96 secondary copper, lead, and zinc plants in 27 States. Less than 10 percent are outside SMSA's while 33 percent are outside the first 57 AQCR's. In 1967, output was 1,160,000 tons of secondary copper, 553,800 tons of secondary lead, and 319,800 tons of secondary zinc.

Emission problems occur primarily from two operations: the melting of scrap and the refining of the molten metal. Particulates are the principal emission, and during the melting operation contain oxides of zinc, lead, and other metals from scrap metal contaminated with paint, oils, dirt, and organic materials.

Available estimates indicate half of the recovery plants are now controlling emissions at a collection efficiency of 95 percent for particulates. The average national level of control is about 49 percent. A further reduction of particulate emissions will be required to an average overall collection efficiency of 95 percent to meet the anticipated process weight rate regulations. This will be accomplished by the use of additional gas-cleaning equipment.

Petroleum Refining. In 1968, there were 269 petroleum refineries in the United States with production capacities from 1,000 to 400,000 barrels per day (one barrel is equivalent to 42 gallons). This industry is expected to experience a 30 percent growth in the next 10 years.

Although there are large numbers of air pollution sources in each petroleum refinery, a few account for the bulk of the emissions. Among these, catalyst regenerators emit carbon monoxide, hydrocarbons, and particulates; combustion processes emit primarily sulfur dioxide; and storage facilities emit primarily hydrocarbons.

Control methods are tailored to specific processes. In general, however, gas-cleaning equipment is employed to meet the particulate process weight regulations. Control of sulfur oxides to a maximum of 500 ppm requires sulfur recovery plants. Hydrocarbon emissions from storage products are controlled by floating roofs. Carbon monoxide emission in exhaust gases is controlled by waste heat boilers. Approximately 90 percent of all organic materials in exhaust gases are recycled for hydrocarbon reduction. The value of materials recycled often sets the cost of control.

These control methods will increase average control efficiencies for particulates from 67 to 82 percent, for sulfur oxides from 37 to 50 percent, for hydrocarbons from 49 to 93 percent, and for carbon monoxide from 47 to nearly 100 percent.

Petroleum Product Storage. Significant gasoline emissions result from the evaporation of gasoline during storage and transfer operations. These emissions occur at (a) petroleum bulk stations and terminals, (b) liquified petroleum gas terminals, (c) crude oil wholesalers, and (d) packaged petroleum products wholesalers.

Anticipated hydrocarbon emission standards will require that storage tanks with fixed roofs will have to be converted to floating roof tanks. Floating roofs reduce evaporative emissions by about 90 percent and are now used on 75 percent of all storage tanks.

Gasoline storage tanks will also require submerged filling inlets, which eliminate about 40 percent of emissions.

The application of these devices to all uncontrolled tanks would result in overall control of 75 percent compared to the present 54-percent control.

Phosphate Fertilizer. Phosphate fertilizers are produced directly from phosphoric acid. There are three types: superphosphate, triple superphosphate, and amonium phosphate. Superphosphate fertilizers are produced by reacting sulfuric acid directly with phosphate rock. There are 157 major superphosphate plants; 33 percent are outside SMSA's and 69 percent outside the first 57 AQCR's.

In triple superphosphate fertilizer production phosphoric acid is used for the reaction with phosphate rock. There are 22 triple superphosphate plants, 91 percent of which are outside the first 57 AQCR's.

There are 54 ammonium phosphate plants which use phosphoric acid, ammonia, and sulfuric acid to produce fertilizers, 80 percent of which are outside the first 57 AQCR's. The phosphate fertilizer industry is expected to undergo a 65 percent growth by 1979, primarily in the production of the higher analysis fertilizers diammonium phosphate and triple superphosphate.

Atmospheric emissions are primarily particulates, mostly fluorides. Particulates are released from rocks during the acid-making process and in the fertilizer-handling operations in sheds. The entire industry is presently controlling gaseous and particulate fluoride emissions; however, fluoride and nonfluoride particulates are still being emitted. In the case of large plants these emissions exceed anticipated process weight regulations. Installation of secondary gas-cleaning systems will be required.

Rendering. There are 920 rendering plants in the United States of two basic types: those that reduce edible materials and those that reduce inedible ones. Rendering of edibles constitutes only a small portion of the industry. A majority of the plants are outside AQCR's.

No estimates are available for the future growth rate in this industry. Odor, dust, and smoke emissions can be controlled, however, when current technology is diligently applied.

Residential Heating Plants. Residential heating plants, especially coal-fired furnaces in single-family homes and apartment houses represent significant sources of particulates and sulfur oxides. In 1967, there were an estimated 58 million residential heating plants. The 100 metropolitan areas contained an estimated 2.2 million dwelling units using coal, 10.9 million using oil, and 17.6 million using gas, in addition to some heated with electricity. Both particulates and sulfur oxides are emitted from coal- and oil-fired plants; however, the sulfur oxide emissions from oil burning do not exceed the maximum limit of the selected regulations for either particulates or sulfur oxides.

Because of the small size of residential heating plants, the most practical control strategy is switching to a different fuel or converting to electric heating. To meet emission regulation it must be assumed that coal will be replaced by other fuels or electricity.

Rubber. The rubber tire and inner tube industry is concerned about particulate emissions (mostly carbon black, but also fine dust, fumes, and smoke) and hydrocarbons from cord dipping.

There is insufficient data available to estimate current emissions. Industry practice is to control particulates with high efficiency gas-cleaning devices to recover this valuable material. These devices will enable tire and tube producers to meet the anticipated process weight rate regulations. Afterburners, used as standard industry practice, are capable of meeting anticipated hydrocarbon standards of 90 percent emission control.

Soap and Detergent. Soaps consist of two general types: sodium and potassium. Synthetic detergents can be divided into five main chemical classes—sulfated fatty alcohols, alkyl-aryl sulfonates, miscellaneous sulfates and sulfonates, cationic agents, and nonionic agents. There are 1,069 plants; 89 percent are within SMSA's and 65 percent are in the first AQCR's.

No appreciable growth is expected during the next 10 years for the soap and detergent industry. Emissions from this industry can be adequately controlled with current technology.

Solid Waste Disposal. In 1967, there were an estimated 12,000 land disposal sites (open-burning dumps and landfills) and about 300 municipal incinerator facilities in the United States. Open-burning dumps emit not only significant amounts of particulates but also carbon monoxide and hydrocarbons. Incineration results in the emission of all three pollutants; however, particulate emissions represent the greatest problem. Municipal incinerators are expected to be increased by 36 percent by 1979.

About 75 percent of the land disposal sites practice open-burning and about 90 percent of the incinerators do not meet the requirements of the New York State Incinerator Standard which was the particulate regulation selected for this analysis. With proper incinerator design and proper control equipment, emissions can be well controlled. The incinerators, however, are considered to comply with the sulfur oxide standard of 1.46 lb of sulfur dioxide per million BTU input, which was used in this analysis.

Steam-Electric Power Plants. The combustion of fossil fuels (coal, oil, and gas) causes a significant portion of the nation's air pollution problem. The two atmospheric contaminants of concern are particulates, and sulfur oxides. Hydrocarbons and carbon monoxide are not emitted in significant quantities when the combustion equipment is operating properly.

During the 30-year period 1938 to 1967, total installed steam-electric generating capacity increased from 26,000 to 210,800 megawatts, and annual net generation has grown from 68.4 to 973.8 billion kilowatt hours. This growth pattern is expected to continue. The growth rate of coal- and oil-fired steam-electric power plants is estimated to be 80 percent during the next 10 years.

The electric utility industry in 1967 had 325 plants deriving 100 megawatts (or more) power from the consumption of coal and/or oil; these are located in 39 states and the District of Columbia. Sixty-five percent are within SMSA's, while 37 percent are in the first 57 AQCR's.

Present control practice for particulates is to use electrostatic precipitators, sometimes in combination with high-efficiency mechanical collectors. The present level of industry control of particulate emissions is estimated at 86 percent. Reduction of sulfur oxide emissions has been accomplished only in a few locations, generally by substituting low-sulfur content fuel for high-sulfur fuel by switching to another type. The cost for desulfurizing the oil to an average of 1.14 percent sulfur content is about $0.40 per barrel.

It is also possible to remove sulfur oxides from stack gas; however, some technical problems must be resolved before this technique receives widespread use.

By 1975, it is anticipated a particulate emission standard similar to the state of Maryland's combustion regulation will be required. This regulation limits particulate emissions on a graduated basis depending upon the heat rating of the boiler. The regulation sets a maximum limit on the quantity of particulates emitted per hour. The requirements also increase in stringency as boiler capacity increases. The sulfur oxide standard for fuel combustion sources will approach 1.46 lb of sulfur dioxides per million Btu input. The impact of these standards will be to achieve 98.7 percent control of particulates, and 64 percent control of sulfur oxides (Table 3-15).

Sulfuric Acid. More sulfuric acid is produced in the United States than any other chemical. In 1967, there were 213 sulfuric acid plants. Of the 213 plants, 62 percent are within SMSA's, and 30 percent are in the first 57 AQCR's. Individual site plant capacities ranged from about 10 to nearly 5,000 tons per day of 100-percent acid; total acid output for 1967 was 27,735,504 tons. A 45-percent growth is expected in the next 10 years.

Most sulfuric acid is produced with the contact process which involves burning sulfur or pyrite to form sulfur dioxide. The sulfur dioxide is catalyzed to sulfur trioxide and then absorbed in weak acid. The sulfur dioxide that remains unconverted and acid mist which escapes from the acid absorption tower are the major pollutant emissions.

Sulfur oxides are not presently controlled and will have to be decreased 86 percent to meet anticipated sulfur oxides standards of 500 ppm. This will require the installation of a second absorption tower to decrease emissions.

To comply with the process weight rate regulation for particulate emissions sulfuric acid producers will have to increase the average collection efficiency of particulates from the present 46 percent to about 67 percent. This will require the placement of gas-cleaning equipment, such as a demister, to lower particulate emission levels.

Varnish. Unlike paint and lacquer production, varnish production involves a cooking process which results in the evaporation and emission of hydrocarbons.

Approximately 4 percent of the feed material is lost through cooking. The composition of the effluent varies. One of the hydrocarbons emitted is acrolein, an irritating compound; organic acids are also emitted. Nearly all (94 percent) varnish-producing facilities in the nation are within the 100 metropolitan areas.

Generally controlled varnish plants employ an afterburner with an operating efficiency of 90–95 percent. Nearly all varnish plants

TABLE 3-16 TAX INCENTIVE STATUS

State	Has Exceptional Program	Has Adequate Program	Has Legislation in the Works	Has No Program
Alabama				X
Alaska			X	
Arizona				X
Arkansas				X
California	X			
Colorado		X		
Connecticut	X			
Delaware			X	
District of Columbia				X
Florida	X (water)	X (air)		
Georgia	X (water)	X (air)		
Hawaii		X		
Idaho	X			
Illinois	X			
Indiana		X		
Iowa			X	
Kansas		X		
Kentucky		X		
Louisiana			X	
Maine				X
Maryland			X	
Massachusetts	X			
Michigan	X			
Minnesota	X			
Mississippi				X
Missouri			X	
Montana	X (water)	X (air)		
Nebraska			X	
Nevada				X
New Hampshire		X		
New Jersey	X			
New Mexico			X	
New York	X			
North Carolina	X			
North Dakota				X
Ohio	X			
Oklahoma			X	
Oregon	X			
Pennsylvania			X	
Rhode Island		X		
South Carolina	X			
South Dakota				X
Tennessee			X	
Texas				X
Utah			X	
Vermont				X
Virginia			X	
Washington	X			
West Virginia	X			
Wisconsin	X			
Wyoming			X	
Guam				X
Puerto Rico			X	
Virgin Islands				X

in California are controlled, but only about 20 percent of the plants located elsewhere are. The overall national level of control is about 19 percent. Anticipated hydrocarbons standards will require a reduction of the organic content of effluent gas streams by 90 percent.

TAX INCENTIVES

In 1969, the Congress provided for a 5-year write-off of pollution control equipment. The previous 7-percent investment tax credit was repealed.

In order to qualify for rapid federal amortization[77] an air or water pollution control facility must first be certified by the state as being installed "in conformity with the State program or requirements for abatement or control of water or atmospheric pollution or contamination." The state certification serves as the basis for EPA certification unless a state does not have an approved implementation plan; then EPA certification is sufficient. Exempted from the *Guidelines*[77] are process changes for pollution control purposes and facilities where their cost will be recovered from profits derived through the recovery of wastes. Facilities which merely disperse pollutants cannot qualify either.

In the majority of states tax incentives are available, also, but records indicate that industry has not taken advantage of available state programs. State officials attribute this to lack of publicity and the amount of effort required for certification.

The state tax incentives vary[78] (Table 3-16). New York, for example, excludes pollution control equipment from property taxes, special levies, and state income taxes.

Most tax incentives take the form of accelerated depreciation, exemption from sales, property, ad valorem, and use taxes, or provide for 1- to 5-years amortization of the cost of pollution equipment for income tax purposes.

REFERENCES

1. Platt, R., Hearings of Subcommittee on Air and Water Pollution, Committee on Public Works, U.S. Senate, on Underground Uses of Nuclear Energy, 91st Cong., Ist Sess. (Nov. 20, 1969).
2. U.S. Congress, Air Pollution Control Act of 1955, Public Law 84-159, (July 14, 1955).
3. Simon, J., "Public Health Reports of John Simon," Sanitary Institute of Great Britain, J. A. Churchill, London, 1887, Vol. I.
4. Royal College of Physicians, "Air Pollution and Health," Pitman Medical and Scientific Publishing Co., London 1970.

5. U.S. Congress, Air Quality Act of 1963, Public Law 88-206, 77 Stat. 392 (Dec. 17, 1963).
6. Muskie, Senator, E. S., "Putting the Legislation to Work: The Job of Government Alone?", *Bull. National Tuberculosis and Respiratory Disease Assoc.* (July–Aug. 1969).
7. U.S. Congress, Federal Water Pollution Control Act, Public Law 80-845, (June 30, 1948).
8. Muskie, Senator E. S., "Role of the Federal Government in Air Pollution Control," *Arizona Law Rev.*, **10**: 17 (1968).
9. U.S. Congress, Federal Water Pollution Control Act Amendments of 1956, Public Law 84-660 (July 9, 1956), 70 Stat. 498.
10. U.S. Congress, Air Pollution Control Act Amendments of 1960, Public Law 86-493 (June 8, 1960).
11. U.S. Public Health Service, Surgeon General, "Motor Vehicles, Air Pollution, and Health," House Doc. 87-489 (June 1962).
12. U.S. Congress, Air Pollution Control Act Amendments of 1962, Public Law 87-761 (Oct. 9, 1962), 76 Stat. 760.
13. U.S. Senate, Subcommittee on Air and Water Pollution, Committee on Public Works, "Steps Toward Clean Air" (Oct. 1964), committee print.
14. U.S. Congress, Motor Vehicle Air Pollution Control Act, Public Law 89-272 (Oct. 20, 1965), 77 Stat. 392.
15. U.S. Congress, Air Quality Act of 1967, Public Law 90-148 (Nov. 21, 1967), 81 Stat. 485.
16. Muskie, Senator E. S., *Proceedings: Third National Conference on Air Pollution,* USPHS Pub. No. 1649, 1967.
17. Johnson, President L. B., "Air Pollution: Message from the President," House Doc. 90-47 (Jan. 30, 1967).
18. U.S. Senate, Committee on Public Works, *Air Quality Act of 1967,* Senate Rep. No. 90-403 (July 15, 1967).
19. Muskie, Senator E. S., "Legislation Dealing with the Environment," *Congressional Record,* 91st Cong. Jan. 23, 1970, Vol. 116, No. 5.
20. Nixon, President R. M., "Message on Environment," House Doc. No. 91-225 (Feb. 10, 1970).
21. U.S. House, Committee on Interstate and Foreign Commerce, *Report on Clean Air Act Amendments of 1970,* House Rep. No. 91-1146 (June 3, 1970).
22. U.S. Senate, Committee on Public Works, *Report on the National Air Quality Standards Act of 1970,* Senate Rep. No. 91-1196 (Sept. 17, 1970).
23. U.S. House, *Conference Report on Clean Air Amendments of 1970,* House Rep. No. 91-1783 (Dec. 17, 1970).
24. U.S. Congress, Clean Air Amendments of 1970, Public Law 91-604 (Dec. 31, 1970), 84 Stat. 1676.
25. Chicago Metropolitan Sanitary District, Metropolitan Sanitary District of Greater Chicago, a municipal corporation, *v* Inland Steel Co. *et al.*, Circuit Court of Cook County.

26. Cahners, Publishing Critical Issue Rep. No. 2, *Cooperation: The Art of Getting Along with Government,* (Nov. 1969), pp. 41–43.
27. National Air Pollution Control Administration, *USDHEW News,* July 31, 1970.
28. Miller, L. A., and Borchers, D. J., "Private Lawsuits and Air Pollution Control," *J. Am. Bar Assoc.,* **56** (May 1970), pp. 465–469.
29. Diamond *v* General Motors *et al.,* California Superior Court, Case No. 947429.
30. Kean Wigoda *et al. v* General Motors, Ford, Chrysler *et al.,* U.S. District Court, Northern District of Ill., Case No. 69c1900.
31. U.S.A. *v* Auto Mfgrs., General Motors, Ford, Chrysler *et al.,* District Court, Southern District of Calif., Case No. 69-75 JWC.
32. Farrell, R. J., "Let the Polluter Beware!" address to Association of General Council, Washington, D.C., Oct. 6, 1969.
33. U.S. Public Health Service, "A Compilation of Selected Air Pollution Emission Control Regulations and Ordinances," Pub. No. 999-AP-43 (1968).
34. Degler, S. E., "State Air Pollution Control Laws," Bureau of National Affairs, Inc., Washington, D.C. (1969).
35. Cluster, R. C., "State and Local Manpower Resources and Requirements for Air Pollution Control," *J.A.P.C.A.,* **19:4** (April 1969).
36. 62 Wash. 2d 834 P 2d 859 (1963) cert. den. 377 U.S. 906. 84 S. Ct. 1166, 12 L. Ed. 2d 177 (1964); 102 Cal. App. 2d Supp. 925, 226 P 2d 587 (1951); 137 Cal. App. 2d Supp. 859, 291. 2d 587 (1955) cert. den. 351 U.S. 990 76 S. Ct. 1046, 100 L. Ed. 1503 (1955).
37. Stumph, T. L., and Duprey, R. L., "Trends in Air Pollution Control Regulations," Hearings on Air Pollution—1970, Subcommittee on Air and Water Pollution, Committee on Public Works, U.S. Senate, Part I, pp. 396–427.
38. Rom, J. J., "Reading Visible Emission" Training Course Manual, NAPCA, Durham, N. C., April 1968.
39. Conner, W. D. and Hodkinson, J. R., "Optical Properties and Visual Effects of Smoke-Stack Plumes," NAPCA Pub. No. AP-30, 1967.
40. Stern, A. C., *Air Pollution,* 2d ed., Academic Press, New York, 1968, Vol. III, p. 706.
41. Air Pollution Manual, AIHA, Detroit, 1968, Part II, "Control Equipment," Ch. 2.
42. American Society of Mechanical Engineers, Standard APS-1 (1966).
43. American Society of Mechanical Engineers, "Recommended Guide for the Control of Dust Emissions—Combustion for Indirect Heat Exchanges" APS-1 (1966).
44. Los Angeles County Air Pollution Control Distirct, "Multiple-Chamber Incinerator Design Standards for Los Angeles County, Oct. 1960.
45. NAPCA, "Interim Guide for Good Practice for Selecting Incinerators for Federal Facilities," Durham, N. C., Jan. 1968.

46. NAPCA, "Bi-monthly Compilation of State and Local Air Quality Standards," Nov. 16, 1970.
47. U.S. Senate Committee on Public Works, Subcommittee on Air and Water Pollution, Hearing on S. 780, 90th Cong., 1st Sess., Pt. 3 (1967), p. 1132.
48. U.S. Senate, Committee on Public Works, "Recommendations regarding the conditional consent of the Congress to various air pollution control compacts," Committee Print, June 17, 1968.
49. U.S. Senate, Committee on Public Works, Hearings on S. 780, the Air Quality Act of 1967, before the Subcommittee on Air and Water Pollution, Air Pollution, 1967.
50. Alfred Watson Memorial Lecture to the Institute of Actuaries, *J. Inst. Actuaries*, BB:**178** (1962).
51. USDHEW, "Air Quality Criteria for Particulates," NAPCA Pub. No. AP-49, Jan. 1969.
52. "Air Quality Criteria for Sulfur Oxides," USPHS Pub. No. 1619, June 1967.
53. Johnson, President Lyndon, Memorandum to the Secretary of Health, Education and Welfare, April 21, 1967.
54. U.S. Senate, Committee on Public Works, Report on S. 780, the Air Quality Act of 1967, Senate Report No. 90-403, July 15, 1967.
55. USPHS, "Air Quality Criteria for Sulfur Oxides," NAPCA Pub. No. AP-50, January 1970.
56. Junge, C. E., *Air Chemistry and Radioactivity*, Academic Press, New York, 1963, p. 382.
57. USPHS, "Air Quality Criteria for Carbon Monoxide," NAPCA Pub. No. AP-62, March 1970.
58. USPHS, "Air Quality Criteria for Hydrocarbons," NAPCA Pub. No. AP-64, March 1970.
59. USDHEW, Errata Sheets for "Air Quality Criteria for Hydrocarbons," AP-64, Sept. 28, 1970.
60. USDHEW, "Air Quality Criteria for Photochemical Oxidants," NAPCA Pub. No. AP-63, March 1970.
61. USDHEW, Errata Sheets for Air Quality Criteria for Photochemical Oxidants," AP-63, Sept. 1970.
62. "Air Quality Criteria for Nitrogen Oxides," Air Pollution Control Office Publication No. AP-84, January 1971.
63. EPA, "Proposed National Primary and Secondary Ambient Air Quality Standards," *Federal Register*, **36**, (21) (January 1971).
64. EPA, "National Primary and Secondary Ambient Air Quality Standards," *Federal Register*, **36**, (84) (April 1971).
65. USDHEW, *Guidelines for the Development of Air Quality Standards and Implementation Plans*, NAPCA, May 1969.
66. Middleton, J. T. "Public Policy and Air Pollution Control," speech at Penjerdel Regional Conference, Swarthmore College, Swarthmore, Pa., June 11, 1969.
67. U.S. House Committee on Interstate and Foreign Commerce Report

on S. 780, the Air Quality Act of 1967, House Report No. 90-728, Oct. 3, 1967.
68. USDHEW, "Control Techniques for Sulfur Oxides," NAPCA Pub. No. AP-52, Jan. 1969.
69. USDHEW, "Natural Emission Standard Study," Report to the Congress, Senate Doc. No. 91-63, March 1970.
70. EPA, "Proposed Standard of Performance for New Stationary Sources," *Federal Register*, **36**: 15704 (August 1971).
71. EPA, "Standards of Performance for New Stationary Sources," *Federal Register*, **36** (247): 24876-24895 (December 1971).
72. EPA, "Standards Proposed for Three Hazardous Pollutants," *Environmental News* (December 1971).
73. USDHEW, Second Report to the Congress on "The Cost of Clean Air," Senate Doc. No. 91-95, April 27, 1970.
74. USDHEW, "Cost of Clean Air," First Rept., Senate Doc. No. 91-40, Oct. 16, 1969.
75. U.S. Department of the Interior, Bureau of Mines, *Minerals Yearbook, 1967*, 1968.
76. USPHS, "Atmospheric Emissions from the Manufacture of Portland Cement" Pub. No. 999-AP-17, 1967.
77. EPA, " Guidelines for Certification of Pollution Control Facilities," *Federal Register*, **36** (189): 19132-19134 (September 1971).
78. Cahner's Critical Issue Report No. 2. *Tax Credits for Pollution Control Equipment Environmental Management* Nov., 1969, p. 44.

4 Air Pollutants

Clean, country air can contain carbon monoxide, methane, ammonia, nitrogen oxide, and ozone. These gases, normally considered pollutants, all come from natural processes in small amounts. Particulates may also be present in the form of precipitation, pollen, volcanic ash, and insects. Unpleasant odors derive from decay. Thus the problem of air pollution must be considered a matter of degree. Man has attained the capability of generating these same and similar substances in massive amounts often far exceeding natural dissipative processes and by so doing has created areas where the atmosphere is uncomfortable and possibly toxic or destructive as well. He has a responsibility to himself and his fellow inhabitants of this planet to apply the same technology which created "the stink" to cleaning it up. This is a considerably complex operation since it requires separate steps of identifying each pollutant, defining safe levels, measuring emissions from industrial processes, monitoring ambient urban atmospheres, and controlling sources of emissions so that safe levels are maintained in surrounding areas. This chapter is concerned with identification of the various common pollutants. The following chapter describes measurement and monitoring procedures, and subsequent chapters will describe the techniques required to reduce emissions to acceptable levels.

Combustion processes account for a greater volume of air pollutants than any other single source. This is due in part to the inherent nature of combustion and in part to its ubiquitous position in human life. It enters into transportation, heating, cooking, power generation, waste disposal, and various industries. Some of these processes are very efficient, converting most of a hydrocarbon fuel to carbon dioxide and water, but others are relatively inefficient, producing large amounts of carbon monoxide. In the latter category is the internal combustion engine, infamous for creating the high CO, hydrocarbon, and nitrogen oxide levels of congested areas. Nitrogen oxides result in varying quantities from all combustion processes using air as an oxidant. They increase as the temperature increases, so efficiency of combustion enters into the pollution problem. The internal combustion engine is inherently inefficient and produces large amounts of CO. The adjustment of the carburetor enters into the amount of the fuel which is actually burned. Fine particulates, hydrocarbon vapors, and aerosols result from the remainder. This also occurs in stationary burners, especially those using residual fuels. Lower-quality fuels can produce immense amounts of fly ash, seen as dark gray clouds emanating from stacks. Also characteristic of poorly refined fuels and coal is contamination by sulfur. Sulfur dioxide is produced during the burning of these fuels, resulting in physical discomfort and harm to exposed persons as well as damage to plants and buildings. To clean up all pollution caused by combustion is impossible, but each pollutant can be reduced to a minimum by various means. The proper choice depends on economical as well as technical considerations.

Other than combustion processes, manufacturing industries probably account for a major portion of the remaining air pollution problem. Chemical industries and refineries invariably release vapors of reactants, products, and by-products, as well as wastes. Other polluters include finishing industries, which account for massive amounts of organic solvents, smelters, food processors, rendering plants, and even dry cleaners. The pollutants produced are as varied as the industries involved, each requiring separate evaluation and control procedures.

Air pollutants fall into two groups—gases and particulates—because of the different effects, testing methods, and control procedures that can be generalized for each. Gases, for example, cannot be seen but irritate the eyes and mucous membranes. Severe gaseous pollution will even cause sickness and long-term physiological damage. Gases are detected either by the continuous measurement of a physical or chemical property or by their collection for intermittent chemical tests. The control of gaseous pollutants requires either a basic change in the source, removal by scrubbing or adsorption, or chemically altering the pollutant—for example, by incineration. Particulate contaminants are often very visible, as in the case of stack plumes, and can be a source of complaints. Their physiological effects are frequently not as bad as those from gaseous pollutants since the nose tends to filter out a large portion; however, lung damage can occur over extended intervals. In addition, toxic metals can accumulate in the body through prolonged exposures. Testing and control procedures of particulates rely on collection by filtering apparatuses with subsequent physical or chemical identification.

Many air pollution problems involve the interaction of particulates with gases such as the reaction of SO_3 and water vapor to form sulfuric acid droplets and the formation of smog. Despite this, present air pollution testing and control technology requires the separate consideration of each component.

Certain gaseous and particulate contaminants are found at such low concentrations as to be undetectable by instrumentation but still remain an air pollution problem because of their odors. These are generally not dangerous; however, they are nuisances and sources of complaints. Eliminating them is often difficult because of the low concentrations involved and the peculiar nature of the human nose. Detection and control techniques of odor problems is entirely different than for gas or particulate pollutants, and therefore odors are often considered as a third class of pollutant despite the fact that physically they must be either gases or particulates.

CLEAN AIR

The earth is surrounded by approximately 500 billion tons of air, of which 20.9 percent is oxygen, 78.0 percent is nitrogen, and 0.9 percent is argon. Excluding water vapor, these gases make up all but 0.04 percent of the atmosphere and three-quarters of the remainder is carbon dioxide. The last 0.01 percent contains the other noble gases, hydrogen, nitrogen oxides, ozone, and traces of many other gases. Table 4-1 summarizes the major gases, and their concentrations on a dry basis. These figures remain essentially constant as long as water vapor is not included, since it varies in a range from as high as 3 percent by volume to less than 1 percent. In ambient air measurements this variation is normally within the reproducibility limits of the calculations and can be ignored. Despite this, samples are often dried before testing.

Water vapor normally cannot be considered as an air pollutant even though it is produced in great quantities in combustion processes. Natural evaporation and precipitation move far greater amounts of water than any man-made process. One exception to this may occur in the near future with the advent of the supersonic transport. This plane may release large amounts of water vapor into the stratosphere causing increased cloudiness and a temperature rise. Water vapor has a synergistic effect upon air pollutants, increasing many deleterious effects. For example, sulfur dioxide, which itself is an irritating and corrosive pollutant, can combine with water vapor and atmospheric oxygen to form highly corrosive sulfuric acid mist. Smog also depends upon water vapor for its formation. Monitoring of pollution in urban areas must consider the humidity and predict its effect on pollutants.

TABLE 4-1 COMPOSITION OF NATURAL AIR

Constituent	*Approximate Concentration*	
Nitrogen (N_2)	78.03	volume percent
Oxygen (O_2)	20.99	volume percent
Carbon dioxide (CO_2)	0.03	volume percent
Argon (A)	0.94	volume percent
Neon (Ne)	0.00123	volume percent
Helium (He)	0.0004	volume percent
Krypton (Kr)	0.00005	volume percent
Xenon (Xe)	0.000006	volume percent
Hydrogen (H_2)	0.01	volume percent
Methane (CH_4)	0.0002	volume percent
Nitrous oxide (N_2O)	0.00005	volume percent
Water vapor (H_2O)	Variable	
Particulate matter	[a]Variable type and quantity	
Ozone (O_3)	[b]Variable	
Formaldehyde (HCHO)	[c]Uncertain	

[a]The types and concentrations of these substances may vary substantially from one region to another and within any period of time from natural conditions.
[b]From ultraviolet radiation and probably thunderstorms, concentrations will vary from 0 to 0.07 ppm.
[c]From biological sources or oxidation of CH_4; possible concentrations not determined.

Pollution monitoring must also take into account meterological and geological considerations to be able to determine the rate of dissipation of pollutants. Emission levels for industries and traffic may be nearly constant all year but seasonal weather variations often make the difference between clean air and pollution emergencies. The effects of prolonged temperature inversions over urban areas have been well publicized.

GASEOUS POLLUTANTS

The concentrations of some gaseous pollutants which have been designated as dangerous would be ignored in standard analytical chemistry or classed as "traces." These levels rarely exceed 0.001 percent by volume of an atmospheric sample and generally lie two or three orders of magnitude below this figure. Volume percent, then, is a cumbersome way of expressing concentration in air pollution work and is rarely used. The most common terms for concentration by volume are parts per million (ppm), parts per hundred million (pphm), and parts per billion (ppb). These are self-explanatory. Table 4-2 compares these units with percent by volume demonstrating the way in which extremely small concentrations can be expressed in whole number values.

TABLE 4-2 COMPARISON OF PERCENT BY VOLUME WITH PARTS PER MILLION, PARTS PER HUNDRED MILLION, AND PARTS PER BILLION

Percent	*ppm*	*pphm*	*ppb*
100.0			
10.0			
1.0			
0.1			
0.01	100.0		
0.001	10.0		
0.000 1	1.0	100.0	
0.000 01	0.1	10.0	100.0
0.000 001	0.01	1.0	10.0
0.000 000 1	0.001	0.1	1.0
0.000 000 01	0.000 1	0.01	0.1

Recently, some authors have preferred the use of the weight of pollutant per unit volume in air pollution measurement. The most common units are micrograms per cubic meter ($\mu g/m^3$). This yields more meaningful information as far as toxicological considerations are concerned, but requires the use of the molal volume of each gas and a conversion to standard temperature and pressure conditions. These units enable rapid estimations to be made of the amount of a pollutant that might be retained in the human body over a given time period.

The most common gaseous pollutants are carbon monoxide, sulfur dioxide, and smog-causing oxidants such as nitrogen dioxide and ozone. These will be discussed individually in the succeeding paragraphs, but it should be kept in mind that synergistic effects and reactions often occur between pollutants and as a result of water vapor, atmospheric oxygen, and sunlight. Carbon dioxide has also been included since it is a poisonous gas and is produced as a technological waste. On the other hand, it is vital to life on this planet, and so is another example of the importance of concentration when defining air pollutants.

Carbon Dioxide

Concern has been expressed over the long-term effects resulting from the steady increase in atmospheric carbon dioxide generated by the use of fossil fuels. Over the past 50 years this increase has amounted to almost 10 percent in some urban areas and the trend is expected to continue or even increase. Another 18 percent rise is expected by the year 2000, and according to some scientists the atmospheric CO_2 could even double in the foreseeable future. The "greenhouse effect" is the predicted result of this increase whereby the average temperature of the earth will rise slowly, eventually melting polar ice, flooding coastal cities, and even changing the balance of the earth. Opposing this effect, however, is another result of the increase in fossil fuel combustion—atmospheric smoke and dust—which blocks the sun's rays partially causing a cooling effect. Which of these factors will eventually dominate is a subject of scientific debate, but a slight cooling trend has been noted since 1940. At this time, discussion of CO_2-imposed disasters remains academic, but more attention has been paid recently to careful monitoring of atmospheric levels the world over.

It is highly probable if a real danger resulting from a CO_2 increase should manifest itself in the near future that scientists will be able to "cancel out" the temperature rise by releasing controlled amounts of fine particulates into the stratosphere. Thus, real concern over CO_2 pollution is needless, especially when the importance of CO_2 to life on this planet is considered. Plants require CO_2 for photosynthesis, so the entire food chain by which man exists is based on it. Also, although it is a waste product of animal respiration and consequently toxic at higher concentrations, a certain percentage stimulates breathing; in fact, the safe level for prolonged human exposure is 15 times the present atmospheric level of 320 ppm. Plant life has been shown to benefit from increases in CO_2 and some greenhouse owners use this to increase yields. So it appears as if an increase in atmospheric CO_2 might even be of direct benefit to mankind by increasing the world's food crop.

Carbon Monoxide

The most ubiquitous pollutant gas in the world today is carbon monoxide (CO). An estimated 102 million tons were generated in the United States alone in 1968 which was equal to all other pollutants generated in that year added together. Almost 60 tons of this CO came from vehicular sources.

Carbon monoxide is a colorless, odorless gas which combines preferentially with the hemoglobin of the blood blocking the body's oxygen transport system. If the concentration is high enough, the body will be rapidly asphyxiated, but at lower concentrations, such as those found in urban environments, an equilibrium becomes established in the bloodstream and the exposed person will most likely have a slight loss in coordination or reaction time which he probably will fail to notice. When removed from the CO-containing environment his lungs will slowly discharge the CO contained, leaving no permanent effects as far as is presently known. So science has largely ignored CO pollution even though levels are rising higher and higher in urban areas as the number of automobiles increases each year. Sulfur dioxide causes immediate complaints of eye irritation. Fly ash darkens the sky and soils the streets, but CO remains unnoticed and considered seemingly harmless.

The primary source of CO is the incomplete combustion of carbonaceous fuels. This occurs where insufficient oxygen or time is allowed for complete conversion to CO_2. The gasoline engine operates by the almost instantaneous combustion of a fuel-air mixture and produces CO in amounts as high as 7 percent in factory-adjusted new cars. This can be readjusted to less than 0.1 percent, but the performance of the engine will be affected. In stationary furnaces, the latitude for adjustment is much broader and CO can be eliminated almost entirely from the emissions. In fact, this is mandatory because of economics. The production of CO results in a loss of two-thirds of the available heat energy. Home heating units are not maintained as closely as industrial furnaces and often produce more CO, but even so, all stationary combustion sources accounted for only 11 percent of the total CO produced in this country in 1968. Other major sources of CO pollution include solid waste disposal and industries such as foundries, petroleum refineries, kraft pulp mills, and steel mills.

CO is emitted naturally by volcanoes and forest fires. Plants generate trace amounts as a waste product from the breakdown of chlorophyll molecules. The hemoglobin molecule in animals is structurally similar to chlorophyll and also produces a molecule of

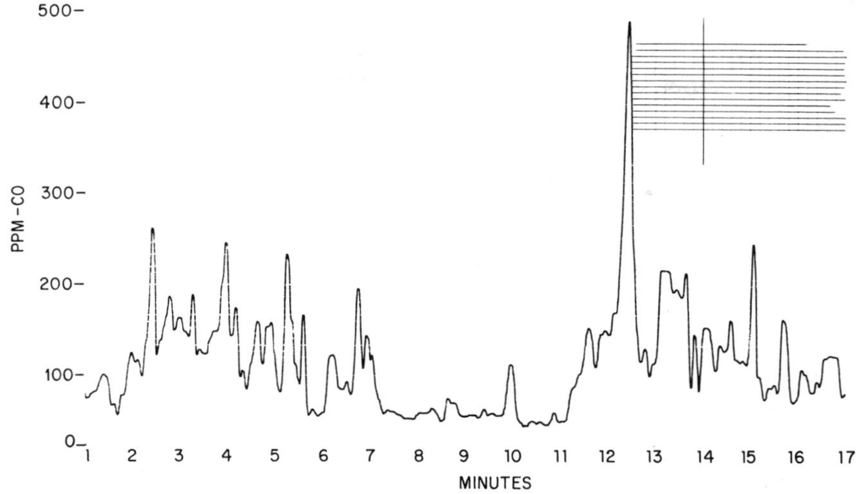

Fig. 4-1 CO measurements taken at toll booth No. 7, Queens's Midtown Tunnel, New York City, on Nov. 9, 1970.

CO when it degrades to urobilinogen. This natural presence of CO in animal metabolism may help to explain how a substance so toxic at high concentrations can seem so innocuous at low ones. Just how innocuous it is, however, is the subject of increasing research because of the great amounts of CO to which people are exposed daily.

The average figures for CO levels in urban communities often reach 50 ppm due to motor vehicles. This is only a small part of the picture for individuals living in this environment. Figure 4-1 shows the instantaneous CO levels at a busy toll booth in a major city. Each peak represents a passing vehicle. The average CO level during the period of the illustration is around 90 ppm, but a person standing at the booth might receive a dose several times as concentrated, depending on the exact moment he inhaled. At every busy intersection this condition is duplicated, exposing motorists and pedestrians alike. A similar situation is self-imposed by many persons who smoke cigarettes. The smoke containing an average of 400 ppm CO will be inhaled many times during the 5-minute or so burning life of the cigarette and the cycle repeated many times a day. Manufactured gas contains a high concentration of CO because of its fuel value. This has been replaced in many areas of the United States by natural gas, consisting primarily of hydrocarbons, but where it is used, it imposes a hazard through leaks or inefficient burners. Even with natural gas the housewife is exposed to CO through the use of her gas stove, oven, or heater. In summation, every urban resident will be exposed to varying levels of CO during each day of his life in the urban environment, and since the present population trend seems to be approaching total urbanization, it is mandatory that the effects of these CO levels be thoroughly understood.

Exposure to high concentrations of CO will produce the following symptoms: headache, dizziness, lassitude, flickering of eyes, ringing of ears, nausea, vomiting, palpitations, pressure on chest, difficulty in breathing, muscular weakness, collapse, unconsciousness, and finally death. These symptoms will occur in any environment, for example at high altitudes, where the body is deprived of oxygen, because CO does just that. It reacts with the hemoglobin in the blood 210 times as rapidly as oxygen to form carboxyhemoglobin, reducing the oxygen-carrying capacity of the blood. Not only is oxygen displaced by this reaction, but the remaining oxyhemoglobin is bound more tightly, decreasing its delivery capacity. Since this is a reversible reaction, removal from the CO-containing atmosphere will result in gradual elimination of CO until a background level is reached. The half-life is about 4 or 5 hr. Blood, then, has an integrating effect as the body is exposed to varying levels of CO throughout each day. Figure 4-2 demonstrates this for the case of the smoker and nonsmoker. The rise in the CO level of the expired air for each person is apparent as a result of the high CO levels of urban rush hours. The increased average CO level of the smoker's breath is also obvious. The physiological effects of nontoxic levels of CO exposure are a function of the carboxyhemoglobin level in the blood. This reaches equilibrium after about 8 hr of exposure in an atmosphere containing CO and is proportional to its concentration. In pure air, the blood always contains a background of about 0.5 percent of carboxyhemoglobin due to the natural breakdown of hemoglobin molecules, but levels above this are due to external CO. Exposure to 30 ppm CO for 8 hr will result in about 5 percent carboxyhemoglobin. Cigarette smokers have carboxyhemoglobin levels of 5–10 percent while garage workers and traffic police may run as high as 18 percent. The immediate effects of these carboxyhemoglobin levels seem to be the impairment of performance. Low levels, below 5 percent may affect choice discrimination and reduce visual acuity. Above 5 percent, certain psychomotor effects have been noted, but definite long-term effects of nontoxic levels of carboxyhemoglobin have not yet been established.

Ill effects of CO pollution have been exceedingly difficult to pin

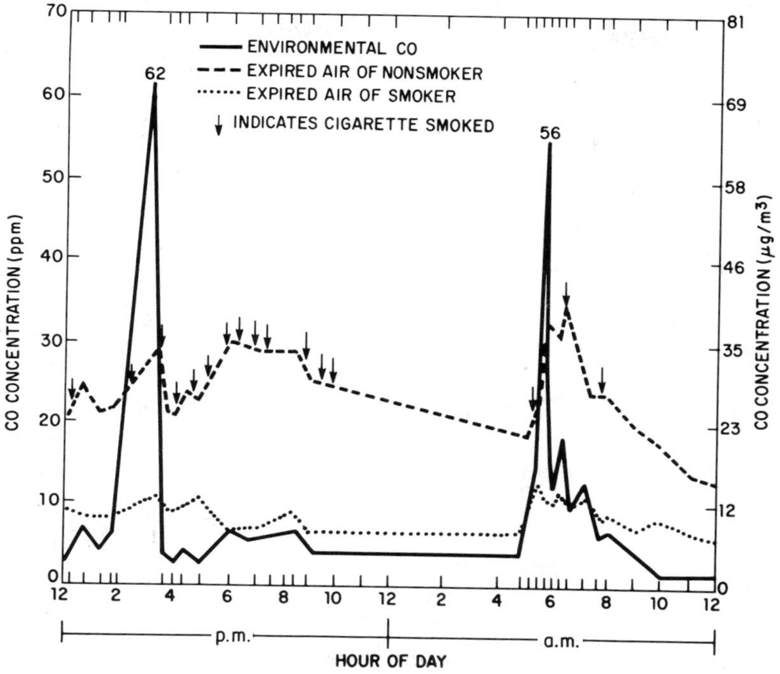

Fig. 4-2 Carbon monoxide levels of environmental air and of expired air of smoker and nonsmoker, Los Angeles and Pasadena, Aug. 1962. (From Goldsmith, J. R. *et al., Excerpta Med. Int. Cong. Series,* **62** (1963), pp. 948–952)

down and therefore have been of little assistance in defining safe levels. Two areas exist, however, which most probably could be used to provide general safe limits. These are the amount of CO which will impair reactions while driving an automobile, and the amount of CO which can be tolerated by patients with chronic heart and lung diseases. The first of these requires massive amounts of statistical information pertaining to the relationship of accidents and highway CO levels. The second may be derived from present research data. For example, a 10 percent carboxyhemoglobin level has been shown to significantly affect oxygen transport. Many individuals in a community have physical impairments equivalent to a 5 percent restriction in oxygen transport and exposure to 35 ppm CO for an extended period would add an additional 5 percent carboxyhemoglobin, producing a dangerous condition. Standards based on figures such as these would require drastic reductions in automotive CO emissions in the near future, and possibly the elimination of the internal combustion engine in the more distant future. Steps have already been taken in this direction in many countries in anticipation of soundly based standards.

Sulfur Compounds

The most infamous and destructive group of air pollutants is based on the sulfur atom. It includes several oxides of sulfur and a number of other compounds that are usually malodorous nuisances. The oxides of sulfur present the scientist with a very complicated situation. Sulfur dioxide (SO_2) is the primary industrial emission in this group. It is a gas and readily amenable to tests on plants and animals to determine its effects. This has been done; however it only yields a small portion of the story because in the atmosphere SO_2 reacts to form sulfur trioxide (SO_3) which readily combines with water vapor to form sulfuric acid mists. These are highly corrosive to man-made structures and have toxicological properties that are related to particle size. Further complicating the picture are various catalytic effects which other particulates have on the formative and destructive reactions of sulfuric acid.

In 1966, 28 million tons of SO_2 were emitted in the United States. Power utilities burning sulfur-containing coal or residual oil were responsible for more than 50 percent of this, and refineries, smelters, coke producers, sulfuric acid manufacturers, and incinerators helped make up the remainder. High concentrations of SO_2 are found in urban industrialized areas with the highest being in northeastern cities where high-sulfur fuels are used extensively for power generation. The range of urban concentrations runs from a high 3.3 ppm 5-min maximum in Chicago to 0.36 ppm, the same maximum measured in Denver. When a 24-hr period is used for averaging, most major cities in the United States have maximum SO_2 levels of tenths of a ppm.

These concentration figures may seem low unless the sensitivity of the body toward SO_2 is considered. The taste threshold is about 0.3 ppm, and it is a very unpleasant experience at 1 ppm. Breathing and pulse rate changes have been noted at this low concentration. A level of 5 ppm SO_2 causes respiratory irritations and even spasmotic reactions in some sensitive individuals.

The sensitivities of plants to SO_2 vary widely. Some trees and shrubs become yellowed by exposures of 0.3 to 0.5 ppm SO_2 for several hours. Lower concentrations over longer periods can cause leaf drop and severe damage to some crops. Alfalfa becomes discolored by an exposure of 1.25 ppm for an hour. It is also possible that damage in the form of growth retardation may occur from very low SO_2 concentrations. Rye grass has shown indications of this.

In the atmosphere, SO_2 can react with oxygen photochemically and catalytically to form SO_3, a highly hygroscopic chemical which will scavenge moisture immediately to form sulfuric acid in the form of the fine drops. The rates of these reactions depend upon the amount of moisture present, sunlight, the presence of other chemicals such as hydrocarbons and nitrogen dioxide, and the presence of particulates. Inorganic sulfate particulates and complex photochemical smog can also result in these reactions. They are so extensive that sulfuric acid and sulfates typically account for from 5 to 20 percent of all suspended particulates in urban air samples and often cause severe reductions in visibility. The ability of SO_2 to cause eye irritation is magnified three or four times when conditions are favorable for the formation of sulfuric acid. Thus, the simple measurement of SO_2 levels is insufficient to predict the toxicological effects of a given environment.

Particulates larger than 5 μ remain suspended in the exhaled air. Those between these sizes are trapped in the lungs and if these are capable of catalyzing the oxidation of SO_2, they will most likely have a drop of sulfuric acid attached. Particles of ferrous iron, manganese, or vanadium salts are examples of active compounds and can contribute to high incidences of lung damage through this mechanism. Many of the air pollution episodes that have occurred in major cities around the world have been the result of high SO_2 pollution along with conditions favorable for sulfuric acid conversion. They have been characterized by sharp rises in mortalities and increased hospital visits for respiratory complaints. The higher death rates probably reflect the additional respiratory burden imposed on elderly and chronically ill persons, but less evident effects to children and healthy adults are undoubtedly present.

The works of man also suffer from the combined destruction of SO_2 and sulfuric acid. Steel structures, power lines, guy wires, fabrics, slate, some building stones, cement and paint are all gradually destroyed by these pollutants. Worse is the damage to irreplaceable statuary, architecture, frescos and shrines that have survived hundreds of years only to be eaten away in the twentieth century.

PHOTOCHEMICAL OXIDANTS

The classical smogs of London and other major cities are caused primarily by the interaction of fog with soot or smoke and sulfur compounds such as SO_2. The automobile has brought about a new kind of smog which has been most often associated with the Los Angeles basin. Known as photochemical smog, it is caused by the action of sunlight on the hydrocarbon and nitrogen oxide components of vehicle exhausts to produce a complex and toxic gas mixture which combines with fog and smoke. It can be very irritating to the eyes and respiratory tract and as a result, it can cause increased incidence of infection and disease. This phenomenon is only beginning to be understood, with most of the research having been done in the past decade.

Photochemical oxidant is the result of a very complex series of reactions, many of which are only suspected at this time. These are illustrated in Fig. 4-3. It can be easily produced in the laboratory, however, by illuminating diluted automobile exhaust with ultraviolet light. Alternately, a stepwise series of events can be enacted, producing the same result. Nitric oxide (NO) and a trace of nitrogen dioxide (NO_2) are mixed with air in a reaction chamber and illuminated with ultraviolet light. This causes an equilibrium

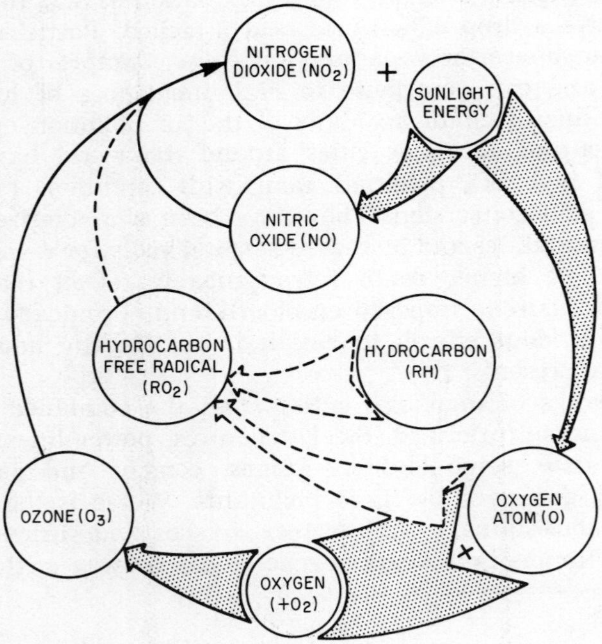

Fig. 4-3 Interaction of hydrocarbons with atmospheric nitrogen dioxide in the presence of oxygen and sunlight. Ref. *Air Quality Criteria for Photo Chemical Oxidants,* U.S. Dept. Public Health, p. 2–7, March 1970. *Used by permission of NAPCA.*

reaction between NO_2 and oxygen, forming ozone. If a hydrocarbon typical of auto exhaust is then passed into the chamber, it quickly converts to a mixture of products including aldehydes and nitrates. Higher concentrations of NO_2 and ozone appear while NO disappears. Also, a new series of compounds called peroxyacyl nitrates (PAN) is generated. The final mixture will contain primarily ozone, NO_2, PAN, and aldehydes. The first two of these are well known as very toxic compounds; the effects of the second two are less well understood.

The components of photochemical oxidant can be divided into primary and secondary substances depending on whether they are produced by vehicle emissions or are the products of atmospheric reactions. Hydrocarbons and NO are primary gases resulting entirely from internal combustion engines. Some aldehydes are also produced this way and a small fraction of NO_2 is formed in the hot exhaust stream. The major portion of aldehydes and NO_2 are secondary substances resulting from the photochemical oxidation of hydrocarbons and NO respectively. Ozone and PAN are exclusively secondary pollutants. The importance of sunlight to these conversions can be seen in Fig. 4-4. They only take place during daylight hours, and the atmospheric concentrations and proportions will be constantly changing throughout each day. This has made the study of photochemical smog all the more complex.

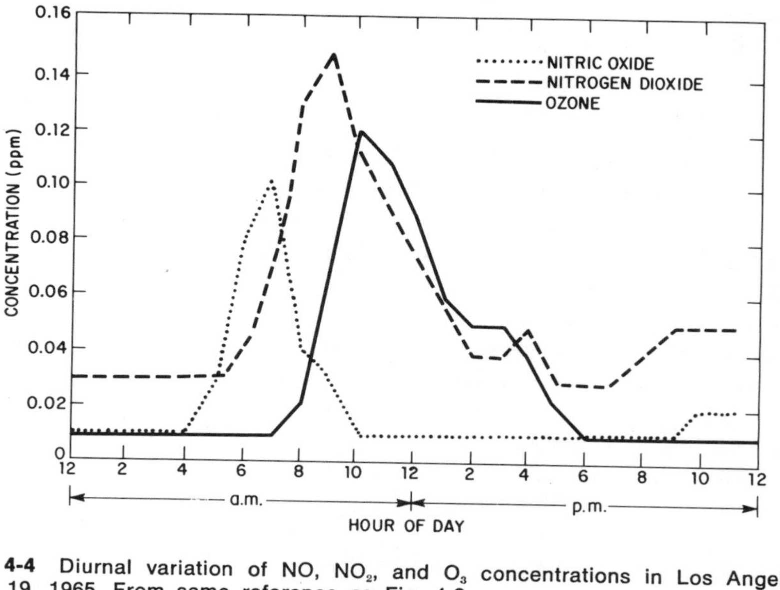

Fig. 4-4 Diurnal variation of NO, NO_2, and O_3 concentrations in Los Angeles, July 19, 1965. From same reference as Fig. 4-3.

In general, the term photochemical oxidant refers to the three compounds: ozone, NO_2, and PAN. Their combined effect is similar but not equal to what would be predicted on the basis of their individual effects. Photochemical oxidant, for example, causes eye irritation at 0.1 ppm, and this is one of its best-known effects. Others include respiratory and sense impairment and increased susceptability to acute respiratory infections. Changes in pulmonary functions have also been noted.

The body is very sensitive to ozone, detecting its odor at as low as 0.02 ppm. Sensory fatigue may follow rapidly, but nasal and throat irritation occur at 0.3 ppm. At 1 ppm, severe restriction of respiratory passages occurs and many persons cannot tolerate higher concentrations. Ozone appears to damage lung tissue by accelerating the aging process, making it more susceptible to infection.

NO is not a hazard at levels found in ambient air, but NO_2 is toxic and causes restrictions in respiratory passages and pulmonary edema. Very little additional information is known about its human effects, but it seems to be similar to ozone.

The cause of the eye irritation of photochemical smogs has yet to be determined. Hydrocarbon products are suspected, but this lack of positive knowledge makes control difficult. Guidelines cannot be established for each component of this smog since they are all interrelated. Also, arbitrarily reducing certain primary substances is foolhardy and wasteful until the exact route to the irritating secondary pollutants is known.

PARTICULATES

Early-rising city dwellers enjoy some spectacular dawns that their country cousins are denied. During these shows, the sun takes on a brilliant red hue as it rises above the horizon, and the surrounding clouds shine in colors from deep violet to fiery oranges and yellows. Unfortunately, the performances are far too short, and the atmosphere quickly takes on its hazy daytime gloom. Most probably, the dawn-watchers would give up their show if the haze would also go away, and this trade off is not totally insane because both are caused by man-made airborn particulates.

Particulates in air pollution studies is a very broad term covering all atmospheric substances that are not gases. Included are ions, molecular clusters, ice crystals, dust, smoke particles, raindrops, pollen, and even insects. This diverse assortment is broken down in various ways depending on the effect to be studied, but common to most considerations is the particle size. Optical and toxicological properties are especially size-dependent. Figure 4-5 compares a wide variety of natural and man-made particulates by their size ranges. It also compares the capabilities of several collection techniques and particle measurement methods. Three size ranges are most useful for the discussion of particulate materials, properties, and effects. These are, in microns: less than 0.1 μ, 0.1–1.0 μ, and greater than 1.0 μ.

The least scientific study has been done on particulates that are less than 0.1 μ in size. In this size range are found ions and Aitken nuclei. These particles are so small that they never settle out of the atmosphere due to Brownian motion caused by collisions with individual gas molecules. When they collide with each other or with other particulates, they adhere, transferring their charge, in the case of ions, to the larger particle. Visible light is in a wavelength range several times longer than 0.1 μ and is not scattered to a noticeable degree. The total weight of all ions and Aiken nuclei at any given time make up a minor fraction of the particulate loading of the atmosphere.

Ions are simply gas molecules or groups of gas molecules that take on a positive or negative electrical charge due to the effect of solar or cosmic radiation. Their lifetimes are inversely proportional to the number of other particles in an area and consequently far less are found in urban areas than for example over the ocean. They readily combine with each other or with solid or liquid particles upon contact. Very little is known about the properties or effects of ions except that the electrical conductivity of the atmosphere is dependent on the number of small, mobile ions present. No physiological or psychological effects to humans have been discovered.

High numbers of uncharged particles less than 0.1 μ are found in polluted urban atmospheres. These have been given the name Aitken nuclei after the developer of a counting technique based on the condensation of water vapor. They are not thought to enter into cloud formation, however. Chemical data are very scarce because of the small sizes, but the numbers of these particles vary directly with pollution cycles and plant photosynthesis cycles. Counts range from less than 1,000 nuclei/cu cm in ocean-air samples to over 300,000 in large urban areas during pollution periods. Possible sources are sea spray, dust storms, volcanoes, forest fires, automobile exhaust, photooxidation reactions, vegetation vapors, and industrial effluents. The mechanism of formation and the effects of these particles are not known.

Particles in the size range between 0.1 and 1.0 μ are formed primarily by the condensation of combustion vapors and products.

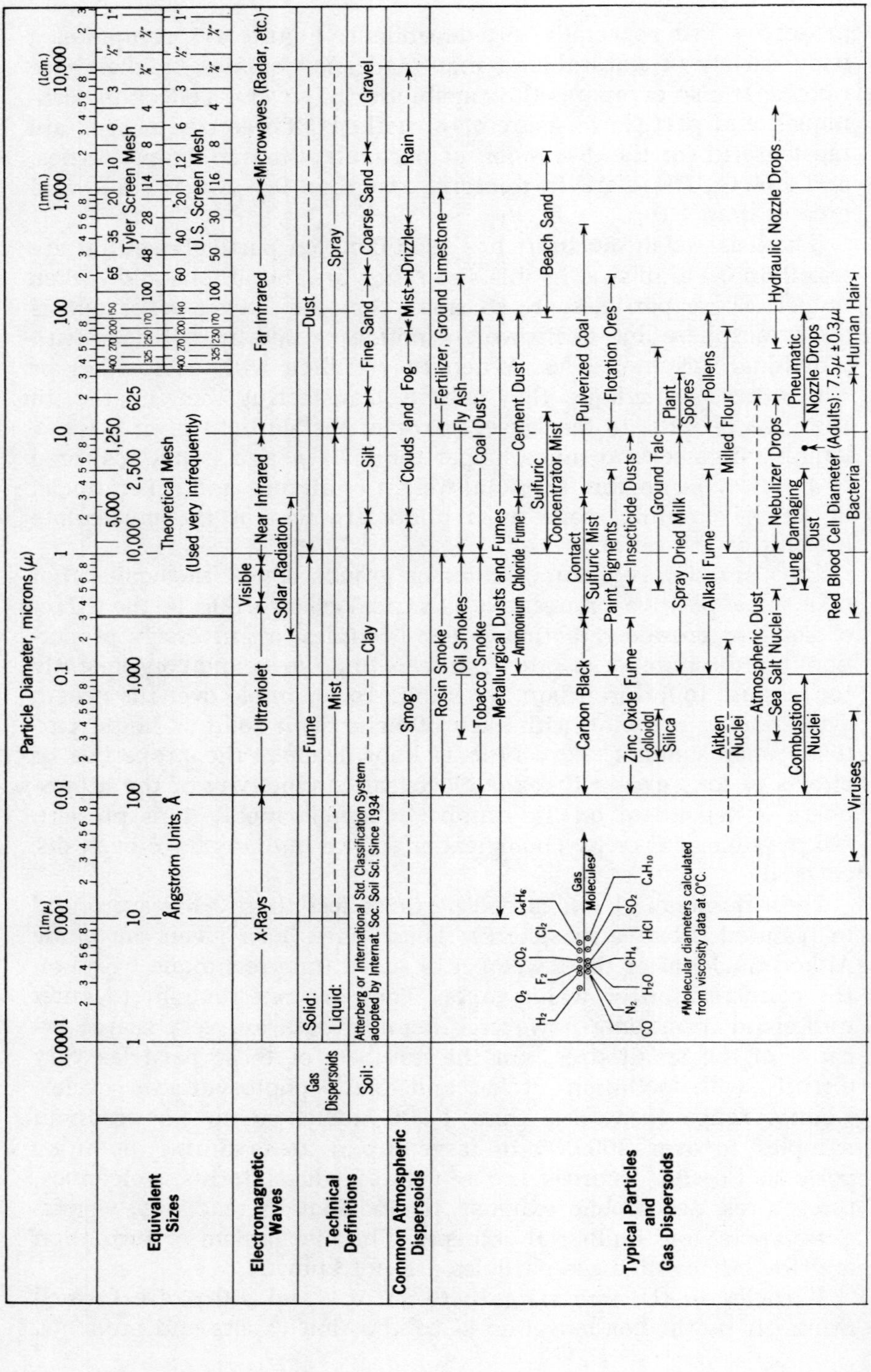
Particle Diameter, microns (μ)
0.0001
(1mμ) 0.001
0.01
0.1
1
10
100
(1mm.) 1,000
(1cm.) 10,000
Equivalent Sizes
Ångström Units, Å
Theoretical Mesh (Used very infrequently)
Tyler Screen Mesh
U.S. Screen Mesh
Electromagnetic Waves
X-Rays
Ultraviolet
Visible
Near Infrared
Far Infrared
Microwaves (Radar, etc.)
Solar Radiation
Technical Definitions
Gas Dispersoids
Solid:
Fume
Dust
Liquid:
Mist
Spray
Soil:
Atterberg or International Std. Classification System adopted by Internat. Soc. Soil Sci. Since 1934
Clay
Silt
Fine Sand
Coarse Sand
Gravel
Common Atmospheric Dispersoids
Smog
Clouds and Fog
Mist
Drizzle
Rain
Typical Particles and Gas Dispersoids
Rosin Smoke
Oil Smokes
Tobacco Smoke
Metallurgical Dusts and Fumes
Fertilizer, Ground Limestone
Fly Ash
Coal Dust
Ammonium Chloride Fume
Cement Dust
Sulfuric Concentrator Mist
Beach Sand
Carbon Black
Contact Sulfuric Mist
Pulverized Coal
Paint Pigments
Flotation Ores
Zinc Oxide Fume
Insecticide Dusts
Colloidal Silica
Ground Talc
Plant Spores
Spray Dried Milk
Alkali Fume
Pollens
Milled Flour
Aitken Nuclei
Atmospheric Dust
Sea Salt Nuclei
Nebulizer Drops
Hydraulic Nozzle Drops
Combustion Nuclei
Lung Damaging Dust
Pneumatic Nozzle Drops
Red Blood Cell Diameter (Adults): 7.5μ ± 0.3μ
Viruses
Bacteria
Human Hair
Gas Molecules
*Molecular diameters calculated from viscosity data at 0°C.

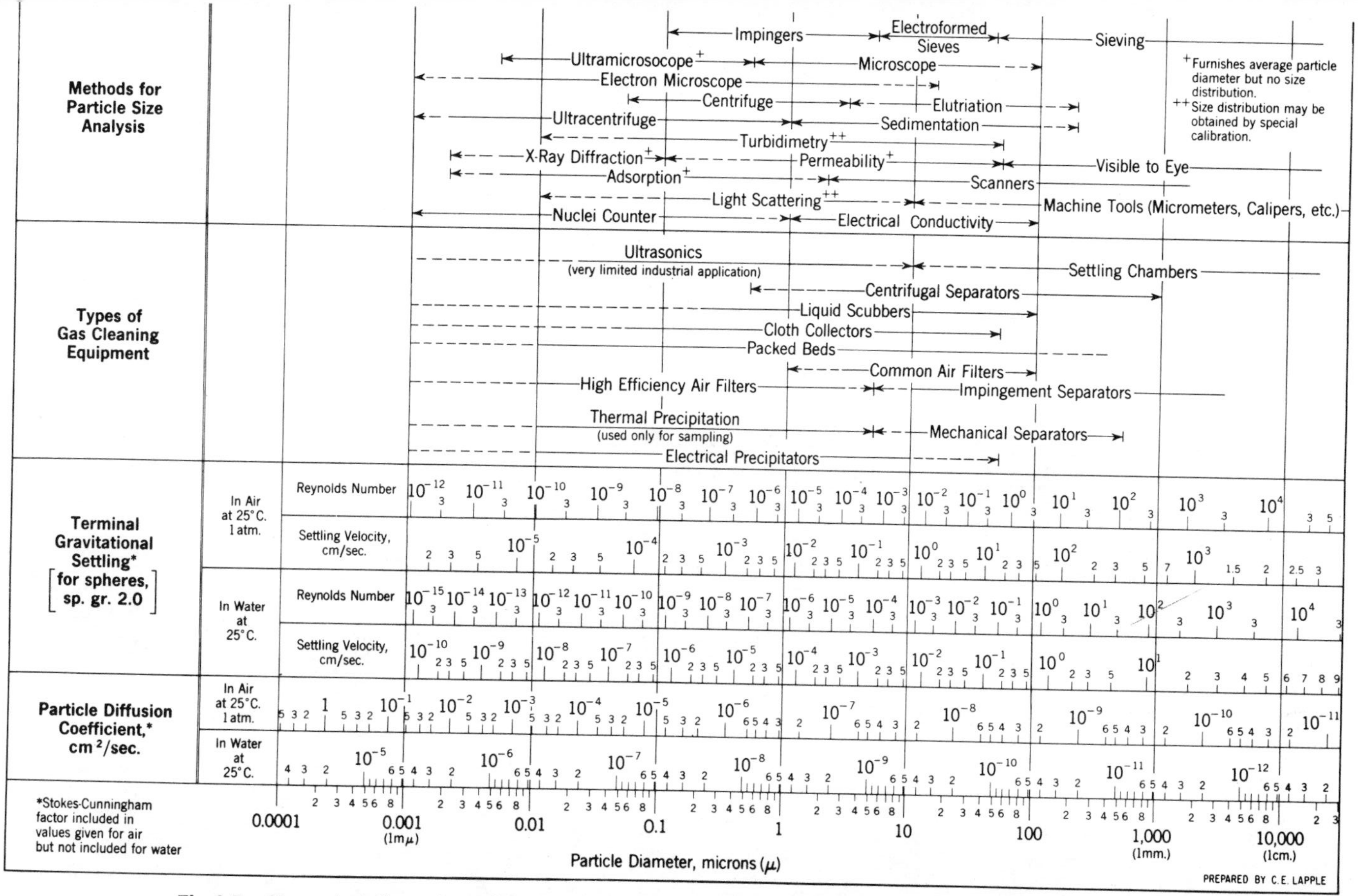

Fig.4-5 Characteristics of particles and particle dispersoids. *(Courtesy Stanford Research Institute)*

Other sources include atmospheric dust and dried ocean spray. They are too heavy to be affected by Brownian motion but settle so slowly that they will remain in air masses for months at a time. The visable spectrum has wavelengths within this range; consequently, these particles are the cause of haze and visibility reductions. Atmospheric reactions, particle collisions, and humidity enter into these considerations and the result is a dynamic system varying according to several factors.

Above 1.0 μ, particles have definite settling velocities and especially above 10 μ. Their origins include the agglomeration of combustion product particles, ashes, dusts, and various products of grinding and wearing operations such as materials pulverized by vehicles and pedestrians. Also, in this category are raindrops, snowflakes, pollen, and insects.

The toxicology of particulates requires consideration of both the chemical composition and the size. Particles down to about 0.5 μ are retained by the nose, while those below this have a good chance of being deposited in the trachea or the lungs. Toxic metals, especially lead and nickel, in the form of dusts and compounds have been given the most publicity in this regard. Inhalation of fine particulates resulting from the automobile combustion of tetraethyl lead has been blamed for the high body levels of lead found in some urban residents. Miners, foundry workers, and machinists must also be aware of this hazard. Radioactive fallout is another form of particulate hazard that is compounded for mankind through the increased chances for intake with food animals and plants. Toxic organic particulates include pesticides and polynuclear aromatic hydrocarbons. Lastly, hay fever sufferers are very much aware of natural particulates in the form of pollens.

REFERENCES

I General

"Air Pollution—The Problem and the Risks," *Soc. Automot. Engrs. J.*, **76**:5 (May 1968), pp. 47–52.

Altschuller, A. P., "Composition and Reactions of Pollutants in Community Atmospheres," *Bull. World Health Org.* (Geneva), **40**:4 (1969), pp. 616–623.

Goldsmith, J. R., "Los Angeles Smog," *Sci. J.*, **5**:3 (March 1969), pp. 44–49.

"Health Effects of Air Pollution," *World Health Org. Chron.* (Geneva), **23**:6 (June 1969), pp. 264–274.

Stern, A. C. (ed.), *Air Pollution.* 2nd ed. Academic Press, New York, 1968. Vol. 1.

Sullivan, J. L., "Air Pollution—Causes and Control," *Occup. Health Rev.* (Ottawa), **20**:3-4 (1969), pp. 9–23.
Viland, C. K., "Air Pollution and Atomic Power," *Mines Mag.*, **60**:1 (Jan. 1970), pp. 17, 19–20, 22–23.

II Gases

Friedlander, S. K., and Seinfeld, J. H., "A Dynamic Model of Photochemical Smog," *Environ. Sci. Technol.*, **3**:11 (Nov. 1969), pp. 1175–1181.
Landau, E., Smith, R., and Lynn, D. A., "Carbon Monoxide and Lead—An Environmental Appraisal," *J. Air Poll. Contr. Assoc.*, **19**:9 (Sept. 1969), pp. 684–689.
Robinson, E., and Robbins, R. C., "Gaseous Nitrogen Compound Pollutants from Urban and Natural Sources," *J. Air Poll. Contr. Assoc.*, **20**:5 (May 1970), pp. 303–306.
Rohrman, F. A., Steigerwald, B. J., and Ludwig, J. H., "SO_2 Pollution: The Next 30 Years," *Power*, **3**:5 (May 1967), pp. 82–83.
Slonim, N. B., and Estridge, N. K., "Ozone—An Underestimated Environmental Hazard," *Environ. Health*, **31**:6 (May/June 1969), pp. 577–578.

III Particulates

Charlson, R. J., "Atmospheric Visibility Related to Aerosol Mass Concentration. A Review." *Environ. Sci. Technol.*, **3**:10 (Oct. 1969), pp. 913–918.
Charlson, R. J., and Pierrard, J. M., "Visibility and Lead," *Atmos. Environ.*, **3**:4 (July 1969), pp. 479–480.
Critchlow, A., and Maguire, B. A., "Airborne Dust in Man's Environment," *Environ. Eng.*, No. 35, (Nov. 1968), pp. 12–20.

5 Analytical Techniques in Air Pollution

The detection and monitoring of air pollutants is a highly specialized branch of analytical chemistry dealing with the qualitative and quantitative analysis of a gaseous mixture containing many different trace materials. These may be gases, vapors, odors, particles, and aerosols. No single procedure could ever be expected to give a complete analysis of so complex a system, but several methods are available which can detect a number of pollutants simultaneously. More important, however, is the location of the air to be analyzed, its degree of pollution, and the final uses of the data. For example, entirely different techniques are required for the monitoring of community air as opposed to studies at pollutant sources such as industrial sites where concentrations of certain individual pollutants may be extremely high. Also, data generated to show regional patterns based on meteorological and industrial interactions will require different techniques from data generated to show seasonal changes and annual trends. Further, each general class of pollutant requires a different approach in its sampling, measurement, and even in the standards developed for acceptable levels.

In general, gaseous air pollutants and some of the finest aerosols are measured by relatively standard analytical techniques adapted for extremely low concentrations. These include absorption in solutions for subsequent reaction and measurement, and instrumental techniques. Concentration is read in parts per million or micrograms per cubic meter. Particulates require either a filtering apparatus for total measurement or a separation device based on the kinetic energies of different particles. Microscopic or elemental examinations are often done in conjunction with these sampling methods. Measurements are given in micrograms per cubic meter or, in the case of dust falls, particles per square centimeter per unit of time. Odors can only be evaluated through the reactions of human beings to them and these will vary according to the individual. The consensus of a group of persons is necessary to give any sort of reproducibility.

GASEOUS AIR POLLUTANTS

Gaseous air pollutants are the most familiar to the lay public and have been studied scientifically for the longest time. For this reason, the techniques available for their study are highly sophisticated and very accurate. Standard dry and wet chemical tests have been published for practically every known industrial chemical which might be found in the air, and in addition, the more common pollutant gases such as those generated by the automobile or power generating stations can be detected by instrumental techniques. Two instruments, the gas chromatograph and the mass spectrometer, have been adapted to the detection of a wide range of gaseous pollutants at one time.

Table 5-1 lists threshold limit values for chemicals which might be found in urban or industrial environments as recommended by the American Conference of Governmental Industrial Hygienists. All of these substances can be detected by specific published techniques. Some are particulates, which will be reviewed later. Since vapors can be defined as the gaseous portion of a gas-liquid equilibrium below the boiling point of the liquid, they will be considered as gaseous for the purposes of this discussion.

The noninstrumental techniques for the detection of these substances all require the contact of a known volume of contaminated air with a specific reagent which can either be in a liquid or a solid form to produce a measurable reaction. The most common reaction is a color change which can be measured either by visual comparison or with a spectrometer. Interfering substances must either be removed before the reaction or compensated for in the measurement step. All these techniques have an inherent averaging time during which a sufficient sample is being collected for the reaction, and this must be taken into account in the evaluation of the results.

TABLE 5-1 THRESHOLD LIMIT VALUES (PARTIAL LIST)[a] ADOPTED BY THE AMERICAN CONFERENCE OF GOVERNMENTAL HYGIENISTS FOR 1970

		Threshold Limit Values	
	Substance	*ppm (v/v)*	*mg/m³*
	Acetone	1,000	2,400
	Acrolein	0.1	0.25
	Acrylonitrile—Skin	40	70
	Ammonia	50	35
	Amyl acetate	100	525
	Aniline—Skin	5	19
	Antimony dust	—	0.5
C	Benzene (benzol)—Skin	25	80
	Benzyl chloride	1	5
	Berylium	—	0.002
	Bromine	0.1	0.7
	n-Butanol	100	300
	n-Butyl acetate	150	710
	Carbon dioxide	5,000	9,000
	Carbon disulphide—Skin	20	60
	Carbon monoxide	50	55
	Carbon tetrachloride—Skin	10	65
	Chlorine	1	3
	Chlorobenzene (monochlorobenzene)	75	350
C	Chloroform	50	240
	Chromate dust (as CrO_3)	—	0.1
	Copper fume	—	0.1
	Cyclohexane	300	1,050
	Cyclohexanol	50	200
	Cyclohexanone	50	200
	DDT—Skin	—	10
C	*o*-Dichlorobenzene	50	300
	p-Dichlorobenzene	75	450
	1,2-Dichloroethane (ethylene dichloride)	50	200
	Epichlorohydrin—Skin	5	19
	Ethyl acetate	400	1,400
	Ethyl alcohol (ethanol)	1,000	1,900
	Ethylene chlorohydrin—Skin	5	16
	Ethylene oxide	50	90
C	Formaldehyde	5	6
C	Hydrogen chloride	5	7
	Hydrogen cyanide—Skin	10	11
	Hydrogen fluoride	3	2

[a]These values represent the highest concentrations in which it is believed that nearly all workers can be exposed repeatedly without adverse effect. The values for substances marked by the letter "C" should never be exceeded and those marked "Skin" can be absorbed through the skin, mucous membranes, and eyes.

TABLE 5-1 (Continued)

	Substance	Threshold Limit Values ppm (v/v)	Threshold Limit Values mg/m³
	Hydrogen sulfide	10	15
	Ketene	0.5	0.9
	Lead	—	0.2
	L.P.G. (liquified petroleum gas)	1,000	1,800
	Mercury-metal—Skin	—	0.1
	Mercury-organic compounds—Skin	—	0.01
	Methyl acetate	200	610
	Methyl alcohol (methanol)	200	260
C	Methyl chloride	100	210
	Methyl chloroform (1,1,1-trichloroethane)	350	1,900
	Methylene chloride (dichloromethane)	500	1,740
	Nitric oxide	25	30
	Nitrobenzene—Skin	1	5
C	Nitrogen dioxide	5	9
	Ozone	0.1	0.2
	Perchloroethylene (tetrachloroethylene)	100	670
	p-Phenylenediamine	—	0.1
	Phosgene (carbonyl chloride)	0.1	0.4
	Pyridine	5	15
	Stibine	0.1	0.5
C	Styrene monomer (phenylethylene)	100	420
	Sulphur dioxide	5	13
	1,1,2,2-Tetrachloroethane—Skin	5	35
	Toluene (toluol)	200	750
C	Tolylene-2,4-di-isocyanate	0.02	0.14
	Trichloroethylene	100	535
	Turpentine	100	560
C	Xylene (xylol)	100	435

Each analytical procedure has an ideal concentration which it can detect with the highest accuracy. This is dependent on the practical degree of cleanliness which can be achieved, the sensitivity and reproducibility of the measuring step, and on the reagents and reaction products both of which are almost invariable at low concentrations. For this reason, the volume of sample gas must be varied to bring the results into the required range. For air pollutants of very low concentrations, it is very possible that the long sampling times required would make the tests useless for the measurement of highly variable sources. Alternative, more sensitive tests must then be found to give better estimations of the peak concentrations which might be encountered.

SAMPLING METHODS

The method of sampling a contaminated atmosphere is dependent on the type of source being sampled and the pollutants present, as well as the sensitivity of the testing method. The most common means of sampling is to connect the testing device to a vacuum source which will suck a known amount of air directly into the reaction chamber. For more complex systems, however, where several tests are being run on one airstream, it may be necessary to pass the sample air through the pump first and blow it into the testing devices. Care must then be taken that the pump, the tubes, and any other parts which contact the sample do not change it in any way.

The best pumps operate on a rotary principle which gives an even flow of gas. Diaphragm pumps give a pulsed flow but are necessary where more pressure is required. Other methods of producing a vacuum include aspirators, hand pumps, and hand squeeze bulbs. These latter two are only used for the so called "grab" samples taken in connection with portable detection units designed to allow rapid tests to be made at field locations.

Grab samples can also take the form of volumes of air trapped at various locations for subsequent analysis at a laboratory. Their advantages include the mobility allowed in collection, and the ability to subject the samples to complex, nonportable analytical instruments. Also, large samples can be collected in very short times and then analyzed by slow methods, thereby shortening the averaging time considerably. Grab samples cannot be used where decomposition or reactions may occur in the sample.

Several types of grab samples commonly used are evacuated vessels and plastic bags. The former might be made of glass or stainless steel and have valves which are opened to take in the sample. They are necessarily small, but can be calibrated to hold a precise volume. Techniques have even been developed whereby reagents are sealed into the evacuated flask so that a specific test can be run at the sampling site. Nonevacuated rigid containers with a valve at each end can be used in a similar manner to the evacuated flasks by simply flushing sample air through them with a hand pump and then closing both valves to trap the sample.

The plastic bag or balloon allows the collection of much larger grab samples. As much as 10 cu ft can be easily collected providing enough sample for a series of laboratory tests. The bags can be

Fig. 5-1 Gas sampling system (Scott Model 301). *(Courtesy Scott Research Laboratories, Inc.)*

made of suitable plastic sheeting including polyethylene and polytetrafluoroethylene. Filling is done with a high-volume pump similar to a vacuum cleaner, or alternatively, the rim of the bag can be afixed to the rim of a rigid container and the interstitial area evacuated if it is not desirable to pass the sample through a pump (see Fig. 5-1).

A similar grab technique involves freezing or condensing all components in an air sample boiling above a certain temperature. The frozen mixture is considerably more compact than the original air sample and can be vaporized for analysis at any time desirable. For example, if liquid nitrogen (–196°C) is used as the condensing medium, all constituents boiling above this temperature including oxygen will be trapped. Normally, a warmer bath such as dry ice-acetone (–78°C), or any of several fluorocarbons would be used. In any sample taken at less than O°C, a primary trap to collect water is necessary to avoid clogging the flow system.

Where a pump is being used to blow a sample gas into a detection device, it must be constructed so that only inert materials come into contact with the sample. This also applies to the tubing, valves, fittings, and flow measuring devices which may also precede the reactor. Glass is by far the best material for these purposes from a chemical point of view, but it is impractical for pump linings and valves. It is also fragile and difficult to work with as tubing, and it cannot be used for hydrogen fluoride detection. Various types of plastics are most commonly used with the specific type, again dependent on the sample. Polytetrafluoroethylene is the best choice, being inert to all known pollutants, but it is expensive and slightly more difficult to handle than polyvinylchloride for example. Polytetrafluoroethylene also is a suitable lining for diaphragm pumps. Lastly, stainless steel is the ideal material for fittings and valves, filter holders, moisture traps, and even tubing, where feasible. Brass can also be used for the more common pollutants with the exception of the sulfur and nitrogen oxides.

The volume of the atmospheric sample being tested must be measured accurately in a form usable in the final results. In general, this means on a dry basis, converted to 760 mm pressure and 10°C. Where hand pump or syringe is used, the volume is simply the displacement of the pump. Continuous pumps require either a rate measurement of their output or a quantitative measurement. Rate measurements can be made in two ways: restricting the flow of the system by means of a calibrated orifice or by measuring the flow with a rotameter. Either method is inexpensive and easy to use. They give instantaneous flows, however, and must be checked often to insure stability. Once a flow is established, the volume of sample can be varied by changing the sampling time. Calibrated orifices and also capillary tubes, which work on the same principle, can be purchased or made to yield relatively constant flows at constant temperature regardless of inlet pressure, once a minimum pressure differential has been exceeded. These are convenient for permanent sampling units once they have been calibrated. Rotameters are more flexible in their application and are widely used in air pollution work. These consists of a tapered glass or plexiglass tube with a scale printed on the outer wall. A movable ball or cylinder on the inside of the tube will float on an airstream passing through the tube when it is in a vertical position at a level proportionate to the rate of flow. They are either marked with air flow rate units or have an arbitrary scale to be used in conjunction with a calibration chart based on air.

Quantative volume measurements are more accurate than flow rates and are independent of sampling time. Three types of meters are available all of which are relatively bulky and expensive. The wet test meter is commonly found in air pollution laboratories and has an accuracy of ±0.5 percent. Its operation involves the displacement of water from compartments of a specially designed wheel, so a humidity correction is required in its use. The dry test meter is similar to the commercial gas meter and is more portable for use in field operations. Its accuracy is around ±1 percent. Also available is a dry cycloid-type meter which is more accurate, but more expensive, than the other two.

Quantitative volume meters can be used for the calibration of rotameters and orifices, but in turn, they should be calibrated occasionally themselves. The easiest method is the one recommended by the ASTM (D1071-55) which involves the siphoning of water into a tared container followed by weighing the water and calculating the volume of air it displaced.

Dry Chemical Tests

A wide variety of analytical procedures can be grouped under the classification of dry chemical tests. These are for the most part qualitative for specific pollutants, but often relatively accurate quantitative estimations can be made. A piece of catgut, for example, will indicate the presence of water vapor in the air by its degree of twist. In the same manner, dry chemical tests provide simple and rapid checks and are most useful when the danger of toxic gases is present. They include primarily indicating papers and tubes, but some unique methods must also be added.

Indicating papers are best exemplified by the common litmus or pH types. They consist of a piece of absorbent material with a reagent dried on the surface and are usually supplied with a series of colored spots for comparison after exposure to the sample. Preparation and storage of the papers is often critical, and only a few have really proved to be useful. These include papers for the detection of bromine, hydrogen cyanide, hydrogen fluoride, hydrogen sulfide, and sulfur dioxide.

An example of an indicating paper is the one used for the detection of hydrogen sulfide. This gas is a toxic by-product of sewage plants. It can be quickly estimated with a small hand-held device containing an aspirator bulb which draws air through a clamp holding a piece of filter paper impregnated with dried lead acetate. The bulb is squeezed a sufficient number of times to produce a brown stain on the paper. The intensity of this stain and the number of squeezes required to produce it determine the concentration of hydrogen sulfide in a range from 6 to 200 ppm. The entire test can be completed in less than 5 minutes.

Indicating tubes are more sophisticated than indicating papers and are commercially available for many hazardous gases. They are generally more reliable, more accurate, and easier to store than papers. The tubes are constructed of glass and sealed at both ends. They contain an inert granular material onto which a dry or damp reagent has been absorbed. To use them, the ends are broken off and a measured quantity of sample air drawn through to produce a color in the tube. Either the shade of color produced or the length of the colored band is used to evaluate the concentration of the pollutant. The tubes are supplied with comparison charts and also are assigned expiration dates after which their accuracy cannot be guaranteed. Commercial tubes have a life of from 3 months to 3 years and accuracies of ±30 percent which is sufficient for most safety checks (see Fig. 5-2).

Indicator tubes are available for ammonia, chlorine, carbon dioxide, sulfur dioxide, hydrogen cyanide, hydrogen fluoride, hydrogen sulfide, nitrogen oxides, and trichloroethlene. The carbon mon-

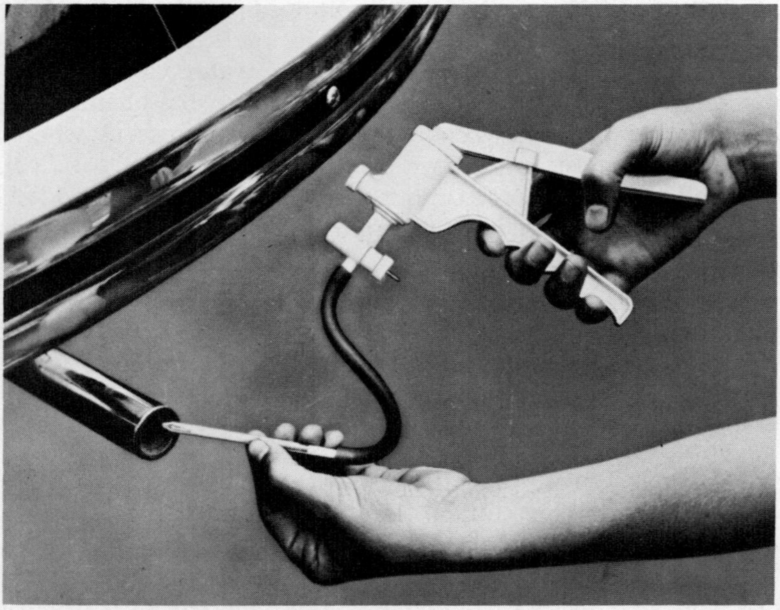

Fig. 5-2 Indicating tube and hand pump for exhaust test. *(Courtesy Edmund Scientific Co.)*

oxide tube is a very valuable indicator. It consists mainly of silica gel. The center band contains potassium palladosulfite while the others are plain. In use, the two ends of the tube are broken off and a constant stream of air passed through. Interfering gases are removed by the plain gel, and a brown band appears if carbon monoxide is present. The length of the band compared with the amount of gas passed through will indicate carbon monoxide levels up to 700 ppm.

Various unique methods which cannot be classified under any general heading have shown value for certain pollutants. Among these are several methods for the detection of ozone which compares the degree of cracking of thin rubber strips. Another is the venerable lead peroxide candle (Fig. 5-3) method for the estimation of long-term sulfur dioxide levels. This consists of a cylinder

Fig. 5-3 Sulfur dioxide monitor. *(Courtesy Research Appliance Co.)*

coated with a dried suspension of lead peroxide which absorbs sulfur dioxide in proportion to the surface area and the concentration in the atmosphere. Hydrogen fluoride and other volatile fluorides are retained by many plants. Among these, Spanish moss has been used to detect fluoride emitters.

Wet Chemical Tests

The basic unit of the wet chemical air pollutant test is the absorption apparatus. It provides a relatively efficient contact between a sample gas and a liquid solvent or reagent to dissolve the component of interest. This will either react to produce a measurable physical change or remain in solution for subsequent reaction and analysis. A well-designed absorber and flow system will absorb 90–100 percent of a gas in a relatively simple and inexpensive analysis procedure.

The absorber may take a great variety of forms, but these can be grouped into two basic types. One is the wet impinger, which directs a stream of air against a surface at sufficient speed to disperse it; the other is the fritted glass scrubber. Both break the incoming air stream into fine bubbles under the solvent, increasing the surface area immensely. The efficiency of the absorber depends on this dispersion and also on the solubility of the gas in the liquid phase.

The ideal liquid phase should be nonvolatile, noncorrosive, stable, nonviscous, nonfoaming, and inexpensive. Water has these qualities and is used for some very polar gases which are soluble in it, such as hydrogen sulfide, hydrogen chloride, ammonia, and methanol. Ethanol is a good solvent for organic esters and aldehydes. Dilute hydrochloric acid is used to absorb primary amines, and dilute caustic for hydrogen cyanide and sulfur dioxide. Many other absorption procedures use a reacting system of a solvent and a reagent. The efficiency of these is no longer dependent on just the dispersion and solubility of the gas since, as it dissolves, the gas is being removed from solution by reaction with the reagent. The faster this reaction, the poorer the solvent that can be tolerated with good results.

The wet impinger is usually associated with fine-particulate collection, but its simplicity makes it useful as a gas absorber also. It consists of a glass tube with a fine orifice (1–2 mm) at one end mounted in a larger tube so that the orifice is 5 mm above the flat bottom. In operation, a gas stream entering the smaller tube is so accelerated by the orifice that aerosols and particles in it are impacted onto the bottom of the larger tube by their momenta.

When liquid is placed in the tube, the gas stream will be dispersed into fine bubbles as it passes through the orifice. Flow rates from 1 to 10 liters/min can be used depending on the design. Impingers are easy to clean and reproducible in operation.

For less-soluble gases, fritted glass scrubbers are required to get sufficient dispersion for adequate dissolution. These are available in many shapes and sizes, but all contain a coarse or extra coarse glass frit separating the incoming gas from the solvent. Air passing through the frit is broken into fine bubbles for maximum solubility. Good dispersions are possible at very low flow rates, but flow rates may vary from frit to frit and cleaning is difficult.

Both wet impingers and fritted glass scrubbers restrict the gas flow in a system, causing a pressure loss which increases as the flow of gas through them increases. Impingers have lower initial pressure losses, but are not at all linear with flow since, as the input pressure is increased, the flow through the orifice approaches the speed of sound—a limiting characteristic. When this flow is reached, the orifice becomes insensitive to further increases in pressure, passing a constant rate of gas. Thus wet impingers have a relatively narrow range of flow rates through which they can be used. Fritted glass scrubbers are more linear in their pressure-loss characteristics and are usable over a wider range of flows. Their limit is often the height of the foam produced before solution carry-over occurs.

The reactions to which the absorbed pollutants are subjected are most often colorimetric, requiring either the addition of a color-developing reagent or merely a waiting period before measurement on a colorimeter. A blank run through the entire procedure with the exception of the exposure to the pollutant is necessary to set the zero point and the readings must be compared with a calibration curve prepared with standard solutions. Other methods of analyzing the solutions include conductivity, titration, polarography, and turbidimetry.

Every gas listed in Table 5-1 has an absorption procedure worked out for it. These and hundreds of others have appeared in journals such as *Analytical Chemistry, Journal of the Air Pollution Control Association,* and *American Industrial Hygiene Association Journal.* Several excellent reviews are also available describing in detail the analytical procedures summarized here and including carefully selected analyses for the more common pollutants. For example, the second volume of *Air Pollution* edited by Arthur C. Stern (Academic Press, New York, 1968) gives exact procedures for eleven inorganic gases, and *Determination of Toxic Substances in Air* edited by Hanson, Reilly, and Stagg (W. Haffer & Sons Ltd.,

Cambridge Eng., 1965) includes procedures for over thirty organic pollutants. For very low concentrations or other special situations, the latest published works should always be sought out through the journals mentioned.

An example of a wet chemical monitoring technique is one often used for determining trace amounts of aldehydes in the atmosphere. These derive primarily from auto exhausts and are thought to cause eye irritation. A 0.05-percent aqueous solution of 3-methyl-2-benzothiazolone hydrazone hydrochloride is placed into each of two fritted glass scrubbers arranged in series. Sufficient air is passed through to give a measurable reaction. The lower limit is around 2 ppb, which would require a sampling rate of 0.5 liter/min for 24 hr. After sampling, the solutions are allowed to stand for an hour to insure the reaction of all dissolved aldehydes before an oxidizing solution is added. This causes a blue color to develop in proportion to the concentration of aldehyde. The color is measured at 630 $m\mu$ for comparison to a Beer's law curve of formaldehyde at several different concentrations. The method is very accurate for formaldehyde and somewhat less for longer-chain aldehydes, but since formaldehyde makes up more than 50 percent of atmospheric aldehydes, the error is not serious.

Various automatic devices can be constructed or purchased to assist in wet chemical atmospheric monitoring. An example would be a unit designed to hold several absorbers with pumps, flowmeters, and timers to allow unattended operation. An occasional visit by a technician to fill fresh tubes and measure the reacted ones would be all the supervision required.

Instrumental Methods

The trend in chemical research over the last several decades has been to automate analytical procedures wherever possible. This has freed chemists from tedious and time-consuming tests allowing more time for basic research. Often, the automatic equipment has been capable of much higher sensitivities and greater accuracies than the manual methods while at the same time achieving results much more rapidly. Many examples of this trend can be found in air pollution research. It is now possible, for example, to collect at one central point simultaneous and continuous data on several pollutants at many different sites. At each location is a series of instruments connected into a data logger which would collect all readings at preset intervals and either record them on magnetic or punch tape or send them to the central office via rented phone lines. A minimum of supervision is required.

Any air pollution instrument will contain a sensing section and a data readout. The sensing section must be connected to a flow system in a monitoring instrument or be equipped to receive discrete samples if it is a laboratory or field emissions instrument. Some monitoring instruments contain their own pumps and flow control devices. Spectrophotometry, chromatography, electrochemistry, and mass spectroscopy are among the means of sensing individual or groups of pollutants. Each of these is capable of responding to several compounds and the instruments must be designed to filter out, absorb, or in other ways to discriminate between the desired pollutant and interfering materials. This process is usually only valid for certain concentrations of pollutant and interfering gases so great care must be taken in designing instrument systems.

The data output from a pollutant measuring instrument usually takes the form of a millivolt signal which can be fed to a strip chart recorder for a permanent, real-time record of the pollutant levels. Instrument outputs can be used in other ways, depending on the needs of each project. For example, data loggers will record instantaneous readings from many instruments taken at predetermined intervals on magnetic tape. This can be fed into a computer for subsequent data manipulations. Care must be taken to insure that the intervals at which data is recorded are truly representative of the atmosphere being tested. Great amounts of data can be handled efficiently in this manner.

Prerequisite to all instrument techniques is the accurate calibration of the output. The amount of drift that can be expected in continuously monitoring instruments must be determined and calibration periods scheduled accordingly. Standard calibration gases are available for most common pollutants, and solutions must be prepared for the organic vapors that are measured. A pure air sample is also necessary to establish zero points. The calibration points should be in the range of the highest concentration expected and include intermediate points to check the linearity of the instrument. Some instruments have built-in electronic calibration points, but standard gases approximating the atmosphere being tested submit the entire system—tubing, valves, electronics—to the calibration.

The most basic form of air pollution instrumentation is an automated version of the absorbers described earlier. Such an instrument contains an absorber designed to continuously contact fresh solution with incoming gas. This mixture is kept in contact long enough to effect reaction and then the solution is passed through a measurement section before going to a waste reservoir.

The measurement is often done by a built-in colorimeter or conductivity cell. Many other measurement techniques are also used; for example, one ozone instrument detects the chemiluminescence of a reaction. Practically any wet absorption test could be adapted to this kind of instrumentation provided it is sensitive to a fairly short atmosphere contact period and its solutions are stable. Also techniques requiring subsequent heating or chemical mixing steps would require more complicated equipment. These instruments are often used for the monitoring of sulfur dioxide, nitrogen oxides, ozone, chlorine, oxidants, cyanides, and sometimes even carbon dioxide.

Some special instruments have been developed for certain gases which are difficult to detect with standard means. Hydrogen fluoride and other fluorides fall into this category and an instrument is available to detect these using a fluorescence quenching technique. Also included here are low concentrations of the nitrogen oxides. These gases are very difficult to measure by any means because of their unstable nature. Two instruments are marketed which use an electrochemical cell to measure both gases. A surprising improvement in accuracy has resulted.

Hydrocarbons can be detected easily over a wide concentration range with a flame ionization detector (FID). This detector, used widely in gas chromatographs, consists of a hydrogen flame into which a calibrated flow of sample air is metered. A potential is applied between the burner and an electrode inserted into the top of the flame. Hydrocarbons in the flame are ionized and carry current between the electrodes in proportion to their numbers. This current is ampified and forms the output of the instrument. Molecules containing several carbon atoms carry more current than methane, but if oxygen, chlorine, sulfur, or other atoms are present, the signal is greatly diminished. Carbon monoxide and methanol give no signal at all. The instrument is only useful where the total concentration of all hydrocarbons is desired, since no discrimination between different types can be made. Also, oxygenated compounds should not be present in large amounts since they will be detected.

Hydrocarbons can also be detected by combustion on a vanadium pentoxide catalyst at 300°C. This is a relatively specific reaction at this temperature, and the heat of combustion measured with thermocouples forms the output. In a similar manner, carbon monoxide can be combusted on hopcalite at 100°C.

Spectroscopy, chromatography, and mass spectroscopy are all laboratory instrumental techniques which have proven very useful in air pollution analysis. Of these, only spectroscopy and chromatography have been adapted to continuous monitoring of specific pollutants or groups of pollutants. The cost of the mass spectrometer will probably prohibit its development in this area for quite some time. Standard scanning spectrometers, chromatographs, and mass spectrometers are most useful in the analysis of complex gaseous mixtures which might include primary and secondary pollutants in various stages of reaction.

Spectrophotometry is the measurement of the amount of radiant energy absorbed by molecules at specific wavelengths. Every compound has a different absorption pattern which gives rise to an identifying "spectrum." This consists of absorption plotted against wavelength and is presented in wavelength ranges from the ultraviolet through the infrared. The peaks in the spectrum correspond to harmonic oscillations of the molecules as the instrument scans a sample with progressively increasing or decreasing wavelengths. If an instrument is set at a single wavelength chosen carefully where a pollutant molecule absorbs strongly and other compounds likely to be present with the pollutant do not, it can be used for monitoring that pollution on a continuous basis. Such an instrument would cost far less than a scanning instrument since only one wavelength is needed which can be isolated with filters instead of a complicated monochromator. Other gases absorbing in the area of the selected wavelength can be used themselves as filters by filling short cells with concentrated mixtures. Another type of instrument, called nondispersive infrared, uses a detector filled with the gas being measured—for example carbon dioxide. When this is irradiated with a broad spectrum of IR radiation, the carbon dioxide absorbs only those wavelengths characteristic to itself. It heats slightly and expands against a diaphragm causing an electrical signal. When an air sample containing carbon dioxide is passed between the IR source and the detector, it absorbs some of the radiation, causing the gas in the detector to cool slightly and changing the electrical signal. The change is proportional to the concentration of carbon dioxide. Such an instrument is illustrated in Fig. 5-4.

Air pollution instruments based on infrared spectrophotometry can be used for the measurement of carbon dioxide, carbon monoxide, sulfur dioxide, ozone, acetylene, ethylene, methane, ammonia, and peroxyacyl nitrates. Those based on peaks in the ultraviolet can be used to detect nitrogen dioxide, sulfur dioxide, ozone, and hydrocarbons. As an example of the type of precautions that must be taken, the infrared carbon monoxide instrument will also respond to carbon dioxide and water vapor. While the former can be filtered out, the latter cannot when low concentrations of CO are being measured, so either the sample must be dried or an alternate detection means used.

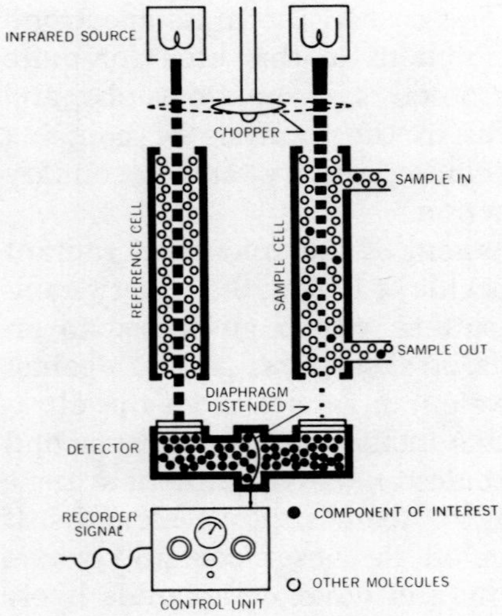

Fig. 5-4 A nondispersive infrared instrument. (*Courtesy Beckman Instruments, Inc.*)

The scanning infrared spectrophotometer is much more precise in its wavelength discrimination than the nondispersive instruments and so can be used in the laboratory to differentiate between compounds with absorptions in the same wavelength region. Analysis can either be done by condensing the higher boiling fraction of an atmospheric sample in a cold trap or by measuring the entire sample in a long path cell. The presence of several complex organic molecules can complicate an analysis considerably, and often it is necessary to separate the component parts first.

The gas chromatograph is one of the most versatile air pollution instruments available, being able to separate complex mixtures containing literally hundreds of components or, with a different column, being able to precisely standardize a calibration gas. Other chromatographs are adapted to continuously monitoring several pollutants at once. In addition, chromatographs are basically inexpensive, simple to operate, and rugged.

Gas chromatography is fractional distillation on a molecular scale. It is done in long columns containing nonvolatile liquid absorbants carried on solid substrates. Sample molecules pass through the columns at rates proportional to their volatilities and their affinities to the absorbent. Practically any combination of volatile materials can be separated under the correct conditions. The gas chromatograph consists of a column in a temperature controlled area, a system for metering inert gas through the column, a sample injection system at the entrance of the column, and a detector at the exit. The signal from the detector plotted against increasing time or temperature forms a chromatogram. Detectors can be chosen for specific sensitivities to certain types of compounds (for example the FID) or can be chosen to detect the widest variety of materials possible. The latter are more useful in basic air pollution work, although more sophisticated instruments may contain dual columns with different detectors on each for the comparison of their outputs.

Atmospheric samples often require the temperature to be increased from below ambient where the lighter gases would come off to several hundred degrees for high molecular weight materials. The column might even need baking out before further samples could be run in case any fractions were retained on the column which would bleed out slowly. In continuously monitoring instruments, a time cycle is used where lighter gases are separated on the column and detected at the exit. Then the flow is reversed so that the heavier ones are back-flushed and detected at the entrance. Hydrogen, carbon monoxide, and hydrocarbons are detected on one specialized instrument in this manner. Chromatographs are also suggested for the detection of sulfur oxides and sulfur compounds, and they are unparalleled for the separation and detection of hydrocarbons. Literally hundreds of components can be resolved from complex petroleum samples.

The mass spectrometer is probably the most expensive laboratory instrument used in air pollution work. It is becoming increasingly important despite this because of its usefulness in identifying small amounts of materials. This is done by destructively ionizing a sample and then accelerating the ions through a magnetic field onto a collector. The time interval between ionization and collection of each ion is proportional to its mass. An oscilloscope is used to read out a spectrum of the masses of all the ions produced from each sample from the lightest to the heaviest. The fragments of each compound are characteristic of its structure, with the largest one usually being the entire molecule after the removal of one electron. For simpler compounds, often this peak is sufficient for identification, but any possible isomers must always be considered. Thus, certain lighter gases in atmospheric samples can be determined by their primary peaks. More complex molecules require careful purification because their fragments are necessary to assist in the identification.

The resolution of a mass spectrometer is its most important characteristic. While the early instruments were of low resolution, the average price-range instruments available today can be described as medium-resolution and are capable of resolving down to about an atomic weight unit. High-resolution equipment is also available which can distinguish between atomic isomers, but the cost and complexity are usually prohibitive to most laboratories. Consequently, the medium-resolution instruments are the most common and their drawbacks must be considered. For example, several important gases have nearly equivalent molecular weights and cannot be distinguished. These include carbon monoxide, ethylene, ethane, and formaldehyde. Also, high molecular weight hydrocarbons and oxygenated compounds such as aldehydes and ketones yield isomeric fragments. While the total organic content of samples containing these can often be determined, individual constituents cannot. An exception is chlorine compounds, identified through the fragment containing the chlorine atom.

The use of mass spectrometry usually requires condensation of large volumes of sample to accumulate sufficient material for analysis. Often, compounds present in trace amounts are lost in this process. Some analytical chemists have coupled mass spectrometers to gas chromatographs forming an extremely powerful identification tool. The ability of the gas chromatograph to separate complex mixtures coupled with the mass spectrometer's speed and accuracy providing an exact molecular weight may solve many air pollution problems. In summation, it must be said that the use of mass spectrometry in air pollution is in a developmental stage, but has a promising future.

PARTICULATE ANALYSIS

The complete analysis of the particulates in an atmospheric sample is as complex as the analysis of the gaseous phase, but in addition, requires the determination of morphological properties. Particle size, for example, has no equivalent in gas analysis. This study is further complicated by the nonhomogeneity of a particulate-containing atmosphere, meaning that no definitive tests can be run on a continuous or instantaneous basis. In fact three distinct steps are required for a complete analysis: collection, size classification or morphological identification, and chemical analysis. On the other hand, a complete analysis is rarely necessary, and certain basic procedures can be readily followed to monitor criteria such as particulate loadings or the presence of hazardous materials.

Most particulate analysis is done with batch samples. The exceptions to this are several subjective procedures involving the visual evaluation of stack plumes by trained observers. Also, confined areas of high particulate concentrations can be monitored continuously by smoke photometers. These are used primarily for stack monitoring and usually involve a light source and a photocell opposite it. The output of the photocell can be recorded or used to activate an alarm at a preset particulate level. Neither this procedure nor visual comparison can tell anything about the size range of the particulates or their chemical composition. Nevertheless, both methods are very valuable in applications where rapid estimations of gross particulate levels are required.

The batch collection systems for particulates are usually filters through which a known volume of air is pulled. Other means also used are impingers, electrostatic precipitators, and cyclones. The choice of one collection system over another is a question of the information desired, since each method has advantages in certain areas. The primary question is the size range of the sample.

Total Particulates

The workhorse of air pollution tests is the high-volume sampler (Fig. 5-5). This is nothing more than a vacuum cleaner piped through a piece of filter paper. It is usually placed directly at a sampling site and run for 24-hr periods. The flow rates through the filter both before and after sampling are used to calculate the

Fig. 5-5 High-volume air sampler. *(Courtesy Precision Scientific Co.)*

volume of the sample, and the particulate weight is measured by weighing the filter paper before and after. Glass fiber paper is preferable to cellulose fiber paper since it allows particles to enter into its fibrous structure, avoiding a rapid build-up on the surface, which causes a large pressure drop. Analysis is more difficult with glass fiber filters since they cannot be ashed and there is some danger of adsorption of trace materials when the sample is solvent-extracted. The principle advantage of the high-volume sampler is its ability to sample a large volume of atmosphere, not only giving an accurate measure of the total particulate loading, but also supplying sufficient sample for any desired number of chemical tests. The sample it collects represents all the airborne particles in an atmosphere which exceed 0.5 μ.

The high-volume sampler is often used in conjunction with a tape sampler (Fig. 5-6) so that variations in particulate levels can be detected over shorter averaging times. For example, a 24-hr high-volume sample might be compared with 24 hourly spots from a tape sampler to yield the proportion of the daily total of particulates generated each hour. Sampling is done by pulling air through a circular spot on a strip of paper or glass fiber tape for a preset time interval. When this period is over, the tape is advanced to a clean area and the cycle repeated. The result is a strip of spots

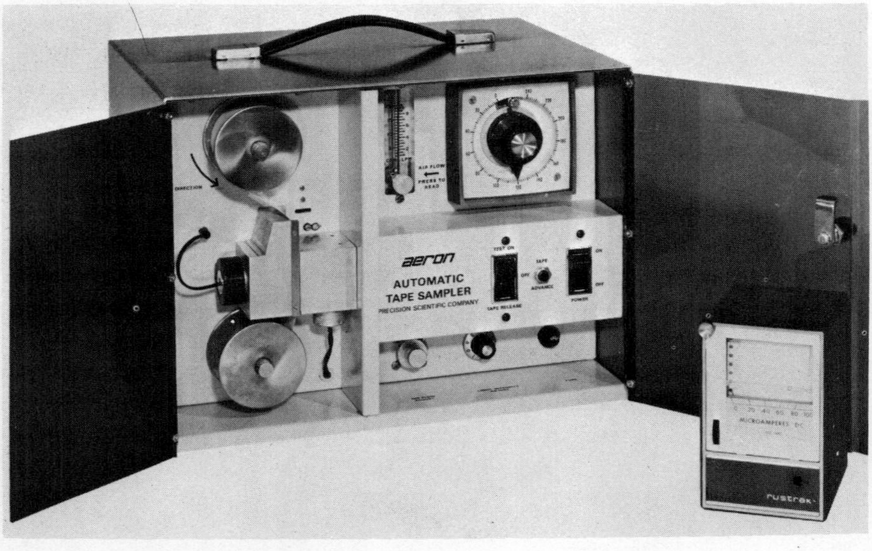

Fig. 5-6 Tape sampler. *(Courtesy Precision Scientific Co.)*

each corresponding to equal consecutive intervals of time. To evaluate them, a transmission of reflectance photometer is necessary, since the actual amount of particulate on each spot is too low for accurate weighing. A scale must be established for the photometer using blank tape as zero and a black spot as maximum. Several indices have been proposed including a soiling index and a coefficient of haze. Chemical tests can also be run on each spot to show variations in inorganic pollutants such as lead or nickel.

The wet impinger is another method of taking small samples of total particulates. These can be used for chemical, morphological, or even gravimetric analyses. Sampling is done in the same way as gas samplings with this device, but considerable errors can arise from long tubing which may trap particles either by electrostatic attraction or by impingement at bends and restrictions. Care must also be taken to insure against any solvent action by the liquid in the impinger, particularly if samples are being collected for microscopic examination.

A very gentle way to collect particles out of an atmosphere is by thermal precipitation. This technique uses the tendency of particles to be repelled by a hot wire. By mounting the wire close to a flat surface such as a microscope slide and then flowing the sample stream very slowly through, specimens suitable for direct microscopic examination can be prepared. The particles are collected so gently that no agglomeration occurs nor are bacteria destroyed.

Other methods of sampling particulates include centrifugal and electrostatic samplers. These devices operate in manners similar to their larger industrial cousins which are used for the removal of particulates from industrial process stacks. Centrifugal samplers are only efficient for particles larger than 5 μ, and while electrostatic samplers are useful for the collection of particles from 0.01 to 10 μ, they tend to volatilize organic containing particles.

Classified Particulates

There are many particulate collection methods which retain only certain size ranges. Others can differentiate several size ranges. These are very useful in the study of pollution effects which are a function of particle size. Haze, for example, is caused only by particles in a size range from about 0.2–1 μ. Particle size is a major variable when considering the effects of pollutants on humans. Particles of 5 μ and above are usually trapped by the nose while smaller ones down to about 0.02 μ will reach and deposit in the lungs.

The simplest size fraction to collect and measure is dust fall.

This includes particles from about 5 μ to over 200 μ which are heavy enough to settle out rapidly by gravity. Smaller particles tend to be more affected by air motion and the smallest ones are kept aloft by Brownian motion. A spherical 10 μ particle will drop in still air at a rate of 30 cm/sec while one that is 1 μ in diameter will fall at a rate of 0.003 cm/sec. Dust fall is measured by simply reweighing a tared container after it has been exposed to the atmosphere for a period of time. The weight of the sample collected is usually expressed in tons per square mile per month and is a measure of the "dirtiness" of a community. Additional tests are often run on the sample to determine the origin of the particles, or to detect the presence of bacteria, for example. Care must be taken in the design and placement of the collectors to avoid the loss of sample through the action of wind or rain. A sticky coating can be applied to the walls to trap particles and high sides provide sufficient volume to hold rain until it can evaporate.

Membrane filtration can be used to collect the smaller particles in an atmospheric sample down to less than 0.1 μ. Many pore sizes are available and accurate classification is possible. The filters may be dissolved in solvent, ashed, or simply made transparent with mineral oil for microscopic examination.

Dry impingers are capable of depositing particles in distinct size ranges on surfaces such as microscope slides. Several units are connected in series, each containing an orifice which is aimed at the surface. As a particulate containing atmosphere is pulled through the train, its velocity increases at each successive orifice, impacting smaller and smaller fractions of particles on the surfaces as their speeds exceed the ability of the airstream to deflect them. Particles down to about 2 μ can be collected this way, and often a fine filter is used to terminate the impinger train.

Another design of dry impinger uses a stack of tightly fitting metal pans each with holes in its bottom. The holes are precisely drilled and their diameter decreases from the top pan to the bottom. A glass or metal plate raised slightly off the bottom of each pan collects the impacting fraction from the gas stream as it is pulled through the stack. Petrie dishes containing nutrient gels can be used in place of the plates for bacteriological studies. As many as nine separate size fractions can be collected in this manner.

Physical Examination of Particles

Particulate samples collected in dust-fall containers, electrostatic precipitators, dry and wet impingers, and thermal collectors can be counted by microscopic examination. The total number of particles or the number in a specific size range per volume of air can then be calculated. Some collectors are designed for direct examination, but the others require the sample to be washed off and dispersed in a known volume of water or alcohol. An aliquot of this is placed on a microscope slide having a counting grid on the bottom and raised sides designed to trap a calibrated amount of liquid.

Membrane filters can be dissolved in acetone and the released particles counted in the above manner. The filters can also be attached whole to counting slides and rendered transparent. While this method may not be as accurate because of uneven distribution, it avoids any possibility of agglomeration, is more rapid, and provides a permanent record. Filters are available with counting grids printed on them.

Measuring the size of particles requires the definition of some standardized dimension that can be applied to all kinds of shapes. Direct observation either by light or electron microscopes must be used for an absolute measurement, but even these only show a profile of each particle. Thus, any measurement technique depends on a random orientation of particles. This is not always true and the judgment of the observer is very important. Special slides and eyepieces are available with imprinted calibrated scales. One such scale has a series of circles of increasing diameters. The observer compares the areas of the circles with the particles to arrive at a "projected diameter" of each. Several statistical diameters are also used which require measuring the randomly oriented particles in one direction according to the movement of the microscope bed. These different diameters cannot be compared against each other, nor do they correspond with classifying devices.

A skilled microscopist can provide much more information than simply the number and size of particulate samples. Often, he will be able to tell exactly what is contained in the sample and the most probable origin. Considering he needs only one particle which may weigh as little as 10^{-15} g, to do this, this is the most sensitive analytical technique known. The tools the microscopist may use in his evaluation start with a microscope of sufficient power to enlarge the particles to a recognizable size. The illumination of the sample is equally important since it must be adjusted to bring out distinguishing characteristics. For example, polaroid filters can be used to show birefringence in crystalline materials. Melting points and mixed melting points can be run on an adjustable hot stage. In addition, reactions can be run, solubilities determined, and other physical properties obtained through microscopic techniques.

The experience of the microscopist in evaluating materials is probably his strongest asset. This combined with a comprehensive collection of slides and photographs of known materials can solve many problems quickly. This is particularly true if some idea of the source of the contamination is known. By locating where the sample was taken and the wind pattern, often the exact plant emitting a pollutant can be determined.

Chemical Analysis

While an atmosphere with a high particulate content may cause discomfort, it is the presence of certain elements and compounds that is actually dangerous. Lead compounds, for example, or carcinogenic organic molecules might be present in atmospheres which appear clean. On the other hand, dense, billowing stack plumes often consist of large, nontoxic particles which settle out rapidly. The determination of toxic materials in the atmosphere follows established qualitative analytical procedures. Tests for inorganic materials usually involve the search for specific elements so destructive sampling and preparation techniques can be used. Organic samples, however, must be handled with great care to avoid decomposition.

Inorganic Analysis. Several instrument techniques are available for the detection of specific elements in a sample. These include emission and absorption photometry and polarography. Proper selection can give the analyst a very selective measure of a single element or a spectrum of all elements present in a sample. Volumetric and gravimetric chemical procedures have been practically eliminated in air pollution work, but a family of very specific reagents have been developed which yield color reactions to indicate the presence of certain elements. These reagents can be used singly or combined into more comprehensive procedures such as the ring oven technique.

The ring oven is an inexpensive device which enables a small amount of sample to be divided up evenly for a large number of separate indicator tests. This is done on a small piece of filter paper clamped in a circular, heated ring. A particulate sample dissolved in an appropriate solvent such as hydrochloric acid is run slowly into the center of the circle of filter paper. As it migrates outward toward the wall, the solvent evaporates, leaving a thin homogeneous ring on the paper. Although the ring contains the same amount of material that was in the original sample, it has been spread out over a much larger area.

A family of color tests have been developed which can be run on segments of the ring. Each is specific for a certain element and, generally, the results can be stored as a permanent record. The reagents may be in the form of a solution into which the paper is dipped or which is sprayed on, or crayons can be used. The latter are rubbed on the ring and the spot melted to produce a reaction. The resulting colors can be compared to standards for identification and quantitative estimations. A monograph by H. Weisz, the developer of the technique, (*Microanalysis by the Ring Oven Technique*, Pergamon Press, Oxford) lists thirty-five elements that can be identified.

Spectrographic techniques for the detection and measurement of elements entail substantial equipment costs but make up the expense rapidly in the time saved running the analyses. Three systems have been used: flame absorption, flame emission, and spark emission.

Emission spectroscopy has been used for many years as a rapid means of determining metals present in air pollution samples. The energy levels of the atoms present in a sample are raised to their excited states very rapidly with a jolt of energy from a flame, arc, or spark. On returning to their ground states, the atoms emit characteristic frequency radiations which are recorded on photographic film. Electric sparks and arcs have more energy than the hottest flames and excite a larger number of elements. The flame is useful when a specific group of elements such as the alkali metals is to be studied without interference from other elements. These methods are usable on dry particulates without special preparation and are very rapid. However, certain interactions between elements will occur which must be understood by the analyst.

For a large group of metals, atomic absorption spectroscopy provides a convenient and sensitive quantitative technique. This relatively recent development enables individual elements to be analyzed in solution down to very low concentrations. Again, a flame is used, but only to vaporize the elements. A radiation source specific for one element is beamed through the flame and the amount of energy absorbed is a measure of the concentration of that element. Separate sources are required for each element to be analyzed and these must be calibrated with standard solutions. Quantitative tests can be run on large numbers of samples in rapid succession with this apparatus.

Where the quantitative measure of two or three elements is desired in a large number of tests, some laboratories have used polarography. This is a measurement of the electrical potential of ions in a solution in comparison to a standard electrode. The electrolyte must be chosen with care for the specific group of elements to be tested and some unity is required to interpret the results. Excellent sensitivity is possible from very small volumes of solution and the sample is not destroyed. An experienced operator is required for the proper maintenance of the equipment and interpretation of results.

Organic Particulates. While the number of inorganic elements that might be found in the atmosphere is less than fifty, thousands of organic compounds may be present. These include pesticides, combustion condensation products, industrial emissions, and vehicle emissions. Many of these compounds have reacted in the atmosphere and are highly unstable, so great care must be taken in all stages of analysis from sampling and separation to detection. With the exception of pesticides and polynuclear aromatic hydrocarbons, little work has been done in identifying organic particulates. Most of the work that has been done consists of extensive analyses of certain areas rather than the development of routine tests. This is not the fault of analytical chemists but is indicative of the infancy of this field. Clear-cut guidelines on the danger of many organics have not been established, making routine tests impractical.

Sampling atmospheres for organic particlates is done in the same manner as for inorganic and total particulates although care should be taken to protect the sample from too much heat or sunlight. The high-volume sampler is very useful due to the low volumes of most organics. In very poor atmospheres, the deposit on the sampler is likely to be oily or waxy. No attempt should be made to dry it, however, to avoid any possibility of decomposition. Smaller samples may be practical when a specific compound or type of compound is being sought, so classifying collectors can be used to isolate the "lung damaging" fraction of particulates, for example.

A general analysis of the organic material in a sample requires a fractionation scheme to divide the components into chemically similar groups. One such scheme uses benzene, ether, or hexane for the initial extraction and then by a series of subsequent extractions using acidic, basic, polar, and nonpolar solvents divides the original sample into four fractions: strong acid, weak acid, neutral, and basic. The acid fractions have been found to contain fatty acids and high molecular weight phenols. Polynuclear nitrogen heterocyclics have been found in the basic fraction, but it is the neutral fraction which has been studied most extensively. This can be further separated by chromatography into an aliphatic, aromatic, and oxygenated fractions. Most of the organic pollutants found in the air will fall into these categories.

The aliphatic fraction contains the long-chain hydrocarbons resulting from petroleum products and combustion. Unsaturated hydrocarbons, terpenes, and substituted aromatics may also be found. Polynuclear aromatic compounds have attracted the most attention due to their carcinogenic properties and are found in the aromatic fraction. Fourteen different species have been positively identified. They can be separated by paper or thin-layer chromatography and detected by ultraviolet absorption or gas chromatography. Esters, aldehydes, steroids, epoxides, peroxides, and ketones will all be found in the oxygenated fraction. These have not been studied extensively since they do not seem to be dangerous at low concentrations.

Pesticides such as DDT, Chlordane, aldrin, and malathion will also go into the oxygenated fraction, but these are usually studied independently without prior fractionation. Tests have been developed through extensive study for all the commercial pesticides for use on biological, water, and air samples. The most versatile system uses gas chromatography, an ignition tube, and a microcoulometric detector. The sample is separated on the chromatograph and each fraction ignited in the combustion tube. Each pesticide will form a characteristic ratio of HCl, H_2S, H_3P among its combustion products. These are detected by the microcoulometric cell.

It is seldom necessary to conduct detailed analytical studies of organic particulates. For general studies, the total organics is sufficient. Injection of fractions into animals can be used to indicate carcinogenic properties, although these tests are not directly correlatable to humans.

ODOR MEASUREMENT

At worst, odors are nuisances. They represent trace contamination of the air by molecules which can be detected by the human nose. No attempt is usually made to identify the offending molecules since they are most often found in complex mixtures. Instead, odors are described in terms such as: fruity, flowery, burning, or nauseous. These are all subjective human responses and often vary from person to person, but they are also the only type of odor evaluation that is possible today. No instrumental means exists which can be used reliably in odor detection or measurement. Human responses must be put on as scientific and reproducible a basis as possible.

Various factors must be taken into consideration for odor evaluation. People react differently to different odors. Some are rather insensitive while others are overly sensitive. Enjoyable smells such as bread baking may become objectionable to persons subjected to them daily. Also, the nose is easily fatigued. In the presence of strong odors, nothing may be detected while the same odor diluted may be very objectionable. Some odors cancel out others. For example, strong odors will make weaker ones undetectable. Certain combinations of odorous substances will be much weaker when mixed together than individually.

While some odors are worse than others, all odors emanating from industrial sources can be considered bad. For this reason, odor evaluation is solely concerned with intensity. The "threshold value" is the only value that can be determined from an odor of unknown constituency. This is the concentration below which an odor is not detectable by the average person. For many pure compounds, threshold values have been determined in terms of their concentrations. Ammonia, for example, must be present in a concentration of 47 ppm before it will be noticed. Sulfur dioxide, on the other hand, can be one hundred times more dilute and still be noticed. Highly toxic trimethyl amine can be detected at concentrations as low as 0.00021 ppm.

Complex odors from industrial processes signal a mixture of compounds. Even when a single compound such as styrene is being produced, leakage from pumps, relief valves, and vents will add vapors of the starting materials as well as reaction by-products to the ambient air. Sewage settling ponds, sanitary land fills, and rendering plants create even more complex odor mixtures which would be nearly impossible to analyze. The industry or organization interested in evaluating its effluents as a preliminary step to control must, therefore, determine an odor threshold value for each possible source without regard to the actual makeup of the odor.

Odor threshold values can only be determined by the average reaction of a group of people. The larger the group the better, from a statistical point of view, but it is more important that the members be selected for near normal odor responses and that they be trained to react to odor intensity rather than quality. Panels consisting of three to twelve members are effective, but even one well-trained person can be used to monitor critical areas around a plant.

An odor evaluation program at an industrial complex would consist of the following steps: identification of odor sources, collection of samples at source concentration, and determination by a panel of threshold values for each odor. Once these values are known for each emitter, the distance over which the odor will be detectable can be calculated based on the emission rate, the height of release, and the local air turbulence. Comparing this to the proximity of local residential areas will clearly define the need for any corrective action.

The odor evalution panel must naturally operate in an odor-free room. Samples are prepared in an adjoining room out of sight from the panel members and are delivered to them in random order. Evaluation can be by passing a quantity of sample into the nose or by inhaling from a larger container or mask. Sample preparation consists of mixing a quantity of source gas with calibrated amounts of odor-free air according to the dilution factor desired. This can be done simply with syringes or specialized devices can be constructed for mixing gases and feeding them to a face mask or hood. Provision can even be made to include a calibration standard by which the tester can check his own sensitivity. This is critical to odor eveluation since a strong odor can easily cause fatigue, invalidating further tests. A period of readjustment is necessary after each test during which the individual breathes only deodorized air, and it is up to the person running the test to guard against supplying samples too quickly or too strong.

Members of the odor evaluation panel sample each mixture and assign it a numerical factor according to the following scale:

0 = Odorless	3 = Moderate odor
1 = Threshold odor	4 = Strong odor
2 = Faint odor	5 = Very strong odor

The statistical average of responses to various concentrations of odorous compounds has been plotted. It yields a linear correlation between these factors and the log of the concentration. For example, a moderate odor (3) might be described as faint (2) after being diluted ten times. The threshold concentration of an odor is the point at which further dilution produces no reaction by panel members. For odors of unknown original concentration, this point can be measured by the dilution factor necessary to reach it. This is often given in the form of "odor units" for purposes of calculation, defined as the number of dilutions necessary to bring a sample to the threshold value. For example, if 1 cu ft of a stack effluent must be diluted with 9 cu ft of clean air before the threshold is reached, it is said to contain 10 odor units per cubic foot. This factor multiplied by the flow rate of the stack in cubic feet per minute yields the odor units per minute emitted by the stack. Discharges of 50,000 to 100,000 odor units per minute are not uncommon in industrial plants, but the height of the stacks and the local air turbulence may make them undetectable a few thousand feet away.

Literally on the other side of the fence are local air pollution groups interested in locating sources of noxious odors in their communities. This task is far more difficult, since they do not have concentrated samples of each effluent to dilute. Complaints may be received about odors at faint or moderate levels, but on evaluation, local air movements might make them undetectable temporarily. Often, all an odor patrol can accomplish is to detect the odor and note the wind direction and velocity. Different smells must be described to some extent to differentiate between different sources, and a log of numerous incidents is often necessary before a source can be pinpointed.

Industries do well to cooperate with local groups in locating odors for often it is a shortcut to clearing up the specific process or equipment causing the problem. A plant survey is unnecessary if the odor causing the complaints can be identified outside the plant.

CONCLUSIONS

From the vast number of sampling and analytical techniques described in this chapter it can be easily seen that most industrial facilities have neither the equipment or expertise to do many of the analyses required to evaluate their own waste problems. Large industrial plants and refineries can afford to purchase expensive analytical equipment because they have other process applications for its use, but the smaller industry cannot justify such expenditures. It must, however, make analyses of its plant effluents if it expects to take proper corrective action for pollution abatement.

In the last few years a new industry has developed in the United States. It grew out of the analytical service laboratories which previously performed chemical and physical analyses on water, minerals etc. It offers similar services in the air and water pollution field. While it is necessary for the plant engineer to carefully review the credentials of such an organization before contracting for its services, by careful selection he can usually obtain complete data on his air pollution problem at a fraction of the cost of employing his own company's material and manpower to do the same job. He can also receive the benefit of expertise in a number of related areas when employing a topflight organization. Finally, most local, state, and national pollution regulatory agencies will more willingly accept the report of a bona fide consulting laboratory than one developed by the company's own staff.

REFERENCES

I. General

American Foundrymen's Society (Des Plaines, Ill.), "Sampling and Analysis for Pollutants," in, *Foundry Air Pollution Control Manual*, 2nd ed., 1967, Ch. 6, p. 29–32.

Clayton, G. D., "Determination of Atmospheric Contaminants," *Am. Gas Assoc. Proc.* (1953), pp. 892–923.

Feldstein, M., "Analytical Methods for Air Pollutants," in *Progress in Chemical Toxicology*, New York, Academic Press, 1963, Vol. 1, pp. 317–338.

First, M. W., "Sampling and Measurements," *Environ. Res.*, **2**:2 (Feb. 1969), pp. 88–92.

Henderson, J. S., "Ambient Air Sampling Instruments, Program Costs and Applications," *Am. Paper Ind.*, **51**:10 (Oct. 1969), pp. 13–14, 16, 18.

Marchesani, V. J., Towers, T., and Wohlers, H. C., "Minor Sources of Air Pollutant Emissions," *J. Air Poll. Contr. Assoc.*, **20**:1 (Jan. 1970), pp. 19–22.

Stern, A. C. (ed.), *Air Pollution*. 2nd ed. Academic Press, New York, 1968, Vol. 2.

II. Gases

Altschuller, A. P., "Atmospheric Analysis By Gas Chromatography," in J. Giddings and R. A. Keller (eds.), *Advances in Chromatography*, Marcel Dekker, New York, 1968, Vol. 5, pp. 229–262.

Brandt, C. S., "Fluoride Analysis," *Int. J. Air Water Poll.* (London), 7 (1965), pp. 1061–1065.

Hanson, N. W., Reilly, D. A., and Stagg, H. E., *The Determination of Toxic Substances in Air*, W. Heffer & Sons, Cambridge.

Kotnik, G., and Scheck, H. F., "Monitoring SO_2 in Stacks," *Instrum. Tech.*, **1699** (Sept. 1969), pp. 52–55.

LaMantia, C. R., and Field, E. L., "Tackling the Problem of Nitrogen Oxides," *Power*, **113**:4 (April 1969), pp. 63–66.

Van Houten, R., and Lee, G., "A Method for the Collection of Air Samples for Analysis by Gas Chromatography," *Am. Ind. Hyg. Assoc. J.*, **30**:5 (Sept.-Oct. 1969), pp. 465–469.

Wartburg, A. F., Pate, J. B., and Lodge, J. P. Jr., "An Improved Gas Sampler for Air Pollutant Analysis," *Environ. Sci. Tech.*, **3**:8 (Aug. 1969), pp. 767–768.

III. Particulates

Bardswick, W. A., and While, F. T. M., "Trends in the Instrumental Assessment of Industrial Dustiness," *Can. Inst. Min. Metall. Bull.*, Oct. 1969.

Ettinger, H. J., and Posner, S., "Evaluation of Particle Sizing and Aerosol Sampling Techniques," *Am. Ind. Hyg. Assoc. J.*, **26**:1 (Jan-Feb. 1965), pp. 17–25.

Gruber, C. W., Pritchard, W. L., and Schumann, C. E., "Recommended Standard Method for Measuring Wind-Blown Nuisance Particles," *J. Air Poll. Contr. Assoc.*, **20**:3 (March 1970), pp. 161–163.

Hemeon, W. C. L., Haines, G. F., Jr., and Ide, H. M., "Determination of Haze and Smoke Concentrations by Filter Paper Samplers," *J. Air Poll. Contr. Assoc.*, **3**:1 (Aug. 1953), pp. 22–26.

Jungreis, F., and West, P. W., "Microdetermination of Lead by the Ring-Oven Technique Applicable to Air Pollution Studies," *Israel J. Chem.*, **7**:3 (1969), pp. 413–416.

Lippmann, M., " 'Respirable' Dust Sampling," *Am. Ind. Hyg. Assoc. J.*, **31**:2 (March-April 1970), pp. 138–159.

McCarthy, J. F.: "Tracing Pesticide Residues to Parts per Billion," *Res./Dev.*, **21**:1 (Jan. 1970), pp. 18–22.

Mercer, T. T., "The Interpretation of Cascade Impactor Data," *Am. Ind. Hyg. Assoc. J.*, **26**:3 (May-June 1965), pp. 236–241.

Mueller, P. K., "Characterization of Particulate Lead in Vehicle Exhaust —Experimental Techniques," *Environ. Sci. Tech.*, **4**:3 (March 1970), pp. 248–251.

Sharkey, A. G., Jr., Shultz, J. L., Kessler, T., and Friedel, R. A., "Determining Organic Contaminants in Air and Water," *Res./Dev.*, **20**:9 (Sept. 1969), pp. 30–32.

Tabor, E. C., "High-Volume Air Sampling," Preprint, Purdue Univ., Lafayette, Ind., 1968, 25 pp. (Presented at the Seventh Indiana Air Pollution Control Conference, Lafayette, Oct. 15–16, 1968.)

Trani, R. R., and Kaelble, E. F., "Particle Size Distribution Measurements Below Five Microns," *Anal. Chem.*, **33**:9 (Aug. 1961), pp. 1168–1170.

West, P. W., and Sachdev, S. L., "Air Pollution Studies. The Ring Oven Technique," *J. Chem. Educ.*, **46**:2 (Feb. 1969), pp. 96–98.

IV. Odors

"Odor Panel Smells Trouble," *Mod. Mfg.*, **1**:7 (Dec. 1968), p. 164.

Reckner, L. R., and Squires, R. E., "Diesel Exhaust Odor Measurement Using Human Panels," *Soc. Automot. Engrs. Trans.*, **77**:3 (1968), pp. 1694–1705.

Sullivan, D. C., Adams, D. F., and Young, F. A., "Design of an 'Odor Perception and Objectionability Threshold,' Test Facility," *Atmos. Environ.*, **2**:2 (March 1968), pp. 121–133.

Vogh, J. W., "Nature of Odor Components in Diesel Exhaust," *J. Air Poll. Contr. Assoc.*, **19**:10 (Oct. 1969), pp. 773–777.

6 Designing a Plant Air Management System

INTRODUCTION

Designing a plant air management system requires the evaluation of all possible corrective action and treatment methods to assure an efficient and economical solution to the problem. A thorough investigation must be made of the complete system, including the source as well as the exhaust to the atmosphere. The evaluation of a pollution problem and the design of a control system involves seven basic steps, namely:

1. The plant survey
2. Testing and data collection
3. Establishing design criteria
4. Evaluating the pollution control system
5. Economic evalution
6. Selecting the control system
7. Engineering, design, and construction

The control program, consisting of these seven basic steps, is usually executed in three phases: evaluation, engineering study, and engineering-construction. Figure 6-1 illustrates the basic air management control program.

The initial or evaluation phase is undertaken to define the pollution problem. A plant survey provides design information to describe the manufacturing process and identify the pollution sources. A testing program is established to determine the prop-

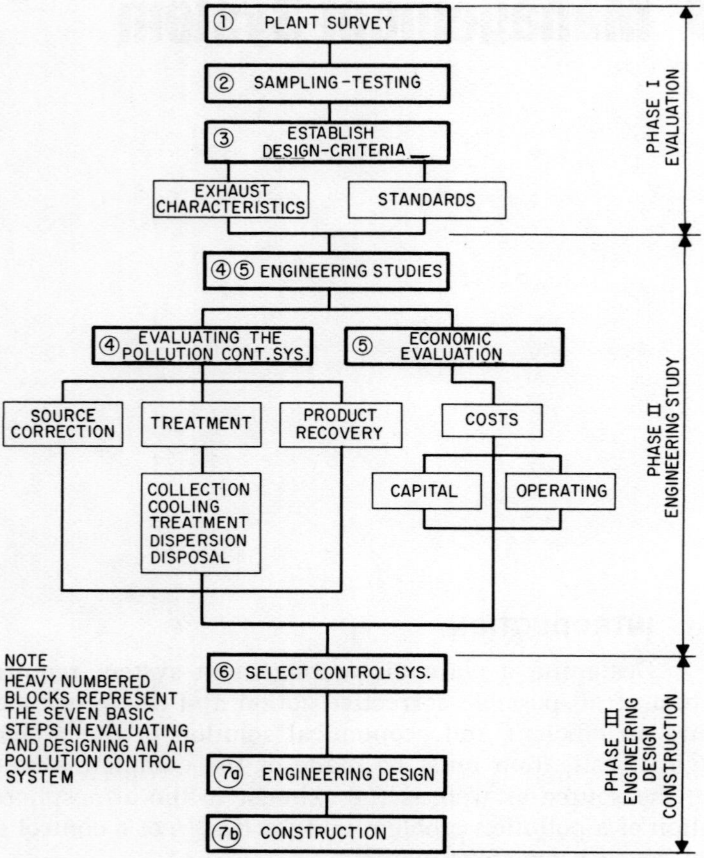

Fig. 6-1 Air management control program.

erties of the exhaust. Part of the field testing could include evaluating pilot-scale treatment equipment to obtain more definitive data for designing the control system. The control regulations and data collected are used to develop the basic design criteria; these establish the characteristics of the discharge and the required exhaust quality.

Next, a detailed engineering study is performed to evaluate suitable control systems for corrective action at the source, recovery of the exhaust pollutants as usable materials or treatment. When source correction or product recovery is not feasible, treatment of the exhaust is required. Developing a treatment system involves consideration of five factors: collection, pretreatment, selection of the control device, dispersion of the final exhaust, and disposal of the collected pollutant. In some systems, one or more of these elements is not required because of the characteristics of the control equipment. The result of the study will be a combination of corrective and treatment alternatives which will meet the system efficiency, dictated by the design criteria, and established by regulations and process requirements.

The final phase of the project includes engineering, design, and construction of the treatment plant, involving principles encountered by the plant engineer in the construction of new plants or the expansion of existing facilities.

Details of the seven steps involved in executing the evaluation and engineering study phases of the air management control program will be emphasized in this chapter. The principles discussed can be applied in investigating a control system for a new or existing plant.

STEP 1: PLANT SURVEY

Preliminary data suitable for developing a pollution control program must be obtained at the plant site. Specifically, the plant survey provides sufficient basic information to:

1. Define the pollution problem;
2. Establish possible solutions;
3. Evaluate the control regulations and community attitudes influencing the design criteria;
4. Develop an overall pollution control program and;
5. Obtain plant site and engineering data necessary to design the control system.

These objectives can be achieved by examining the plant site and obtaining information describing the local meteorology, topography, surrounding community, and applicable control regulations.

Plant Site

The plant site survey should begin with an understanding of the manufacturing process. A simplified diagram should be prepared describing the sequence of manufacturing operations, detailing stream flows and conditions.

Preliminary information to define the pollution problem can be obtained by inspecting the storage and manufacturing facilities. First, an inventory is made of the raw materials and finished products stored so that potential pollutants can be identified. This includes a tabulation of the quantities of all materials stored, classifying them according to emission source and quantity lost to the atmosphere, as well as to volatility, dust potential, and toxicity.

The potential pollutants are next traced to the manufacturing facilities to determine emission source, cause, and possible remedy. This involves a detailed analysis of the problem. Data should be collected to determine whether source correction is worth considering in the design stages. This includes evaluation of:

1. The condition of the equipment;
2. Operating procedures;
3. Maintenance and housekeeping practices and;
4. Possible overloading of the equipment.

In addition, information must be collected to design a treatment or recovery system. Each of the pollution sources found in the survey are inspected, the pollutants identified, and data obtained to define the problem. Principally, the data required are the exhaust rate, concentration, and characteristics of each of the pollutants emitted. The emission frequency and the peak pollution loads establishes the magnitude of the problem. The nature of the problem is determined by whether the pollutants create an odor, nuisance, or other harmful effect to the surroundings.

Finally, any existing air pollution control systems should be described, detailing both the exhaust treated and the control device. The number, size, and location of existing exhaust systems could influence the selection of the control device. The type, capacity, and efficiency of existing control devices should be detailed, observing their suitability for additional pollution control.

Sampling Points and Plan. A test program must be initiated with a minimum of interference to the manufacturing operation. The actual sampling program will be developed away from the plant site; therefore, it is important that some basic information be obtained during the survey. This includes:

1. The required tests
2. The available sample points
3. Accessability of the sample points
4. Interference of sampling with the manufacturing operation
5. The number of sampling stations required
6. The number of personnel required for sampling
7. The facilities available for field testing and shipping the samples to the laboratory

A sample diagram should be established at the plant site, detailing sample points and testing requirements.

Available Plant Facilities. The plant will have physical characteristics which will limit the installation or operation of some treatment systems. These include available plant area, location of the treatment plant in relation to the manufacturing facilities and available plant utilities. These factors will influence the selection and economics of the treatment method.

In addition to selecting a suitable area for the equipment, a check must be made of available plant utilities. Water, power, compressed air, gas, steam, and a condensate system could be required. An important utility often overlooked is the secondary waste treatment facility. In many treatment systems, a liquid or solid waste is discharged. Some preliminary information must be obtained from the plant as to available facilities to stabilize and discharge these wastes. This includes:

1. Available on-site waste treatment facilities
2. Compatibility of the waste with existing waste treatment systems
3. Permissible discharges to the municipal sewer
4. Solid waste disposal facilities
5. Charges for outside waste disposal services

Meteorology and Topography

Data must be obtained for estimating the stack dispersion characteristics to the local surroundings. Required meteorological data includes recorded wind observations and frequency of inversions. In addition, data describing the weather characteristics in the area should be obtained from the local U.S. Weather Bureau Office. An investigation of the terrain surrounding the plant site will also assist the engineer in evaluating the dispersion characteristics from the plant. Flat unobstructed sites are favored for optimum dispersion of the exhaust. Any discontinuity in the terrain can result in conditions which can cause pollution fallout. A building or natural obstruction in the vicinity of a stack can cause a disturbance in the plume rise, causing air motions favoring a portion of the plume to fall to ground level. When the stack is near natural topographic irregularities, the exhaust can be trapped in a cavity causing poor dispersion. This situation is typical when a plant is located in a valley.

Community

The plant survey should include an investigation of the community surrounding the plant, classifying the region as residential, rural, or industrial. Problems resulting from pollution emission could be an offensive odor, visible emissions, or a toxic gas. Obviously, the emission of a toxic pollutant cannot be tolerated. An odor would cause more of a problem in a residential area than in a remote location. In many residential areas, the mere sight of a steam plume from a stack has caused complaints and problems to a plant.

In rural regions, damage to vegetation or animal life has been found to cause severe problems. The treatment plant may have to be designed for extremely low pollution levels to protect these sensitive elements of the environment, as well as man. If the plant under study is surrounded by other manufacturing plants, the survey must include a description of the surrounding industries. The contribution of pollution by the plant under survey should be related to the emissions by neighboring sources and individual responsibility established.

Local Control Regulations

Part of the plant survey should include a review of local control regulations as well as establishing a contact with the responsible control authority. Many communities have local or regional regulations which will greatly influence the basic design considerations. These regulations should be obtained directly from the local control authorities.

The data collected during the plant survey will enable the engineer to develop a control program. The purpose of the survey is not to solve the problem, but to obtain basic data relating to the process, problem, environment, and community to establish design criteria. The information should be obtained in an orderly manner and tabulated for easy reference. Table 6-1 details a Plant Survey Form which could be helpful for this purpose. Having completed the plant survey, the information obtained is evaluated and the need for testing or data collection is determined.

STEP 2: TESTING AND DATA COLLECTION

The information obtained from the plant survey may not adequately define emission characteristics. In many cases, exhaust characteristics are obtained by calculating material balances, plant inventory and operating records, or from nameplate information. The result is that these data represent average conditions, not accounting for load variations, product diversity, or changing operating conditions. The uncertainty of the loading makes selecting the equipment or predicting treatment efficiencies extremely difficult.

TABLE 6-1 AIR POLLUTION CONTROL SURVEY

PLANT VISITED: .... LOCATION: ....
PERSONNEL CONTACTED: .... DATE: ....
SURVEY ENGINEERS: ....

I. *Plant Site*

A. Description of Manufacturing Process:
Reference Drawing or Sketch ....

B. Products Manufactured:

C. Plant Capacity:

D. Plant Operation: .... hr/day; .... days/week

E. *Raw Materials*

Mat'l.	Quantity Used, lb/day	Storage		Pollution Potential				Hazard
		Tank No.	Capacity	Volatile?	Dust?	Toxic	Odor	

F. *Pollution Sources*

Source	Operating		Ex-haust SCFM	Batch? Contin-uous?	Dura-tion, Hours	Pollutants				Source Correction	Product Recovery	Type Treatment
						Gaseous		Aerosols				
	Press.	Temp.				Type	lb/hr	Type	lb/hr			

1. Segregation of streams feasible?
2. Feasibility of substituting raw materials?
3. Value of recovered pollutants?
4. Possible improvements to existing exhaust systems?

G. *Existing Pollution Control Equipment*

1. Type: ....
2. Supplier: .... Model No.: ....
3. Capacity: .... Type: ....
4. Description of device: ....
5. Pollutants treated? ....
6. Spare capacity? ....

H. *Available Plant Facilities*

1. Plant area available for proposed control system
Location: .... Indoors .... Outdoors ....
Physical Dimensions: ....
2. Utilities
Air: .... SCFM @ psi
Steam: .... lb/hr @ psi
Condensate return unit:
Gas: .... cfm of ....

TABLE 6-1 (Continued)

Oil: .... gal/hr of ....
Process Water: .... gpm @ .... psi & .... °F
Cooling Water: .... gpm @ .... psi & .... °F

3. Waste Facilities
   a. Plant waste treatment facilities?
   b. Municipal waste system?
   c. Plant sanitary waste system?
   d. Other systems?

I. Sampling and Test Plan
   Sketch ....

Sample Point	*Required Tests*	*Frequency*	*Sample Pt. Accessible?*	*Pollutants Stable*	*Shipping Requirements*

II. *Meteorological Data*
   1. Prevailing wind direction
   2. Wind speed
   3. Maximum and minimum atmospheric temperature
   4. Frequency of inversions, precipitation, fog, etc.
   5. Address of local U.S. Weather Bureau Office

III. *Topography*
   A. Open Site
   B. Discontinuity of terrain
      1. Nearest building or natural obstruction
      2. Describe terrain surrounding plant

IV. *Community*
   A. Region surrounding plant residential? Rural? Industrial?
   B. Description of area surrounding plant?
   C. Description of surrounding vegetation or farm life.
   D. Neighboring manufacturing and/or process plants.
   E. Describe stack and stack emissions of neighboring plants.

V. *Regulations*
   A. State
      1. State control agency:
      2. State control contact:
      3. Applicable regulations:
   B. Local
      1. Local control agency:
      2. Local control contact:
      3. Applicable codes or standards:

The exhaust pattern and characteristics can be determined by stack tests.

First, the sampling stations are established. This is determined by the number of plant emission sources. A sampling program is developed to obtain sufficient samples to establish the exhaust properties. The length of the sampling period will be influenced by the manufacturing and product variations.

Next, the required tests must be specified. The testing must include determination of:

1. The quantity of pollutants emitted
2. The composition or dilution of the pollutants in the exhaust
3. The total flow volume
4. The exhaust temperature and pressure

Finally, the test equipment is selected and procedures outlined. Some basic process characteristics must be known before suitable test equipment can be selected. These include the nature of the pollutants, processes, and operations. The pollutants must be classified as either a dust, mist, or vapor. If the pollutants are aerosols (dust or mist), the relative size of the particles should be approximated. A gaseous pollutant should be classified as organic or inorganic and its condensing, absorption, adsorption, or combustion characteristics estimated. The pollutant concentrations should be approximated.

The nature of the process and operation will greatly influence the testing program. The temperature and pressure as well as any exhaust characteristics which will be detrimental to the testing equipment must be noted. The frequency of sampling the exhaust will be determined by the nature of the manufacturing operation. Another important consideration is the stability of the collected pollutants in the testing and storage equipment. With the information available from the plant survey, the testing apparatus can be selected and a testing schedule established.

Detail testing and sampling techniques were described in Chapter 5.

Pilot Plant Study

Many times, the control problem is of such a nature that equipment selection cannot be made on the basis of static stack tests alone. In such cases, it is advantageous to obtain design data using pilot-scale models of control devices. Under these conditions, continuous performance data can be obtained and the selection of equipment becomes a problem of equipment scale-up. Materials of construction can also be evaluated during the pilot studies.

When developing a pilot study, certain exhaust variables which influence the equipment performance must be carefully considered.

These include:

1. Pollution load
2. Characteristics of pollutants
3. Pollution concentration
4. Temperature of the exhaust
5. Density of the pollutant
6. Volume of the exhaust

These exhaust properties must be related to the test equipment characteristics, such as:

1. Pressure drop or energy requirements
2. Flow velocity
3. Critical physical characteristics:
   Cyclone: Physical dimensions
   Electrostatic precipitator: Collecting area/gas volume
   Fabric filter: Ratio of gas volume to filtering area
   Wet collectors: Physical dimensions
   Adsorbers: Physical dimensions, bed depth, and adsorbent
   Scrubbers: Liquid/gas ratio, physical dimensions
   Incinerators: Residence time, turbulence, and temperature

The relationship between the exhaust variables and the equipment characteristics determine the system performance and provide the basis for scale-up to the commercial unit.

STEP 3: ESTABLISHING DESIGN CRITERIA

The design criteria is developed from information obtained from the plant survey and data collection phases of the program. The data is collated and the characteristics of each of the plant emissions determined, with particular emphasis on establishing the variations in flow characteristics and volume. Where possible, exhaust streams are combined so that the number of control systems is minimized. The relative location and treatment compatibility of the individual streams must be considered in combining the exhausts.

The required treatment is determined by establishing the permissible limits for the individual pollutant emissions. These limits are established by federal, state, and local regulations. Having defined the exhaust characteristic and required emission quality, the control system performance can be calculated. Table 6-2 summarizes the steps involved in developing the design criteria. Next, the engineer must evaluate the pollution control system to establish suitable corrective or treatment methods.

TABLE 6-2 DESIGN CRITERIA FORM

	EXHAUST						Combined Stream			Treated Exhaust		
Stream No.												
Location												
	Min	*Avg*	*Max*	*Min*	*Avg*	*Max*	*M*	*A*	*Mx*	*M*	*A*	*Mx*
Particulates and Aerosols: lb/hr												
0–5 μ												
5–10												
10–20												
20–40												
40 μ												
Total												
Gaseous: lb/hr (Specify Components)												
Volume: cfm												
Temperature: °F												
Pressure: in. water												
Collection Efficiency												
Required Quality (ppm or μg/m³)												
Recommended Materials of Construction:												

STEP 4: EVALUATING THE POLLUTION CONTROL SYSTEM

After the design criteria have been established, the preliminary design is initiated. Developing a pollution control system involves an engineering evaluation of the factors illustrated in Fig. 6-2, namely:

Source correction
Collection system
Pretreatment of the exhaust
Evaluating suitable control equipment
Dispersion of the exhaust
Disposal of the collected waste

Each of these factors influence the whole system and must be carefully examined.

Source Correction

Source correction requires a detailed review of the manufacturing operations to reduce both the quantity of pollutants and the total volume exhausted from the pollution source. The purpose is to reduce the size and required performance of the treatment

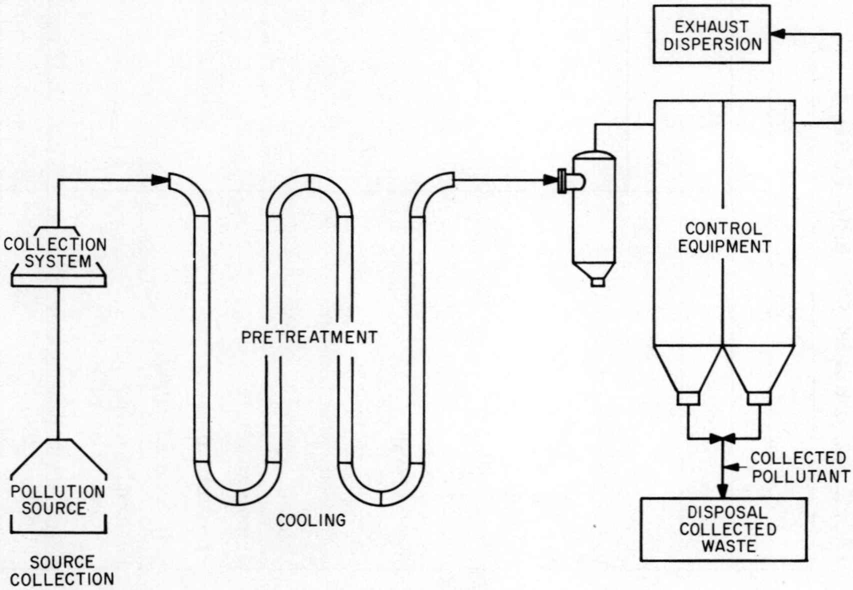

Fig. 6-2 Pollution control system.

system or eliminate its need. The process should be studied to consider the possible improvements to the manufacturing process or recovery of the pollutant as reclaimed product.

The manufacturing process is investigated to consider changes which could minimize or eliminate the quantity of pollutants emitted. This could involve a change in raw materials or equipment. Replacing toxic, volatile, or dust-producing raw materials with low-polluting substitutes could solve the pollution problem. Changes could be made to the manufacturing equipment or operating procedures. This might be as simple as improving a hood and exhaust system or placing covers on open process kettles. In some cases, major revisions to the manufacturing operation or equipment may provide the most practical solution to the problem.

A change in operating procedures might also eliminate excessive pollution emissions. Housekeeping procedures can be established which avoid open storage of volatile liquids or dusty waste materials. Operating process equipment at design or optimum conditions can reduce the emissions. Overloading an existing operation results in inefficient equipment performance and in many cases higher pollution levels.

External exhaust of pollutants can sometimes be eliminated by recovering the chemicals emitted for process reuse. It can be assumed that product recovery has been considered in the initial plant design. However, there is a new factor to be added to the economic evaluation—the cost of treatment.

The total volume or dilution of the pollutants should be minimized to reduce the size of the proposed treatment system. A check should be made of the plant exhaust system for excess air volumes, either by design, poor maintenance, or operation. In addition, the volume of hot exhaust can be reduced by cooling.

The feasibility of altering an existing ventilation system to segregate the stream containing the emission from the total exhaust should also be considered. In some cases, the cost of altering an existing system can be easily offset by the savings of reduced size treatment equipment.

The more expensive the treatment equipment, the more justified is source correction.

Collection System

The design of a new, or the revision of an existing, exhaust system, suitable for the pollution control equipment requires the specialized talents of an experienced engineer. The American Society of Heating, Refrigerating, and Air-Conditioning Engineers

(ASHRAE) *Guide*[1] and similar mechanical standards[2] detail design principles useful for this purpose and the reader is referred to them for design details. Some basic factors important in estimating the ventilation requirements, selecting the hood, and designing the transfer system are worth reviewing.

The ventilation volume is influenced by the rate of air displacement and the explosive limits of the pollutants. The primary purpose of the air flow to the hood is to capture the contaminants. The release of contaminants is accompanied by a displacement of air. This displacement could be a result of expansion of hot or pressurized exhausts, air motion created by manufacturing equipment in motion or ejection of airborne particles. Drafts and air currents near the displaced air tend to create a turbulent effect, dispersing the contaminants. The exhaust ventilation must exceed the air displacement rate, and overcome the effects of turbulence, to assure capture of the contaminants.

Ventilation volume is dependent upon the quantity of pollutants emitted, the characteristics of the contaminants, the cross-sectional area of the source, and the height between the release and the capture points. Obviously, the greater the quantity of release, the greater the exhaust required. In the case of toxic or explosive contaminants, the exhaust volume is determined by dilution requirements as well as capture volume. Protection of personnel from toxic concentration of contaminants and safety requirements of the plant must be considered. The exhaust may have to be diluted to concentrations considerably below the toxic threshold or the lower explosive limits of the pollutant.

Finally, the type of enclosure influences the required exhaust flow. The type selected will depend on the nature of the operation, with consideration given to the equipment, manufacturing operation, and the requirements of the personnel. Commercial hoods are usually classified into three broad categories: enclosures, receiving hoods, and remote hoods. Enclosures surround the emission source either partially or totally. Receiving hoods are located so the pollutants are emitted into the hood. Remote hoods are those designed to collect the emissions with an exhaust volume greater than the air displacement rate. Remote hoods are the most difficult to design and the least desirable in an air pollution control system.

A minimum flow volume is preferred in a collection system designed for pollution control. For this reason, complete enclosure of the source is preferred. If enclosure is not practical, the hood should be located as close to the source as possible and shaped to control the area of contamination.

The principles of duct system design and exhaust fan selection are detailed in mechanical standards used in the design of building ventilation systems. However, some factors important to the design of pollution control systems are worth mentioning. These include the following:

1. Transfer systems for particulate matter must be designed for transport velocities adequate to prevent settling of solids in the duct.
2. Inlets to the treatment devices should be designed to provide uniform flow distribution, avoiding overloading of any section or creating high turbulence conditions.
3. Systems containing condensible fumes should be insulated to prevent fallout of liquid, or at least designed for the collection of the liquid.
4. The exhaust fan operating curve should be checked with the characteristics of the abatement equipment to assure design compatibility.

The exhaust system could affect the operation of the treatment equipment; therefore, it must be carefully designed.

Pretreatment of the Exhaust

The temperature, and often the humidity, of high-temperature exhausts must be adjusted to conditions that can be tolerated by the control equipment. In most cases, the exhaust must be cooled to temperatures below 500°F. The exception being exhausts fed to incinerators. The temperature limits of common control equipment are listed in Table 6-3. Three methods, shown in Fig. 6-3, are generally considered for cooling a hot exhaust—namely, cooling by dilution with cool air, quenching with a water spray, and the use of cooling columns.

Cooling by dilution is inherent in systems where emissions are released to an exhaust hood. A volume of room air, depending on the design of the hood, will mix with the hot exhaust, resulting in cooling of the hot stream. In some systems, air is deliberately injected into the exhaust duct, prior to the treatment device, for the purpose of cooling the hot gas stream. However, dilution increases the flow volume and is generally restricted to small treatment systems. The required quantity of dilution air (A) to reduce the exhaust temperature to acceptable limits (tm), can be estimated by energy balance. Equations (1a) and (1b) summarize the energy balance relations.

$$\text{Cooling of exhaust} = \text{heating dilution air} \qquad (1a)$$

$$EC_{pe}\,(t_e - t_m) = AC_{pa}\,(t_m - t_i) \qquad (1b)$$

TABLE 6-3 CONTROL EQUIPMENT CHARACTERISTICS(4)

Control Equipment	Pollutant	Optimum[a] Size Particle (microns)	Optimum Concentration (grains/cu ft)	Temperature Limitations, (°F)	Approximate Pressure Drop (in. w.g.)
Mechanical collectors:	Aerosol				
Settling chamber		> 50	> 5	700	< 0.1
Cyclone		5–25	> 1	700	1–5
Dynamic precipitator		> 10	> 1	700	Fan
Impingement separator		> 10	> 1	700	< 4
Bag filter		< 1	> 0.1	500	> 4
Wet collectors:					
Spray tower		25	> 1	40–700[f]	0.5
Cyclonic		5	> 1	40–700[f]	> 2
Impingement		5	> 1	40–700[f]	> 2
Venturi		< 1	> 0.1	40–700[f]	1–60
Electrostatic precipitator		< 1	> 0.1	1000	< 1
Gas scrubber	Gaseous		[c]	40–100	< 10
Gas adsorber			[d]	40–100	< 10
Direct incinerator				2000	< 1
Catalytic combustion			combustible vapors	1400	> 1

[a]Minimum particle size collected at approximately 90% efficiency under usual operating conditions.
[b]Space requirements: S, small; M, moderate; L, large.
[c]Dependent on solubility of pollutant.

Where, E = exhaust flow, lb/hr
C_{pe} = average specific heat of exhaust (t_m to t_e), Btu/lb-°F
t_e = temperature of the exhaust, °F
t_m = temperature of the mixture, °F
A = dilution air flow, lb/hr
C_{pa} = average specific heat of the air stream, (t_i to t_m), Btu/lb-°F
t_i = ambient temperature of dilution air, °F

		REQUIRED UTILITIES							
Efficiency	*Space[b] Requirements*	*Power*	*Water*	*Steam*	*Fuel*	*Solid Waste*	*Liquid Waste*	*Collected Pollutant*	*Remarks*
< 50	L	X				X		dry dust	good as precleaner
50–90	M	X				X		dry dust	low initial cost
< 80	M	X				X		dry dust	
> 80	S	X				X		dry dust	
> 99	L	X				X		dry dust	bags sensitive to humidity, filter velocity & temperature
< 80	L	X	X				X	liquid	1. waste treatment required
< 80	L	X	X				X	liquid	2. visible plume possible
< 80	L	X	X				X	liquid	3. corrosion
< 99	S	X	X				X	liquid	4. high-temperature operation possible
95–99	L	X				X		dry or wet dust	sensitive to varying conditions & particle properties
> 90	M-L	X	X				X	liquid	same as wet collector
> 97	L	X		X			X	solid or liquid	adsorbent life critical
> 98	M	X			X			none	operating costs prime consideration
> 98	L	X			X			none	contaminants could poison catalyst

[d]Adsorber: concentrations less 2 ppm non regenerative system; greater than 2 ppm regenerative system.
[e]See Chapters 8 and 9 for specific requirements.
[f]Limited by materials of construction when sprays not in operation.

Quenching the hot exhaust both cools and humidifies the airstream. The increased moisture content could be detrimental to the operation of some devices, such as bag filters. Where increased humidity does not present any problem, a quench column can be an effective method of cooling a hot stream. The minimum water requirements can be estimated using equations (2a) and (2b). The actual water use is dependent on the spray nozzle design.

$$\text{Cooling of exhaust} = \text{heat absorbed by water} \qquad (2a)$$

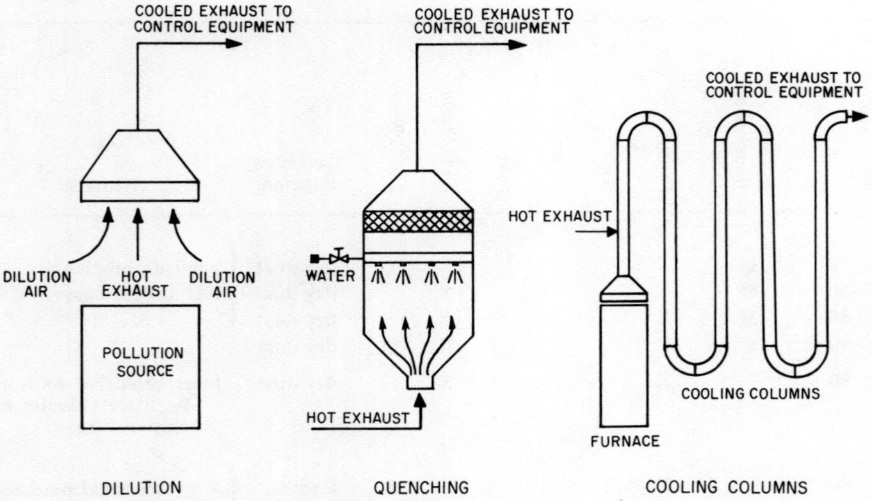

Fig. 6-3 Exhaust cooling methods.

$$EC_{pe}\ (t_e - t_m) = W\ (H_e - h_w) \tag{2b}$$

Where, W = minimum water flow, lb/hr
H_e = heat content of water vapor @ t_m, Btu/lb
h_w = heat content of inlet water, Btu/lb

Cooling columns are essentially vertical duct sections added to increase the surface area and therefore the heat transfer to the surrounding atmosphere. They can be either natural or forced convection types. A limitation in the use of cooling columns is the requirement of large floor area. The required surface area can be estimated as follows:

Heat loss exhaust = heat transferred to atmosphere

$$EC_{pe}\ (t_e - t_0) = UA\Delta t_m$$

Where t_e = temperature of exhaust, °F
t_0 = cooled exhaust temperature, F°
t_a = ambient air temperature, °F
A = duct surface area, sq ft
U = overall heat transfer coefficient, Btu/sq ft, °F

$$\Delta t_m = \frac{t_e - t_0}{L_n\left(\dfrac{t_e - t_a}{t_0 - t_a}\right)}$$

Methods of estimating the overall heat transfer coefficient are detailed in heat transfer texts such as Kern's *Process Heat Transfer*.[3] It should be noted that the overall heat transfer coefficient includes the transfer coefficient for radiation and convection to the atmosphere.

In addition to the methods mentioned, many operations utilize recovery heaters where the heat content of hot exhausts is reclaimed for heating the cold air supply. The exhaust temperatures are thereby maintained at levels which can be tolerated by the control system.

Each of these cooling methods has a definite application and the method selected will be dependent on costs and limitations imposed by the control equipment.

Evaluating the Control Equipment

The selection of a suitable control device requires a careful consideration of the characteristics of the pollutants and the equipment. In many cases, poor system performance can be traced to the selection of equipment not suitable to pollutant characteristics.

Basically, control equipment can be classified into two major categories: Those which remove aerosols and those which remove gaseous pollutants from an airstream. Equipment available for each of these two broad categories include:

Aerosols (including particulates)

- Settling chambers
- Cyclones
- Impingement separators
- Centrifugal separators
- Bag filters
- Wet collectors
- Electrostatic precipitators

Gaseous Pollutants

- Gas scrubbers
- Adsorbers
- Incinerators
  - Direct Incineration
  - Catalytic combustion
- Condensers

Equipment details can be found in Chapters 8 and 9. Equipment selection involves consideration of the pollutant, airstream and equipment characteristics, as well as the required cleaning efficiency.

Pollutant Characteristics: Aerosols (Sprays, Mists, Dust, and Fumes). The nature of the pollutant influences the equipment choice. Aerosols, using a broad definition, fall into four classifications: namely, sprays, mists, dust and fumes. In common usage, aerosols refer to solid or liquid particles of microscopic size (smoke, fog, or mist). Liquid droplets suspended in air, greater than 10 μ in size are defined as sprays, those less than 10 μ, as mists. Similarly, submicron solid particles suspended in air are referred to as fumes, those greater than 1 μ, as dusts. In addition, size distribution, concentration and physical properties are important factors to be considered in selecting the equipment.

The concentration of the exhaust influences the efficiency of a control system. Generally, the efficiency of the system and the mechanical performance of the equipment decrease with increasing dust loading, the principal reasons being the increased duty, reentrainment of the dust, and difficulties in handling the collected dust. Exhausts to collection equipment are normally maintained at loadings less than 10 grains/ft^3. When exhaust loadings are greater than this figure, precleaners are used. There are exceptions to this rule; some equipment, such as bag filters with automatic cleaning devices, can tolerate loadings as high as 80 grains/ft^3. However, many times the principles of economics override, since it may be less expensive to install a precleaner than overwork the principal control equipment.

Size distribution will greatly influence system performance. The important consideration is the weight-size distribution, not the mere presence of submicron dusts. The greater the weight percent of submicron particles, the more difficult is dust separation. Similarly, the particle density and shape influence the separation. The mechanical separation becomes more difficult as the difference in the properties of the two phases diminish. As the size of the submicron particles decrease, the solid phases approach the molecular characteristics of the airstream.

The operation of collection equipment is influenced by the abrasiveness, adhesiveness, resistivity, and hygroscopic characteristics of the pollutants. Abrasive particles could cause excessive wear and require special precautions. Materials which are adhesive or hygroscopic in nature cause operational problems which could be detrimental to system performance.

Pollutant Characteristics: Gaseous. Gaseous pollutants can be chemically classified as organic or inorganic. The selection of a treatment system is dependent on the chemical nature of the pollutants, as well as concentration, and characteristics such as condensability, corrosiveness, combustibility, toxicity, adsorbtivity, and solubility.

The concentration of the pollutants influences the performance of a system. Generally, the higher the pollution loading the greater the treatment efficiency of the system, although the quality of the exhaust will be lower. For example, a system with an inlet concentration of 100 ppm may have a treatment efficiency of 90 percent, while an inlet condition of 10 ppm could result in a reduced efficiency of 70 percent; yet the exhaust of the two streams will be 10 ppm and 3 ppm respectively. This is the significance of reducing the emission quantity. The concentration cannot be considered independently from the required efficiency. The allowable emission rate may be the controlling factor.

Properties such as condensability, combustibility, adsorbtivity, and solubility are significant since they determine the type of device which can be used—i.e., condenser, incinerator, adsorber, or scrubber. The toxicity of the contaminant is important since this property will affect the degree of cleaning required, as well as the manner of disposing of the collected pollutant. The corrosiveness of the contaminants influences the equipment selection, cost, and the materials of construction. A corrosive vapor is normally not collected in a wet system since moisture may increase the corrosion problem. All of these factors must be considered in selecting equipment.

Airstream Characteristics. The airstream characteristics influence the size and efficiency of a control system. Temperature and flow rate are the most critical properties of the exhaust. The temperature limits the type of device that can be used for control. Cooling the exhaust to temperatures which can be tolerated by the control device is an important factor in most control systems.

The flow rate determines the size of the control equipment since most equipment performs effectively over a definite flow range. Velocities through control equipment beyond recommended design values affect the treatment efficiency and may impose excessive wear. Hence recommended flow velocities must be obtained from the equipment manufacturers, since they are a function of the design. Generally, the efficiency of control devices depending on mechanical separation is increased with increased velocity up to a point. Control devices which depend on contact time or chemical separation are more effective at reduced flow velocities, although very low velocities may cause laminar flow and poor mixing.

Properties such as viscosity and density may be critical in the selection of the equipment and influence the design of the system.

Equipment Characteristics. Equipment characteristics such as pressure drop, utility requirements, and floor space must be considered in designing a control system. Table 6-3 details the performance characteristics of common control devices. Some limiting considerations in selecting equipment are worth mentioning:

(1) Settling chambers and cyclones are basically low efficiency separations, which can remove large size particles, at high dust loadings. For these reasons, they are effectively used as precleaners.

(2) Bag filters are effective control devices for removing submicron particles. However, they must be utilized at temperatures less than 500°F, depending on the filter material, and at definite filtering velocities, to prevent destruction of the bags.

(3) Electrostatic precipitators are also effective in removing submicron particles, but they are sensitive to varying flow conditions and particle loading. A change from design conditions could drastically reduce collection efficiency. Instrumentation to adjust for varying conditions is generally expensive.

(4) Wet collectors or gas scrubbers are capable of removing both particles and vapors from an airstream. However, when used for hot exhausts, a visible stream plume is often emitted which could create a nuisance. In addition, the contaminants are discharged in a liquid blowdown stream, requiring suitable waste treatment disposal facilities.

(5) Gas absorbers are effective in removing select gaseous pollutants from an airstream. However, certain contaminants, such as dust, could affect the life of the adsorbent. Disposal of the condensed pollutant or the adsorbent could present a problem.

(6) Incineration of dilute combustible pollutants or exhausts at ambient temperatures could be a problem. The burning of dilute streams will not be self-supporting. Exhausts at ambient temperatures must be preheated to combustion temperatures of 1200–1500°F. In both cases, auxiliary fuel costs can be high. Catalytic combustion and heat recovery units reduce the incineration auxiliary fuel costs but add capital expense. Catalysts are also subject to poisoning and short life.

These limitations must be considered in evaluating the control equipment.

Materials of Construction. An important factor in specifying the abatement equipment is selecting materials of construction which will minimize corrosion and erosion. The engineer will be guided by the design of existing manufacturing equipment, emitting exhausts to the treatment system. When this information is not adequate, he must rely on information in the literature or the experience of the equipment manufacturers. The properties of the pollutants and the exhaust characteristics, such as temperature, moisture content, and composition, are extremely important in materials selection. When the problem is difficult to solve, a competent corrosion engineer should be consulted.

Cleaning Efficiency. The final consideration in selecting the treatment device will be the required system efficiency. The latter is related to the loading and the permissible exhaust concentration by the relation:

$$\text{System performance} = \frac{\text{loading (ppm)} - \text{permissible exhaust (ppm)}}{\text{loading (ppm)}}$$

The permissible exhaust concentration is dictated by the control regulations and is related to the toxic or nuisance effect of the pollutants, both of which are beyond the control of the plant engineer.

Unit loading will greatly influence its performance, since high pollution loadings require a high removal efficiency to meet stringent emission standards. Generally, the efficiency of a treatment system can be increased by reducing the emission rate, concentrating the exhaust loading, or using a multistage cleaning system.

Selection Guide. The general guides discussed here are intended to assist the reader in the selection of control equipment. However, for any general rule, there will be specific exceptions. There can be no substitute for field testing to obtain performance data. Guides can be offered to permit the preliminary evaluation of treatment devices. These guides, illustrated in Fig. 6-4, can be summarized as follows:

(1) The selection of equipment for removing aerosols from air streams depends on the properties of the particles.

- (a) Low concentration mists can be collected with an electrostatic precipitator, a wet collector, a high density fibrous filter, or a packed bed mist eliminator.
- (b) Sprays can be removed with mechanical separators or coarse-packed beds.
- (c) Fumes can be collected with high-energy wet collectors, bag filters, or electrostatic precipitators.
- (d) Dusts can be removed from an airstream with wet collectors, inertia or impingement separators.
- (e) Coarse particles can be removed from exhausts with settling chambers or cyclones.
- (f) A primary cleaner, cyclone, or settling chamber should be considered for exhaust loadings greater than 10 grains/ft^3 to reduce the load on the principal abatement equipment.

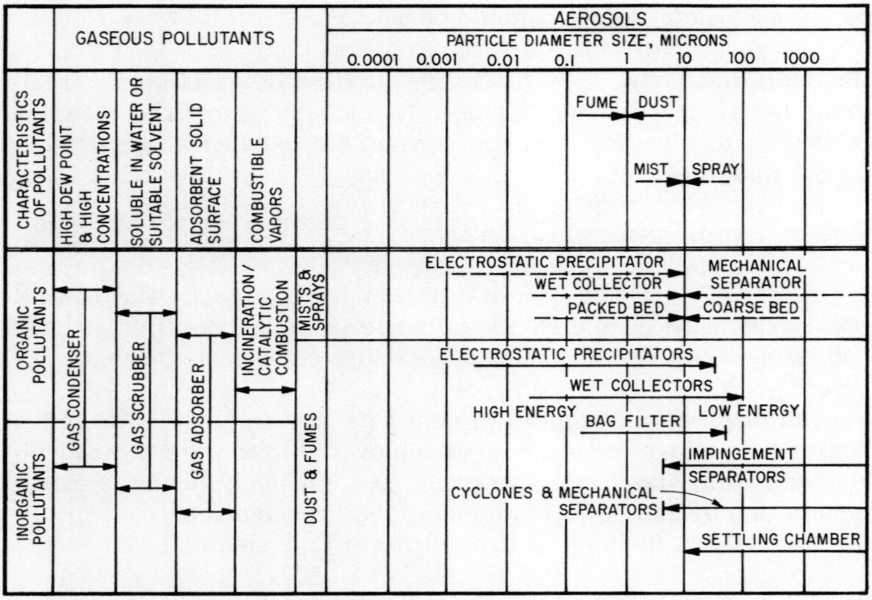

Fig. 6-4 Treatment equipment application.

(g) Abrasive particles can be collected with wet collectors or bag filters operated at reduced flow velocities. Electrostatic precipitators can be used, but the unit must be sized to reduce erosion wear.

(h) Adhesive or hygroscopic materials are best collected with wet collectors. Bag filters, electrostatic precipitators and cyclones are not effective control devices.

(i) Dusts with low resistivity characteristics are effectively collected with electrostatic precipitators.

(2) The selection of control equipment for gaseous pollutants is dependent on the chemical nature of the contaminant. Some guidelines for organic pollutants can be summarized as follows:

(a) Vapors present in high concentrations and with high dewpoint temperatures can be removed by condensation either by direct cooling or compression.

(b) Highly soluble organics can be removed from the exhaust by scrubbing with liquid or a suitable solvent.

(c) Pollutants which have molecular weights higher than the normal components of air can be removed by adsorption. Generally, adsorption is practical for removal of organic vapors which can easily be evaporated from the adsorbent at low pressure steam temperatures.

(d) Combustible vapors can be burned in direct or catalytic units.

(e) Vapors which are highly explosive or flammable, and soluble in water, are best collected in wet systems.

(3) Inorganic gaseous pollutants can be removed from airstreams by condensation, scrubbing, and with adsorbents such as silica gel, alumina, or activated carbon. The use of adsorbents for inorganics has limited application so that scrubbing and sometimes condensation remain the principal treatment methods.

Acid vapors and mists are readily removed from exhaust systems by scrubbing. The efficiency of the unit is increased by using an alkaline scrubbing liquid.

The final selection of control equipment should be based on previous experience or field testing, along with a detailed evaluation of the specific process conditions discussed.

Dispersion of the Exhaust

The small quantities of pollutants exhausted from the control system should be emitted at heights which assure that negligible concentrations reach the surrounding plant area.

Generally, the factors influencing dispersion are wind velocity, effective height, the quantity of pollutants emitted and atmospheric conditions. These variables are illustrated in Fig. 6-5. The maximum ground level concentration is decreased with increasing wind velocity and decreasing emission rates. In addition, the concentrations reach a maximum under inversion atmospheric conditions. Here the pollutants settle at distances close to the source emission. Dispersion of stack exhausts is discussed in detail in Chapter 9.

Disposal of the Collected Waste

All air pollution control systems generate either a dust or a liquid waste. The exception is an incinerator which, when properly operated, stabilizes the organic pollutants to the corresponding combustion products. Disposal of the system waste represents a major problem. Consideration must be given to storage, transportation, and stabilization of the collected material. The value of the material, either for direct utilization or conversion to a usable product, must be determined. If reclamation is not practical, alternate methods of disposal must be considered.

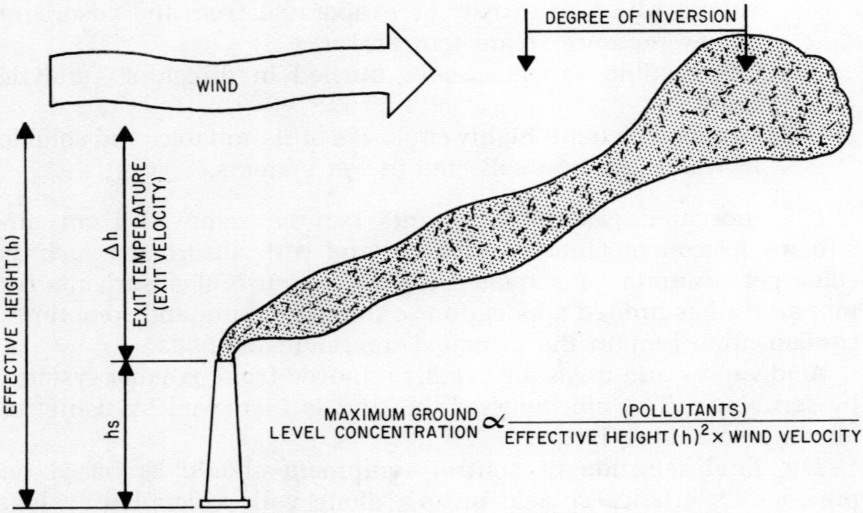

Fig. 6-5 Factors influencing dispersion of exhaust pollutants.

Solid wastes can be hauled away, buried in a sanitary landfill, or destroyed in an incinerator. The method selected is a matter of economics. Generally, hauling costs become prohibitive with large quantities or long transportation distances. In both these cases, on-site treatment usually is the most economical disposal method; the exception being local industrial disposal facilities shared by neighboring manufacturing plants.

Disposal of liquid wastes presents an equally difficult problem, except that the alternatives are more varied. Because liquid wastes are generally greatly diluted, hauling large quantities any great distance is not economical. If the contaminants are not toxic to domestic sewage organisms, industrial wastes can be discharged to municipal sewage systems with only a slight increase in sewage disposal charges. If these disposal methods are not satisfactory, then on-site treatment systems must be installed. Organic wastes can be treated by biological treatment or activated carbon adsorption systems. Consideration should be given to using concentrated organic wastes as a fuel. Organic biodegradable substances can be discharged to existing on-site domestic sewage plants, and the combined wastes treated.

Inorganic wastes will require either physical or chemical treatment facilities. Acid or caustic wastes must be neutralized. Wastes containing solids can be treated by screening, clarification, flotation, or flocculation. Some inorganic materials can be treated by chemical precipitation, oxidation, or reduction. Many inorganic and organic substances can be stabilized in aerated ponds.

When liquid wastes are to be discharged into a receiving stream, treatment is extremely important. The engineer must not exchange an air pollution problem for a liquid waste problem!

STEP 5: ECONOMIC EVALUATION

After completing an engineering evaluation of the pollution control system, an economic evaluation must be made of the proposed corrective or treatment alternatives. The basis for the economic study involves principles commonly employed in evaluating plant improvements—a balance of capital and operating costs. Estimating procedures are detailed in Tables 6-4 and 6-5.

TABLE 6-4 AIR POLLUTION CONTROL SYSTEM ECONOMICAL EVALUATION FORM

System: ______________________		
Fixed Costs (Table 6-5: Section B):		$ ________
Annual Operating Costs:		
Utilities:		
Fuel	$ ________	
Electricity	$ ________	
Water	$ ________	
Steam	$ ________	
Compressed air	$ ________	
Chemicals	$ ________	
Labor and supervision	$ ________	
Maintenance	$ ________	
Disposal collected pollutant	$ ________	
Total Operating Costs:		$ ________
Chemical Reclaim value:		$ ________
Total Annual Costs:		$ ________

First, an estimate is made of the capital costs for each of the proposed systems; this includes the land, site development, equipment, installation, accessories, and overhead costs as detailed in Table 6-5. The fixed annual costs can be estimated as indicated in Section B of Table 6-5—the major item being the depreciation value over the useful expected life of the equipment.

TABLE 6-5 CAPITAL COSTS ESTIMATING FORM

A. *Total Costs*	
Land @ $ ____________ /Acre	$ ________
Site development	________
Equipment & material costs	
Exhaust system	________
Control equipment	________
Auxiliary equipment	________
Instrumentation & controls	________
Electrical	________
Piping	________
Concrete	________
Structural	________
Painting	________
Insulation	________
Equipment installation	________
Control building	________
Laboratory equipment	________
Secondary waste treatment facilities	________
Subtotal	________
Engineering—construction fees & overhead	________
Start-up	________
Total	$ ________
B. *Fixed Costs*	
Annual depreciation of capital costs (Section A) over useful life; ________ years	$ ________
Capital charges	________
Interest	________
Taxes	________
Insurance	________
Total Annual Fixed Costs	$ ________

Next, the total annual costs are estimated for each of the proposed systems, as detailed in Table 6-4. This includes both the annual fixed and operating costs. Any chemical reclaim value is credited to the system and deducted from the annual costs.

The critical items in this type of evaluation are:

1. Expected useful life of the equipment.
2. Capital costs of the system.

3. The costs of the utilities and chemicals required to operate the system.

The optimum design for an air pollution control system can be obtained only if cost and treatment effectiveness are properly balanced.

STEP 6: SELECTING THE CONTROL SYSTEM

The final stage of the engineering study phase involves selecting the control system. For any pollution problem, there will generally be more than one control solution. The decisive factors which will influence the selection are:

1. Control regulations.
2. Feasibility of source correction or product recovery rather than treatment of the plant exhaust.
3. Optimization of the design to balance cost and the required effectiveness.

These three considerations have been discussed in detail in previous sections. Other factors which may restrict the options are the available utilities and location and space for the control system. The location of the proposed control equipment, relative to the manufacturing facilities and utilities, could have a decisive influence in selecting equipment.

Many times the area available for control equipment is indoors or on a building roof. When the system must be installed indoors, floor space and head room are critical. Plant space is held at a premium and there will be considerable restrictions as to what kind of equipment can be installed. When control equipment must be installed on the roof, structural support of the equipment could present a problem. The roof supports are generally designed for nominal loads and may not be capable of supporting the dead or vibrating loads of the treatment system.

Outdoor installations will require winterization of equipment. Heat loss and freeze prevention must be considered in designing the treatment system. All electrical equipment must be suitable for outdoor conditions. When the installation is a considerable distance from the plant, the exhaust transfer system will be expensive and must be carefully designed.

In addition, the available utilities must be taken into consideration. Water, power, gas, steam, compressed air, or a condensate system could be required, depending on the setup. An important utility often overlooked is the secondary waste treatment facility.

In many control systems, a liquid or solid waste is discharged and treatment or disposal of it is an extremely important factor.

The final selection should take into consideration design and maintenance problems as well as the economics of the system. After the control system is selected, the final phase of the project, engineering design and construction, is initiated.

STEP 7: ENGINEERING DESIGN AND CONSTRUCTION

The engineering design and construction phase of the project will be familiar to the engineer involved in plant improvement and construction. Figure 6-6 describes the factors involved. Initially, a flow diagram is established to describe the treatment method, from which an engineering flow sheet is developed. The engineering flow sheet details the operation, process control, and equipment characteristics. After the engineering flow sheet is developed, the plot plan, engineering specifications, and design drawings are completed and contractor specifications issued. The project is completed with the construction and start-up of treatment facilities.

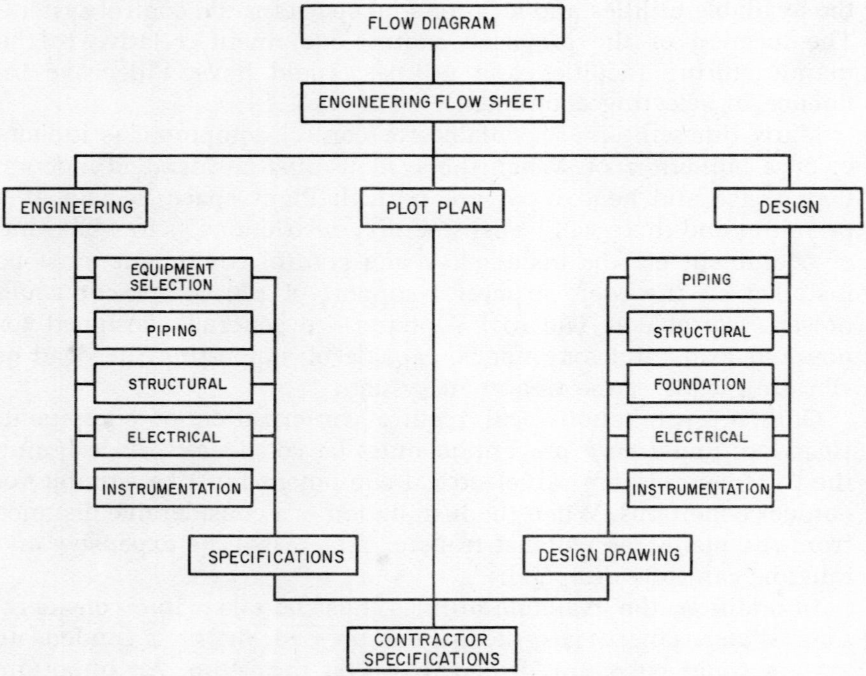

Fig. 6-6 Engineering and design phase air pollution project.

REFERENCES

1. American Society of Heating, Refrigerating & Air Conditioning Engineers, *ADHRAE Guide and Data Book,* 1969.
2. American Industrial Hygiene Association, *Air Pollution Manual,* Brawn-Brumfield, Inc., 1960, Vols. I & II.
3. D. Q. Kern, *Process Heat Transfer,* McGraw-Hill Book Co., New York, 1950.

7 Improvement of Existing Systems

INTRODUCTION

When an air pollution problem occurs in an existing industrial system, the first task may be to recognize it. Although some existing pollution problems may be readily identified, others are not so easily located and often require comprehensive efforts at identification. The smoke from a poor combustion process is an obvious example of a common air pollution source that can be identified by a simple technique—using the Ringlemann Chart. Invisible hydrocarbons and nitrogen oxides are examples of pollutants that are difficult to identify without sophisticated means.

A visible stack discharge is by itself a poor criterion. Steam plumes are readily visible and are often identified by the public as a source of air pollution. This is unfortunate, since money is often expended to eliminate harmless steam plumes while invisible pollutant emissions go unnoticed and unchecked. This then is the first problem—identification of the pollutant emissions from existing systems.

IDENTIFICATION PROCEDURES

The first step in identifying pollution sources is to survey the industrial installation and to tabulate every existing stack, chimney, hood, fan discharge, etc. that vents to the atmosphere.

These should then be examined in detail as to the process raw materials and products that have access to the air. Many systems, such as the ventilating system discharge from an office building or school, may be easily dismissed as nonpolluting. Others may be cleared from being offenders after a simple engineering evaluation. An example might be the stack from a gas-fired boiler supplying the heating steam to an office building or school (although, as will be shown later, this is not axiomatic). After these two groups have been evaluated, a third group may remain on which even the most competent engineer cannot render a judgment without field test data. Field test data should be collected so as to be directed toward two goals. The first is identification of the pollutants; the second, to take the data necessary to assist in selecting proper corrective equipment.

Since we cannot economically test for everything, good engineering judgment also can be effective. The competent engineer will identify the raw materials that enter the process and draw upon established technology to predict what might be present in any discharge to the atmosphere. In the case of the simple gas-fired heating boiler, it would be foolish to test for particulate matter. On the other hand, if the process involves potential toxic materials, such as lead or copper, and if they could be vented to the atmosphere, the discharge should be sampled for the presence of these pollutants.

Table 7-1 shows a typical format for identifying air pollutants

TABLE 7-1 TYPICAL INDUSTRIAL EMISSION SOURCE INVENTORY

Firm Name: XYZ Chemical Co.

Plant Address: 1234 ABC Road, Newark, Delaware, 19711

Representative to Contact: John Doe, Plant Manager

Emission Source Reference No.: 36B4

Plant Location: S.E. Corner of Roof of Bldg. 36

Process or Operation Producing Emission: Direct gas-fired oven drying sand

Type of Emission Point: Stack

Dimensions of Emission Point: 12 inches by 18 inches rectangle

Height of Emission Point Above Ground: 52 Feet

Discharge Temperature: 372°F.

Velocity at Discharge: 3600 ft/min (measured)

Gases in Discharge: CO_2 = 3%; H_2O = 19%; N_2 = 70%; O_2 = 12% (Calculated)

Particulate in Discharge: 0.12 grains per std cu ft (measured)

Other Discharges: None

Fuel Burned: 5400 SCFH of natural gas

from existing sources. Whenever data is estimated, it should be so identified; similarly, test data should be identified. Care should always be taken to assure that test data is representative of actual operating conditions. For example, a pulverized coal-fired boiler equipped with mechanical dust collectors will have its particulate emissions increase with load. A test at light load is therefore of little value if pollution abatement at the maximum firing rate is the ultimate goal.

A special difficulty arises when the emission conditions are not steady state. Many processes are transient in their output and must be carefully evaluated and sampled. It is important that the emission samples, if possible, cover the entire operating cycle for a given process when the conditions are not steady state. Some batch or cyclic processes may be difficult to sample. A relief valve discharge is an example. Whatever the process, good engineering judgments must be made when field testing the pollutants to determine their quantity and identity.

To complete the first step, the physical properties of the pollutant emission must be evaluated to the extent that these proper ties will affect the performance of any corrective equipment. Table 7-2 identifies some of these properties.

If particulate matter is the problem, it is essential that the

TABLE 7-2 PHYSICAL PROPERTIES OF EMISSIONS

A. *Gases*
   1. Volume Rate of Flow
   2. Temperature
   3. Pressure (if not atmospheric)
   4. Analysis (by volume—include all major and minor components)

B. *Solids in Cases*
   1. Volume Rate of Flow
   2. Temperature
   3. Pressure (if not atmospheric)
   4. Analysis of Gases (by volume—include all major and minor components)
   5. Solids (by weight per acutal cubic foot of gas flow)
   6. Solid Analysis (i.e., ash, carbon, HCl mist, etc. by weight)
   7. Solid Sizing (usually percent in each size range from submicron up to approx. 60-μ size)
   8. Solid Specific Gravity (water = 1.0)
   9. Solid Resistivity (ohm-centimeters, required for electrostatic precipitators)

particle sizes be known. If electrostatic precipitators are to be considered, the electrical resistivity of material is important. If scrubbers are a possible alternate, the solubility of gases must be known. If reactions are to be used, the molecular form of the pollutants must be known. For example, if fluorine is present, it is essential to know in what form, if one is to advise a chemical step for its removal.

The purposes of Tables 7-1 and 7-2 are not to make the task of identification so complex that it is considered too formidable for solution. Rather, it is to point out that a systematic approach with good judgment in selecting field test variables is essential for a rational approach to identifying the air pollutants from an existing process. It is only after this has been done can one proceed toward solutions that are anything but hit-or-miss.

SOURCES

To attempt to define or categorize sources of air pollutants is futile in the sense that it inevitably fails to cover those beyond the scope of the knowledge of any individual. The sources must be identified for new processes as well as old. We shall therefore categorize sources by the type of emissions. The emissions to be covered are:

1. Inert solid matter
2. Active solid matter
3. Gaseous hydrocarbons
4. Nitrogen oxides
5. Sulfur oxides
6. Others

These sources can be at high or low temperatures, diluted with air or inert gas, and existing in any combinations.

Inert Solid Matter

Inert solid matter is often referred to as either particulate or fly ash. It includes all manner of stable solid products that do not change their molecular form after a combustion or process condition, and do not react readily with the environment after they are emitted. Examples would be the ash in coal and oil fuels, minerals from a calcining kiln, or metal oxides from an incinerator. While usually chemically inert, their presence may be objectionable, both as a cleanliness factor and from the fact that they may serve to act as a nucleus for formation of droplets of more objectionable pollutants.

Particles of this type may be separated from the gas stream carrying them by any of the methods covered in Chapter 8. Where combustion processes are involved, the chief obstacle to the successful methods outlined may be the temperature of the combustion gases. While steel parts may be heated to about 900°F, without appreciable loss of strength, a temperature of 750°F is a more practical maximum. Special metal alloys may be used up to 1800°F but are costly to fabricate. Ceramic devices could conceivably be used to 3000°F but are not yet commercially available. Bag filters are limited to a maximum temperature of 550°F and then only through the use of special fibers. High-temperature nylon bags coated with Teflon offer promise of extended bag life in the upper temperature range.

The usual practice is to cool the temperature of combustion gases either by direct or indirect heat transfer. Direct heat transfer includes injection of water, steam, or air. Indirect heat transfer includes removal of heat by means of waste heat boilers and process heat exchangers. The cost of the latter must be carefully evaluated economically against the increased size of cleaning equipment required for direct cooling. Table 7-3 shows the relative volumetric changes required for various cooling methods for a combustion gas at an initial temperature of 1800°F.

Table 7-3 is not directly related to size since velocity may be permitted to increase with the injection of water or steam into the gas stream without increasing the pressure drop. Table 7-4 shows the approximate variation in the required size of the equipment if the pressure drop, expressed in inches of water, is kept constant.

It would appear from Tables 7-3 and 7-4 that water injection is preferred to air injection and that both are preferred to steam injection. This cannot be concluded until a final comparison of fan horsepower required to move the gases through the cleaning equipment is made. This ratio is shown in Table 7-3 if sizing is as shown in Table 7-4.

TABLE 7-3 VOLUME EFFECTS OF COOLING 1800°F COMBUSTION PRODUCTS

	RELATIVE VOLUMES		
Method	*750°F*	*500°F*	*250°F*
Indirect cooling	1.0	0.8	0.6
Water injection	2.4	2.0	1.6
Air injection	2.9	3.6	6.2
Steam injection	3.9	5.0	22.5

TABLE 7-4 SIZE EFFECTS OF COOLING 1800°F COMBUSTION GASES

Method	RELATIVE VOLUMES 750°F	500°F	250°F
Indirect cooling	1.0	0.9	0.8
Water injection	1.7	1.6	1.5
Air injection	2.9	4.0	8.0
Steam injection	2.9	4.2	22.8

EXAMPLE. Compare a 6-in. w.c. pressure drop multicyclone separator for an incinerator discharging 1,000 cfs of 1800°F gas for the various methods of cooling possible.

ASSUMPTIONS. Cyclone cost is $1.00 per cfm and heating surface cost $5,000 per million Btu/hr. Maintenance per year is 5 percent of investment. Electricity costs 1¢/kw-hr. Depreciation per year is 10 percent of investment. Management wants 10 percent net return on investment and 8,000 hr/yr operation.

Step 1 Assume indirect cooling to 750°F

$$\text{Specific volume of gases} = 0.0242\ (\text{temp}^\circ\text{R})^1 = 0.0242\ (1800 + 460) = 55\ \text{cu ft/lb}$$

$$\text{Pounds of gas} = \frac{1{,}000\ \text{cu ft}}{\text{sec}} \times \frac{3600\ \text{sec}}{\text{hr}} \times \frac{\text{lb}}{55\ \text{cu ft}} = 65{,}500\ \text{lb/hr}$$

$$\text{Average specific heat} = 0.26^1$$

$$\text{Heat rejected} = \frac{65{,}500\ \text{lb}}{\text{hr}} \times \frac{.26\ \text{Btu}}{\text{lb-}^\circ\text{F}} \times (1800^\circ\text{F} - 750^\circ\text{F}) = 18\ (10^6)\ \text{Btu/hr}$$

$$\text{Investment for indirect cooling} = \frac{18 \times (10^6)\ \text{Btu}}{\text{hr}} \times \frac{\$5{,}000}{10^6\ \text{Btu/hr}} = \$90{,}000$$

Step 2 Determine factors on multicyclone with indirect cooling.

$$\text{cfm} = 1{,}000\ \frac{\text{cu ft}}{\text{sec}} \times \left[\frac{750 + 460^\circ\text{F}}{1800 + 460^\circ\text{F}}\right] \times \frac{60\ \text{sec}}{\text{min}} = 32{,}000$$

Installed cost of cyclone is therefore $32,000 at $1/cfm.

$$\text{Specific volume of gases} = 0.0242 \times (750°\text{F} + 460°\text{F}) = 29.5 \text{ cu ft/lb}$$

Assume the head on the multicyclone is 6 in. of water.

$$\text{The head in fluid flowing is} = \frac{29.5 \text{ cu ft}}{\text{lb}} \times \frac{62.4 \text{ lb } H_2O}{\text{cu ft}} \times \frac{\text{ft}}{12 \text{ in.}} \times 6 \text{ in. water} = 920 \text{ ft}$$

$$\text{Theoretical fan power} = 920 \text{ ft} \times \frac{32{,}000 \text{ cu ft}}{\text{min}} \times \frac{1 \text{ lb}}{29.5 \text{ cu ft}} = 1 \text{ million ft-lb per minute} = 30 \text{ hp}$$

$$\text{Actual fan horsepower} = 30 \div .6 \text{ efficiency (assumed)} = 50 \text{ brake horsepower} = 50 \text{ kw electric input}$$

$$\text{Power cost} = \frac{8000 \text{ hours}}{1 \text{ yr}} \times \frac{50 \text{ kw-hr}}{1 \text{ hr}} \times \frac{\$.01}{\text{kw-hr}} = \$4000/\text{yr}$$

Step 3 Using Tables 7-3 and 7-4, economically compare the various methods of cooling neglecting the cost of labor, water, and steam.

	METHOD OF COOLING			
	Indirect	*Water*	*Air*	*Steam*
Multicyclone investment	\$32,000	\$54,500	\$92,500	\$92,500
Heat exchanger investment	90,000	—	—	—
Annual Operating Cost				
Depreciation (10%)	12,200	5,450	9,250	9,250
Maintenance (5%)	6,100	2,725	4,625	4,625
Power	4,000	9,600	11,600	15,600
Total operating cost	22,300	17,775	25,475	29,475
Return on investment factor*	24,400	10,900	18,500	18,500
Total cost	\$46,700	\$28,675	\$43,975	\$47,975

*For 10 percent net return on investment, a 20 percent charge per year on investment required will cover taxes and other expenses to permit evaluation of the various alternates.

Step 4
Conclusion: Water injection is the preferred method of cooling. Air injection is the second preference.
Caution: The above example is intended to be illustrative only and actual basic data should be used for individual problems.

Active Solid Matter

Active solid matter includes that particulate matter which can be further oxidized or that will react chemically after being admitted to the environment. If the particles are larger than 1–2 μ in size, they can be collected in the same manner as the inactive solid matter covered in the preceding paragraph and then given supplemental treatment. This is often done for some processes. For example, sawdust can be collected from an airstream by the methods given in Chapter 8, and then incinerated.

Unfortunately, many types of active particulate matter appear in submicron size and require more sophisticated forms of removal such as with scrubbers or electrostatic precipitators. One form of submicron active solid matter is smoke, which is formed of submicron particles of carbon. It can be categorically stated that smoke can and should be eliminated by improving the combustion process. Improved burners, furnaces, and/or combustion controls, may be required, but these are all available and it is only poor engineering and management that permits smoke to be formed. Frequently, new equipment can be justified economically as replacing obsolete equipment that produces smoke. This is especially true for industrial process burners, incinerators, and boilers burning liquid and gaseous fuels. Development has lagged on small solid-fuel firing units but, even here, careful engineering, operation, and maintenance can succeed. Smoke is unnecessary and obsolete and should not be tolerated. Careful attention to the combustion process is all that is needed to eliminate it.

The discussion on the elimination of smoke that follows is general in nature and specific cases require individual study. Basically, smoke-free combustion requires adequate time for the combustion process to be completed, good turbulence in mixing the fuel and air, and sufficient temperature for the reaction rate to proceed to completion and produce the complete products of oxidation that are desired.

It is unfortunate that the time required for a combustion process cannot be defined in rational terms although some attempts have been made to define parameters. These parameters include heat-release rates in the furnace, residence times, and flame travel distances. Since the other two requirements of turbulence and temperature have significant effect upon the reaction rate, exact values of these parameters often have little meaning. If it is assumed for the moment that turbulence and temperature are fixed and a furnace is smoking, the following corrective actions can be taken:

(1) Increase the excess air flow to the furnace. While low excess air flames are often desirable, the availability of more excess air provides more oxygen and often results in a shorter flame.

(2) Provide a catalyst to shift or increase the reaction rate to eliminate carbon formation. One useful catalyst at low temperatures is water (or preferably steam). For this reason it is often used in flares and incinerators to control smoking. Another combustion catalyst is manganese which can be readily added to liquid fuels or spayed in a liquid form on solid ones. Such a catalyst can permit low excess air firing in furnaces of limited volume.

(3) Enlarge the furnace. If the flames are leaving the furnace too soon they may be chilled by the heat exchange surface or the atmosphere and produce smoke as a spoon does when put into a candle flame. Increasing the furnace size to accommodate the longer flame will allow the reaction to proceed to completion and thus avoid smoke. Reducing the maximum firing rate can also help.

(4) Change fuels. Some fuels simply burn faster than others. Solid types are the most difficult and, the higher the volatility of the solid fuel, the more likely it is to smoke. Anthracite coal is better than bituminous, for example, as is coke. Liquid fuels are next in line and of these residual oil is harder to burn satisfactorily than the lighter distilled oil. No. 2 fuel oil is easier to burn smoke free than No. 6 fuel oil, and blends of the two, such as No. 4 or No. 5, offer degrees of improvement. The sulfur content has a significant effect on the soot produced and the effect becomes more pronounced as the heat exchanger temperature decreases. For example, high-sulfur No. 6 fuel oil in a small low-pressure heating boiler should be avoided if one is concerned about air pollution. While No. 2 fuel oil is better and as manufactured contains no ash or sulfur, there has been an alarming tendency for this type, as delivered to the customer, to contain sulphur after all. Natural gas is the easiest of all fuels to burn without producing carbon, although it too can produce soot if the flame is quenched too soon or a serious (and dangerous) air deficiency exists.

Turbulence is the second area in which improvements may be made to reduce smoke. In fact, it amounts to one of the most feasible and successful methods. The extent to which the fuel and air are stirred together affects both the rate and completeness of the combustion reaction. The term "well stirred reactor" defines the goal and the attention which must be paid to both ingredients being reacted—fuel and air.

Fuel preparation is important where the purpose is the intimate mixing of fuel and air in the correct proportions. Solid fuels may be hogged, chipped, ground, or pulverized in order to both reduce fuel size and permit more uniform (continuous) feeding. The finer the solid particle, the easier it may be fed and the more intimately fuel and air can be brought together.

Liquid fuels must be vaporized to burn (as must solid fuels) and the various preheating devices and atomizing devices have this as their function. Pool vaporization is often poor although sometimes useful. Wicks are seldom of commercial value. Usually liquids are atomized to fine droplets so that the heat of the flame will complete the vaporization. Heating the liquid may reduce its viscosity and surface tension so that finer droplets are obtained. Some atomizers are better than others. Rotary cup burners and pressure atomizers are mechanical types that have a very limited operating range over which they produce good droplet size. Pressure atomizers generally atomize better than rotary cup atomizers. Wider ranges of flow with good atomization can be achieved with two fluid atomizers where a compressed gas, such as steam or air, is used to impart kinetic energy to the liquid. Sonic atomizers utilize sound waves to break up the liquid into fine droplets over the widest possible range and have the advantage of improving the atomization as the flow decreases.

Even the best of atomizers cannot function well if it is clogged with dirt or carbon particles. Dirt particles should be removed with strainers. Additives can be used to disperse sludge in storage tanks. Liquids remaining in idle burners should be blown out and many burners provide this feature to prevent the heat from cracking the fuel and forming carbon. On multiburner furnaces, idle burner nozzles should be removed entirely from the furnace when other burners are firing. Burners should be cleaned as often as necessary. This can be as frequently as three times per day to as seldom as once a month. Burner tips should also be replaced on a scheduled basis. Although small in quantity, the ash in most heavy oils in a year will erode a nozzle tip to the point where it cannot atomize well. Whenever a liquid burner has functioned well under certain conditions and later starts to smoke under the same conditions, it is wise to examine its atomizing parts for cleanliness or wear.

Even more important than the atomizing or vaporizing aspect of turbulence is the manner in which the air is admitted to mix with the fuel. The more violently and thoroughly the combustion process is stirred, the more complete and smokeless the reaction can be.

In fact, a poor atomizer can be compensated for with a high degree of turbulence. As a rule, this turbulence is achieved in three ways. The oldest way is through the use of refractory baffles, arches, or orifices (chokes) to enforce mixing of the combustion gases while they are still in the flammable state. Another older way is through the use of over-fire air jets which use steam or a high-pressure blower to inject air at high velocity into the combustion zone. The newest and perhaps best way is to utilize means to inject all of the combustion air in a turbulent fashion. Since this requires energy which is not available from natural draft, a forced draft air supply is an essential ingredient. There is unfortunately a tendency to object to high forced draft fan pressures. No. 6 oil can be burned stoichiometrically with forced draft fan pressures of 10 to 20 in. of water, while it can be categorically stated that natural draft can never achieve enough turbulence to give such smokeless performance. In all cases, the turbulence should be such as to promote recirculation and good mixing. Tangential firing and forced vortex firing are two examples of good design which promote recirculation and which have good turbulence.

Where turbulence producers are used they should be maintained. This is readily understood with many of the devices used, such as refractory baffles and over-fire air nozzles, but is frequently misunderstood in connection with the diffuser plates often used on gun type burners. The purpose of these conical shaped discs is to serve as a bluff body and to turbulently retain flame at the tip of the gun to give a stable flame that will not pulsate and smoke. If they perform their job well they will be in contact with flame and may eventually warp and disintegrate. While heat-resistant materials will extend their life, they should be regarded as expendable as burner tips. If the emphasis is on long diffuser life or on minimum forced draft fan pressure, it cannot be on air pollution.

In summary, turbulence can be improved by:

1. Good fuel injection
   a. Good atomizers
   b. Good maintenance
2. Refractory shapes to improve mixing
3. Over-fire air jets
4. Forced draft burners (of good design!)

The final element of good or smokeless combustion is temperature. Too high a temperature is seldom a problem in creating smoke unless the high temperature distills volatiles at too rapid a rate in batch-fed furnaces. Some synthetic materials exhibit this characteristic.

Usually smoke will be produced if the combustion temperature does not exceed 1400°F for most solid and liquid fuels. Methods of increasing temperature that are genarally useful are:

(1) Reduce the excess air if it is too high. Gaseous fuels require at most 10 percent excess air. Liquid fuel requirements vary from 1 to 75 percent depending on the quality of the burner. Solid fuels vary from 15 to 100 percent depending on the fuel and the method of firing. Several hundred percent excess air may be enough to chill any flame and cause smoking. Caution must be taken in reducing the excess air at the burner if its source is infiltration through a leaky furnace. Sealing the leaks or reducing the negative draft is the proper method when this is the case.

(2) Eliminate excessive water from the fuel. Water will mix well with some fuels and depress the flame temperature appreciably where it is present in quantity. A moisture separator in an atomizing steam line may remove significant amounts of water from the flame. Leaking heating coils in tanks should be repaired. Fuel deliveries should be tested to detect abnormal water content and stored fuel protected from moisture condensation and weather.

(3) Reduce the heat absorbtion rates in the furnace. Excessively cold surfaces may be covered with refractory directly or baffles may be installed to prevent line-of-sight radiation to cold surfaces.

(4) Eliminate extreme low-firing rates in furnaces with large cooling surfaces. Most burner manufacturers like to claim high turn-down rates for their burners. If such burners are applied indiscriminately to highly cooled furnaces, the flame temperature at low-firing rates may be too low to prevent smoking.

Once good conditions of time, turbulence and temperature are established in a furnace and smoking is eliminated, the first step has been achieved. To maintain good conditions requires two more steps. The second is adequate combustion controls. These may vary from simple off-on controls to either sophisticated metering or proportioning (positioning) types. The oxygen analyzer can be used as a continuous checking and adjusting device and as lower excess air operation is approached, this or some equivalent instrument must then be used. For fixed fuel firing rate burners with excess air, such as used with incinerators, a temperature device may provide adequate control. Optical pyrometers are especially rapid and suitable for such service.

Once the equipment and controls are in place, the third step of adequate maintenance must be taken. The best design and equipment cannot function if it is not properly maintained.

Usually, incinerators are among the worst smoke offenders. This is often the result of insufficient management direction in supply-

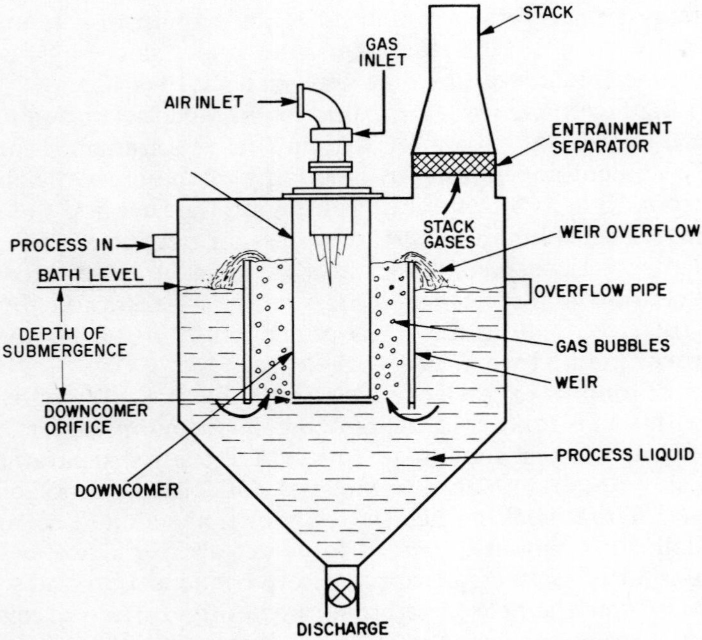

Fig. 7-1 Submerged exhaust incinerator for oxide emission control and recovery. *(Courtesy Thermac Research & Engineering Corp.)*

ing adequate funds to either automate the unit fully or to supervise and train suitable operators. Another form of active particulate emission is soot, usually greatest at times of soot blowing in steam boilers. This is a problem associated with sulfur in the fuel and its control will be discussed more fully in a later section.

Other active solid emissions include metallic oxides that may be toxic to both marine or animal life. When formed in combustion flames, they are usually submicron in size and require careful attention to equipment design to entrap them. Fig. 7-1 shows an incinerator designed to control copper oxide emissions from a liquid fuel. Submerged combustion exhaust is first used to cool the high-temperature combustion gases which then pass to a high-pressure drop scrubber, shown in Fig. 7-2, for final cleanup to an acceptable level of emission.

Of course, these particles could have been removed by electrostatic precipitators, but in this case, the submerged combustion exhaust offers an economical way of water cooling to a temperature below 190°F. Submerged exhaust cooling also has the advantage of requiring only a level control for regulation rather than sprays and temperature control devices. In a vertical configuration, it also

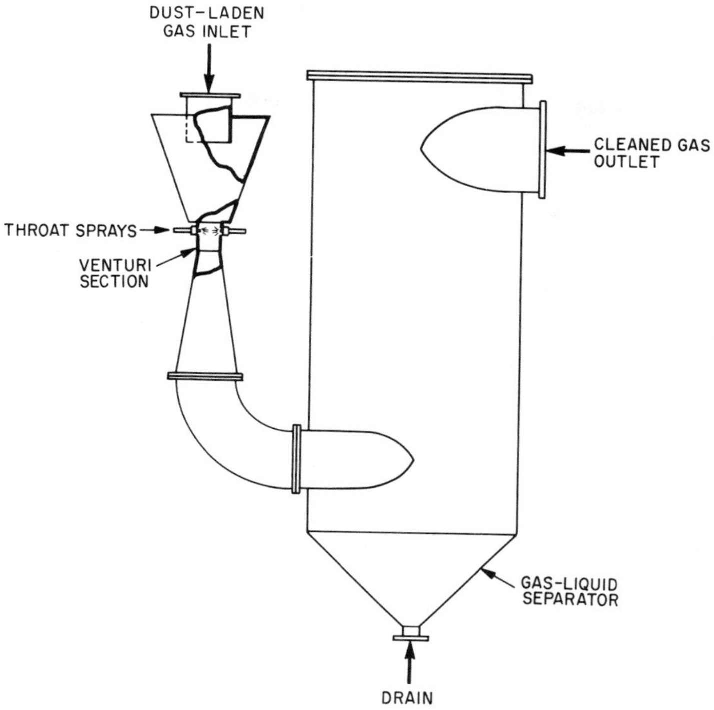

Fig. 7-2 High pressure drop scrubber. *(Courtesy Chemical Construction Corp.)*

allows for easy transition from refractory to metal which may be a difficult problem in horizontal spray systems. While Tables 7-3 and 7-4 clearly show the economic advantages of water injection, they do not show the careful maintenance and operation required. For this reason, air injection may sometimes be used in preference. Submerged exhaust overcomes the more serious problems of water injection at a penalty cost of fan horsepower. In some cases, the 75–80 percent efficiency obtainable in removing submicron particles from gases with submerged exhaust may eliminate the need for further cleanup equipment.

Gaseous Hydrocarbons

Gaseous hydrocarbons may occur from numerous sources. Some come from filling of storage tanks or venting processes with high vapor-pressure liquids; others from the widespread use of solvents. Because they are usually invisible, they represent a very difficult pollutant to identify and control. Where the quantities are large, they are often economically recoverable, either by condensation or activated carbon adsorbtion. If not recovered they may be flared, incinerated, or abated. Each gaseous hydrocarbon vent must be carefully examined to determine if it enters the flammable range. It is desirable to exclude air, if possible, and carry the fumes and gases to a recovery or disposal point without oxygen. Hazards at start-up and shutdown can be avoided by using inert gas purges. If this is not possible, air may be used to convey the gases, if extreme care is used to stay below the lower explosive limit. A maximum value of 25 percent of the lower explosive limit is often specified. For small quantities this is not onerous, but for large ones, it results in sizable equipment and high operating costs.

To reduce operating costs, incinerators and fume-abatement equipment often are designed to do only a partial job, and many pollution regulations specify only a partial abatement requirement. A value of 90 percent is common. As a result, many units are ineffective in odor and air pollution control. It is possible to rationally design units with high performance. Designing to achieve a lower performance is difficult and subject to error. It is this author's opinion that we should design to a high standard of performance and then experiment with the equipment to determine how the operating paramaters can be varied to reduce costs. This is true of both recovery and combustion disposal systems.

Wherever possible, steady state venting is desirable. Intermittent discharging of large quantities results in poor usage of large equipment. Evacuation rates, tank filling rates, and venting should all be controlled to limit instantaneous venting requirements.

Recycling or recovery systems should always be evaluated for economic feasibility. Vapor compressors and condensers are common on tank vents of liquified petroleum and natural gas storage tanks where the heat loss of the storage vessel results in a small but constant boil-off of some gases. Adsorption systems using activated carbon or molecular sieves are also common in film coating operations where the coating operation depends on a solvent. In essence, such systems operate as closed types and prevent the gaseous hydrocarbons from reaching the atmosphere. Other closed systems include conservation vents on tanks, floating tank tops, gas holders, and collapsible fabric tanks.

The methods described in Chapter 9 are also applicable and will be discussed here only from a systems point of view. If recycling or recovery is not feasible, destruction by oxidation is often the answer because it produces nonpolluting gases.

Flares are simple open-end pipes supplied with ignition devices and flameholders. The vented gases burn in the open with atmospheric air. The resulting combustion can be either good or bad, depending on the hydrocarbons involved and their heating value. Flares must be located at a height sufficient to prevent thermal damage to both people and equipment. The height of a flare stack needed to prevent thermal effects on the surroundings can be calculated by the following formula which allows for wind.

$$\text{Height of flare (ft)} = \frac{(Q)^{1/2}}{A} - 20D$$

where: Q = Btu/hr burned in the flare
A = 310 for 20-min human exposure or continuous equipment exposure, or 180 for continuous human exposure
D = diameter of flare in feet

To determine if a vented gas is suitable for flaring, an adiabatic combustion temperature at stoichiometric conditions of about 1800°F is usually required. In other words, if the gas is too low in heating value to provide this calculated temperature, it must be enriched with other fuels in order to be flared. Flaring of combustible mixtures requires special precautions to prevent flashbacks. Flares can be improved in their operation by making them more complicated than simple open-ended pipes with pilots. Possible improvements include the addition of steam jets to reduce smoking, windscreens, refractory cup flameholders, and induced air to the center of the flares. As these are added, however, the weight and cost of the flare increases until it may be less economical than an incinerator. Figure 7-3 shows a typical flare.

Flares have their greatest advantage as low-cost disposal units for large intermittent quantities of combustible gases. Since large flares may disturb residential areas at night, they should not be used indiscriminantly on a continuous basis. If sufficiently high, they may also serve as safe venting points if combustion is lost. If the flare is impractical, burners are required to provide better control of combustion conditions.

In-line burners (Fig. 7-4) provide fuel and/or air to burn the hydrocarbons in an enclosed space. Unlike flares, they can handle any hydrocarbon level if properly designed, but usually are restricted to applications where the maximum concentration of hydrocarbons is less than 25 percent of the lower explosive limit. Their efficiency of combustion will be a direct function of (1) the mixing of the auxiliary fuels and the waste stream, and (2) the final furnace mixed temperature after combustion. If the final temperature is less than 1400°F, or mixing is very poor, (i.e., some

Fig. 7-3 Flare. *(Courtesy John Zink Co.)*

waste gases short-circuiting the flame) complete destruction may not be achieved. Some air pollution codes specify a percentage reduction in hydrocarbons with 90 percent being a common value. Since this performance can only be determined by test, it is wise to obtain equipment with sufficient temperature capabilities to insure operation at 1400°F or higher. Once purchased, a high–temperature unit can be lowered in operating temperature with consequent fuel savings, but the converse is not always true.

If the duct cross section of an in-line burner is too great for a single burner to effectively cover, multiple jet burners or profile plates are used to give the best possible coverage. Gas is the fuel most often used. The duct temperature is usually controlled by adjusting the fuel input.

More sophisticated burner systems, often called fume abaters,

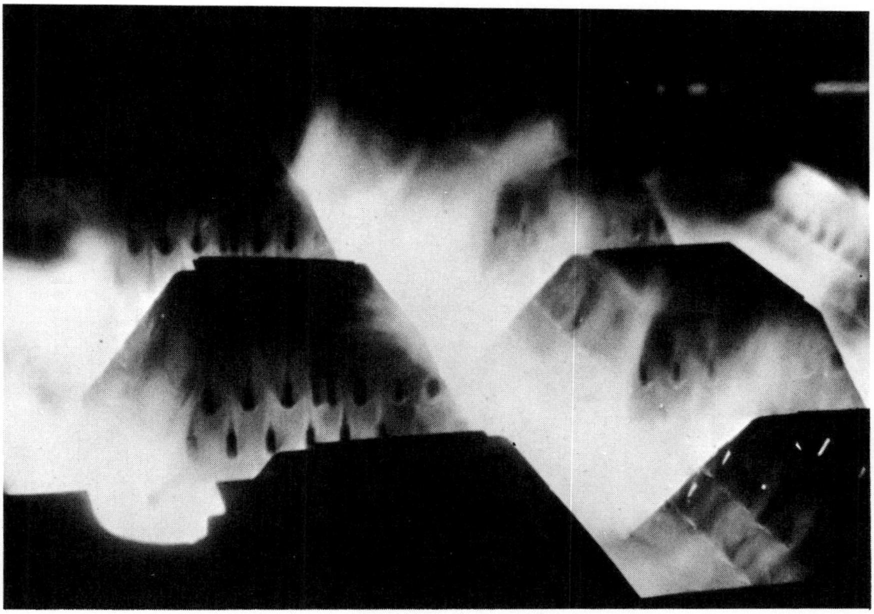

Fig. 7-4 In-line burner. *(Courtesy Maxon Co.)*

are available. They consist of complete disposal at a single point. Whereas the in-line burner is usually installed in a duct with adequate flow characteristics, the fume abater may include its own draft-producing equipment. If fumes are delivered to the unit under positive pressure, it is a forced draft unit; if combustion gases are pulled from the unit, it is called an induced draft unit. In either case, the draft may be produced by either mechanical fans or venturi jets. Figure 7-5 shows a venturi jet arrangement, whereas Fig. 7-6 shows an induced draft unit. The induced draft unit pulls in excess air after the combustion chamber and cools the gases to an acceptable level (800°F or below, usually). This is a disadvantage because the fan size is larger, but one that is often offset by the fact that the entire fume abater is below atmospheric pressure. This reduces odor problems which can occur in a forced draft unit if airtight construction has not been achieved.

If fume abaters operate below the lower explosive limits, they can use the air conveying the fumes as combustion air for the auxiliary fuel required. If the fumes are in the explosive range, the operator has a problem and the usual solution is to dilute the fumes to 25 percent of the lower explosive limit. If the fumes are

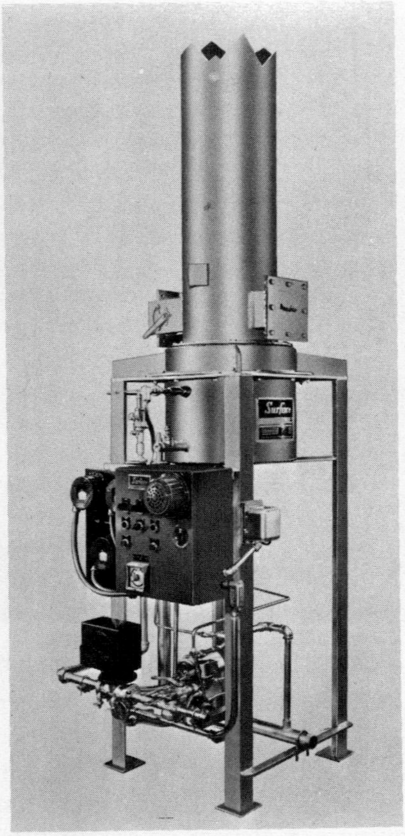

Fig. 7-5 Venturi jet fume abater. *(Courtesy Surface Combustion Co.)*

above the upper explosive limit, the best arrangement is to mix them with combustion air at the burner.

Every fume-abatement system must be carefully evaluated to determine that it does not deliver an explosive mixture to the burner. Average flows cannot be considered adequate information under transient conditions. To prevent upsets, the process must be carefully evaluated and consideration given to the rate that fumes are pumped or forced out of processes. Even so, operator errors can always occur and a flame arrester must be mounted close to the fume abater. No flame arrester is capable of deterring all possible hazardous situations, but the quenching and water bath types are the best. The quenching type has either narrow slots or supported screens which have openings less than the length of the

Fig. 7-6 Induced draft fume abater. *(Courtesy Hirt Industries)*

radical reaction chains formed during combustion. The combustion reaction is stopped by physical interference with the formation of the necessary radicals. Quenching types are best when constructed to withstand heat or detonations but have serious drawbacks because the fine openings plug easily with dirt and have high pressure drop. Another type is the water bath. This type has perforated pipes which cause bubbles to rise through water, thus cooling and interrupting any flame. Disadvantages are pressure drop, prevention from freezing, and a maintenance of liquid level.

Another type of flame arrester consists of bundles of small diameter tubes through which the fumes must pass. This type depends on velocities in the pipes greater than the flame propagation speed with quenching by cooling if the flow reverses. Tubing of ⅜-in. OD and 2 ft long will quench natural gas. Advantages are low pressure drop and relative freedom from plugging. Disadvantages are the inability to stop flame propagation when heated. Larger tubes are sometimes used when consistent flow is assured

EXAMPLE. A 10,000-gal methanol tank is to be vented to a flame incinerator. The methanol is delivered in a 5,000-gal trailer which should be unloaded in 1 hr. Estimate the size of fume abater required to control the vent by direct incineration.

ANSWER. If the maximum tank temperature is assumed to be 104°F, the partial pressure in the space above the level can be 261 mm Hg. This is 30.4 percent methanol by volume in air. The volume of vented vapors is:

$$\frac{5000 \text{ gal}}{\text{hr}} \times \frac{\text{cu ft}}{7.48 \text{ gal}} \times \frac{\text{hr}}{60 \text{ min}} = 11.12 \text{ cfm, say } 12 \text{ cfm}$$

Since vapors should be transported at 25 percent of the L.E.L. and the L.E.L. of methanol is 5.8 percent, the safe transport level is 1.45 percent. To achieve this 240 cfm of air must be added at the tank vent point and the input pumping rate of methanol restricted to (5000/60) = 83 gpm. A fan will have to be sized at (240 + 12) = 252 cfm.

If natural gas is burned and all fumes are to be incinerated (100 percent destruction), the lowest practical operating limit is 1800°F, which corresponds to 150 percent excess air. Since 167 cfm is equivalent to 1 million Btu/hr of fuel, then the burner size will be:

$$\frac{252 \text{ cfm}}{2.5} \times \frac{10^6 \text{ Btu/hr}}{167 \text{ cfm}} = 600{,}000 \text{ Btu/hr of natural gas}$$

Caution: The use of 150 percent excess air results in running at 40 percent of the stoichiometric air/fuel ratio. Since the L.E.L. of natural gas is 46% of the stoichiometric ratio at room temperature, the burner must be fed additional gas until the combustion chamber is warmed up.

Catalytic units are a good method of oxidizing various hydrocarbons if the process is carefully designed and controlled. Unfortunately, catalysis has often been regarded as an art rather than a science and results have frequently been poor. Some elementary facts, if understood, can preclude difficulties. First, catalysis is only a means of speeding up the reaction rate to thermodynamic equilibrium conditions. If at equilibrium carbon dioxide and water are the lowest free energy state of carbon, hydrogen, and oxygen at 500°F, a catalyst is a device to get the hydrocarbon converted to carbon dioxide and water quickly. Catalysis cannot make anything oxidize that does not thermodynamically want to and, if the combustion process proceeds rapidly without catalysis, then catalysis can add nothing.

Secondly, catalysis is dependent upon certain minimum activation temperatures and materials. Oxidation reactions are usually of a heterogeneous nature between solid metals, such as platinum, palladium, silver, or oxides of copper, chromium, vanadium, and silver and gases. The metal platinum is the most active oxidation catalyst and the most widely used. The temperature at which platinum may initiate a reaction can be as low as 350°F for ethylene, or as high as 800°F for methane.

Thirdly, the catalyst metal must be active. This means that the surface area must be large and not fused or filled with other substances. An active platinum catalyst surface will absorb moisture from the tip of the human tongue and stick to it. A catalyst that has been overheated so that the surface area is at a minimum, or that is poisoned by absorbed materials, will not do this. Sulfur and lead are two catalyst poisons often encountered. Spent catalyst can sometimes be regenerated by reheating carefully, but all heterogeneous catalysts usually deteriorate with use.

Thus, a catalyst bed may offer low-temperature oxidation of hydrocarbons with little or no auxiliary fuel cost, but only if both minimum and maximum temperatures are carefully controlled, poisons are avoided, and replacement costs of spent catalyst are not excessive. A typical catalyst unit is shown in Fig. 7-7.

The temperature control problem is a difficult one if the hydrocarbon-to-air ratio varies widely, giving possible high temperatures. The loss of surface at high temperature lowers the catalysis rate at low temperatures and can defeat an otherwise successful installation. Catalyst installations must therefore be carefully controlled and protected from damaging situations. As more understanding

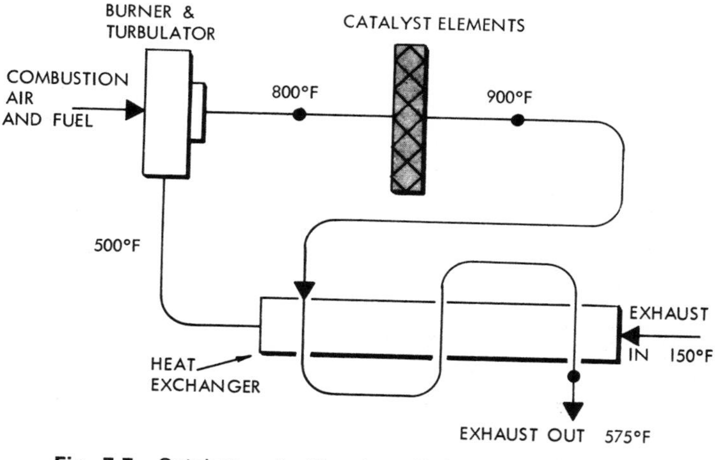

Fig. 7-7 Catalyst unit. *(Courtesy Universal Oil Products)*

grows about catalysis, the use of catalysts should become more widespread in applications to fume incineration for fumes below the lower explosive limit.

EXAMPLE. A mercaptan trace in 3,000 scfm of air must be abated. Compare a catalytic versus a direct flame incinerator.

Since any mercaptan has a very low threshold level for odor, direct afterburning will be similar as with the previous problem. The burner size at 1800°F will be:

$$\frac{3000 \text{ cfm}}{2.5} \times \frac{10^6 \text{ Btu/hr}}{167 \text{ cfm}} = 7{,}200{,}000 \text{ Btu/hr}$$

Assume it has been proven that the mercaptan will oxidize to an ordorless substance at 500°F. Part of the airstream can be diverted around the burner so that the mixed streams have this final bulk temperature. Let this fraction be X and assume that the initial temperature is 100°F. Then since the 3,000 cfm unit burned about 120 cfm of gas,

$$(1 - X)\left[\left(\frac{3000 \text{ ft}^3 \text{ air}}{\text{min}} \times \frac{.075 \text{ lb}}{\text{ft}^3} + \frac{120 \text{ ft}^3 \text{ gas}}{\text{min}} \times \frac{.045 \text{ lb}}{\text{ft}^3}\right) \times 0.26 \frac{\text{Btu}}{\text{lb°F}} \times (1800 - 500)\right] =$$

$$(X)\left[\left(\frac{3000 \text{ ft}^3 \text{ air}}{\text{min}} \times \frac{.075 \text{ lb}}{\text{ft}^3}\right) \times .24 \frac{\text{Btu}}{\text{lb°F}} (500 - 100)\right] \text{ and}$$

$$X = 77\%$$

Thus, the burner size must be only $2(10^6)$ Btu/hr, showing the reduction in fuel costs possible with catalysis from direct flame incineration where all the fumes are passed through the main flame. Of course, if a 90-percent efficient unit is used the fuel savings will not be as drastic.

Nitrogen Oxides

Probably one of the last major air pollutants to be regulated, nitrogen oxides, play an important role in smog production. Fortunately, the formation of nitrogen oxides in combustion reactions is described by very basic relationships that allow it to be controlled. Figure 7-8 shows the basic thermodynamic relationships for the formation of nitric oxide in combustion flames. Combustion reactions are the principal industrial source of nitrogen oxides. Nitric oxide, NO, is colorless. Below 1100°F, this further oxides

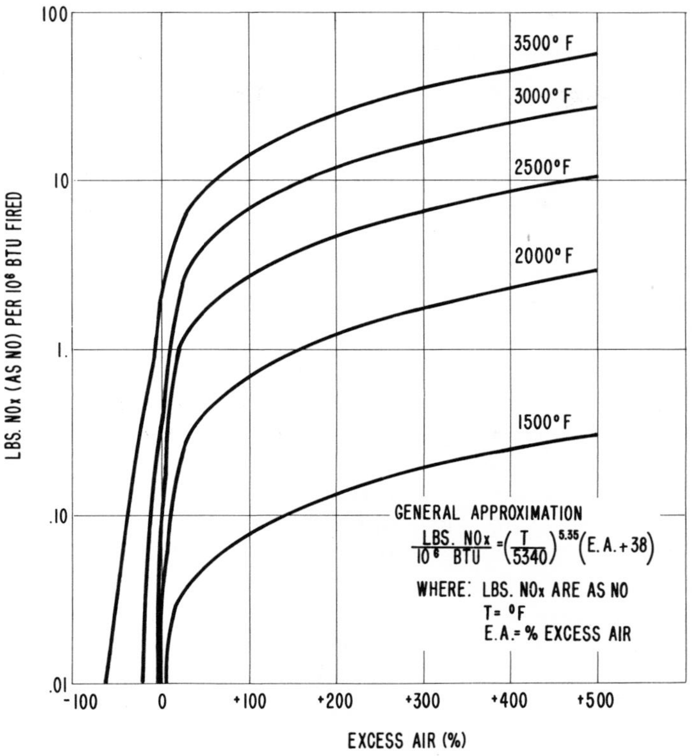

Fig. 7-8 NO formation in flames.

to nitrogen dioxide, NO_2, which has a reddish-brown color and is a noxious gas.

Essentially, only NO is formed in flames and the amount formed is a function of excess air and temperature. Low excess air firing and/or low-temperature firing can control the formation of nitrogen oxides with conventional fuels. Where waste fuels contain nitrogen compounds, additional consideration must be given to providing sufficient residence time to permit decomposition of the nitro-bodies. Rapid quenching tends to prevent the NO from decomposing and should be avoided when nitro-bodies are burned. Similarly, methane and NO_2 or methane and NH_3 can be fed to a catalyst bed at low temperature with good results. The major requirement is that the NO_2 pass through the flame or catalyst bed. If it does not, the temperature may change the NO_2 to NO, making it colorless and dispersing it for reconversion to NO_2. Needless to say, no abatement has been accomplished in such a case.

Sulfur Oxides

Sulfur oxides are considered to be a major air pollutant from combustion processes. Usually, the literature does little to differentiate between SO_2 and SO_3. It is the latter that readily combines with water to form sulfuric acid with all its problems for combustion equipment. SO_3 can be prevented from forming by either removing the sulfur from the fuel or by the use of low excess air firing. As shown in Fig. 7-9, the effect is similar to that for nitric oxide formation except that the temperature effect is reversed. Low excess air, by preventing the SO_2 from oxidizing to SO_3, allows the former to be dispersed over a much wider area before it slowly converts to SO_3. In addition, the elimination of acid dew points in the combustion equipment and stacks also eliminates the formation of acid "smuts" and can eliminate the need for soot blowing.

Numerous schemes for treatment of gases to remove sulfur dioxide are in embryonic development stages. The addition of magnesium dioxide can convert the sulfur dioxide to $MgSO_4$, which can be removed as inert particulate matter. Other means of wet-scrubbing and dry catalytic treatment are under study but at this time there is no commercially developed process available. Low-sulfur fuels are becoming available and offer the industrial plant the current most appropriate method of eliminating SO_2.

Other

Other forms of air pollutants may issue from industrial processes. This is especially true with incinerators. The halogens

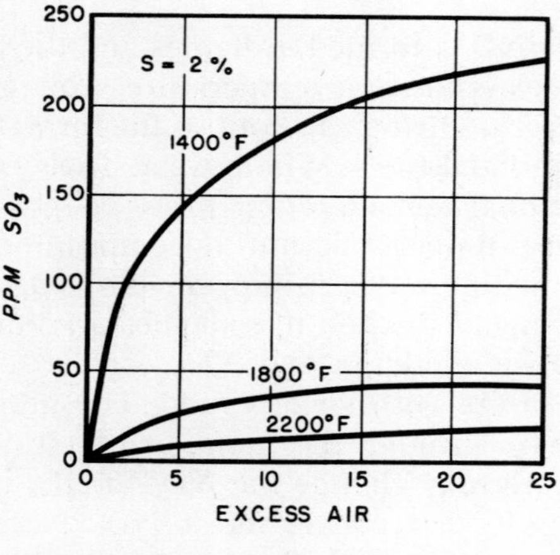

Fig. 7-9 SO_3 formation in flames.

are found increasingly in man-made compounds and form very troublesome acids. Fortunately, the most common (HCl) is readily absorbed in water and can be either reused or neutralized before being released to the environment. Many scrubbers used for HCl absorption, however, emit significant HCl mist to the atmosphere because of entrainment of water droplets in the exit airstream. Care must be exercised to utilize good demisters in such cases. Other halogens, such as fluorides, can be scrubbed with calcium hydroxide, in which case they form a solid precipitate, calcium fluoride.

Although phosgenes and cyanides are often reported as a product of combustion processes, careful burner design and good combustion control will eliminate their formation. They are not normal products of combustion and result only where combustion conditions are very poor. Both gases can be handled similarly to gaseous hydrocarbons should they be vented from industrial processes.

The range of other air pollutants which can issue from industrial processes is almost limitless. Those that are toxic or pungent are readily recognized. The paper industry can readily identify its problems with carbon disulphide as can the gas transmission company in its handling of mercaptans. The challenge is greatest to the industry whose discharges are odorless and invisible. This is not to say that all effluents must be prohibited. The current controversy over whether carbon dioxide is causing the earth's temperature to increase or decrease is a case in point. It is not possible for a narrow unimaginative viewpoint to determine a proper answer to this question—whatever the answer may be. It is possible for each industrial operation to become as knowledgeable as possible about its processes as is feasible, to identify them, determine their pollutive quantity, and assign priorities to correction. Where controversy exists, emphasis should be placed upon sufficient research to obtain better bases for determining if the effluent is good or bad. More properly, do the good (or bad) characteristics outweigh the bad (or good) features? A narrow viewpoint or a short-range one is unlikely to give the correct answer.

REFERENCE

1. R. D. Ross, *Industrial Waste Disposal,* Van Nostrand Reinhold Co., New York, 1968.

8 Selection of Equipment for Particulate Removal

INTRODUCTION

Particulate matter (dust) comes in great varieties of size, grain loading, shape, chemical composition, specific gravity, bulk density, friability, stickiness, resistivity, wettability, cohesiveness, etc. The selection of equipment for particulate removal is further complicated by the quantity and quality of entraining gas; by process variables such as continuous or intermittent operation; changes in gas quantity, dust loading, or moisture content; and by other major process variables such as seasonal variations, soot blowing, and process upsets.

Because of these unlimited variables, the first step in the selection of equipment for particulate removal is the identification of the gas-cleaning problem in terms that can be directly related to the performance characteristics of the four basic types of equipment commercially available for particulate removal:

Mechanical collectors	Wet scrubbers
Fabric filters	Electric precipitators

The questionnaire shown (Fig. 8-1) summarizes the data required by the equipment manufacturer to make recommendations that are both technically feasible and economic. Complete operating data is the foundation upon which equipment selection and performance guarantees are based. The questionnaire is designed to enable evaluation of the four approaches, one against the other.

1. Is This a ☐ New or ☐ Existing Plant
2. Type of Plant and Process
3. Origin of Gas and Dust

Number of Process Units
4. Reasons for Cleaning Gas
   - ☐ Air Pollution Control
   - ☐ Recovery of Valuable Product
   - ☐ Use for Subsequent Process
   - ☐ Protection of ☐ Equipment or ☐ Process
5. Describe Existing or Proposed Precleaners, Gas Coolers or Conditioners (on Reverse)
6. Geographical Location and Altitude of Plant

☐ Indoor or ☐ Outdoor Location
Attach sketch of available space if limiting factor.

7. Power available Phase Volts Hz
8. Gas

	Min.	Max.	Design
Lbs./Hr.			
°F			
CFM			
Inches W. G.			
% Moisture by Volume			

Chemical Composition (on Reverse)

9. Dust

	Design
Inlet Grns./ACF	
Size — % By Weight — 0- 5 Microns	
5-10 Microns	
10-20 Microns	
+20 Microns	
True Specific Gravity	
Bulk Density — Lbs./CF	

10. Efficiency Required — Max. Allowable Dust Loss
Lbs./Hr. Grns./ACF
Applicable Air Pollution Code

11. Design Specifications —
Temperature Pressure
Wind Load Snow Load
Earthquake Factor Min. Hrs. Hopper Cap.
12. Unusual Characteristics of Dust
☐ Poisonous ☐ Combustible
☐ Explosive ☐ Builds up on Metal Surfaces
☐ Difficult to Remove from Hopper
13. Miscellaneous Process Data
Operating Cycle
Interval Between Overhauls
Can Multiple Sources of Pollution be Served by One Collector
14. Is there a previous history within your Company of Air Pollution Control Equipment on a similar service. If so your comments (on reverse) will be appreciated.
15. Miscellaneous
Operating Cycle
Interval Between Major Overhauls
Can Multiple Sources of Pollution be served by one Collector
16. Do you prefer dust to be collected ☐ Dry ☐ Wet. If Wet complete Questions 17-20.
17. Scrubbing Water
Source — ☐ Lake ☐ Municipal ☐ River ☐ Sea ☐ Well
Availability — ☐ Unlimited
☐ Seasonal Limits, if Any (on Reverse)
Maximum GPM
Cost per Million Gallons
Pressure
Chemical Problems— Hardness
— Corrosive
Cleanliness — Parts per Million of Sediment
— Nature of Sediment
— Sediment Seasonal
18. Is a Process Scrubbing Liquid Available ?
Qualities

Quantities
Pressures
Limitations
19. Disposal of Scrubber Water
☐ To Sewer ☐ To Existing Settling Pond ☐ To Proposed Settling Pond ☐ Reuse After Existing Clarifier ☐ Reuse After New Clarifier ☐ Collected Dust is Valuable Enough to Recover.
20. Suitable Materials of Construction
☐ Mild Steel ☐ Stainless Steel ☐ Rubber Lined
☐ PVC ☐ FRP ☐ Other

COMPANY
ADDRESS **ZIP**
LOCATION OF PROJECT
INDIVIDUAL **TITLE** **DATE**

Fig. 8-1 Questionnaire. *(Courtesy Belco Pollution Control Corp.)*

It should always be used when the type of collector cannot be predetermined and then submitted to at least one manufacturer of each basic type.

One of the most difficult problems facing the plant engineer is the assembly of accurate information. For a new plant it is the responsibility of the process licensor or the engineering consultant to specify the operating conditions of all process equipment including that required for particulate control.

Naturally the responsible engineer will include sufficient margins to cover all practical operating contingencies to make certain of meeting process criteria as well as present and future air pollution codes.

For an existing process, the best and safest solution is to measure the gas to be cleaned by pitot tube traverse over the range of operating conditions, as well as to sample the entrained particulate matter by isokinetic techniques. Gas composition, temperature, and moisture content should be measured simultaneously. The dust samples taken should be analyzed to establish pertinent characteristics as well as quantity.

An increasing number of industrial firms, faced with meeting a wide variety of air pollution codes, have established in-house capability for making gaseous and particulate determinations on existing processes. The established methods for these determinations are summarized in "Determining Dust Concentration in a Gas Stream"[19] and "Determining the Properties of Fine Particulate Matter"[20]—both publications of the American Society of Mechanical Engineers. The approximate cost of required apparatus is as follows:

Gas quantity	$ 200
Proximate gas analysis	400
Dust sample	1,000
Dust sizing	3,000
Dust analysis	5,000

Plant engineers lacking this in-house support should engage a consultant to make these measurements. There are a growing number of firms now available to serve you promptly and economically. While such tests may cost several thousands of dollars, it is poor economy to install an air pollution control system costing tens or hundreds of thousands of dollars based on inadequate data.

Before such consultant specialists were available, it was customary for the manufacturers of air pollution equipment to offer such services free, or on a basis wherein the cost would be deducted from the price of equipment purchased subsequently. Such an approach to the establishment of design data, while an apparent economy, is probably a false one. Ultimately the purchaser of air pollution control equipment must pay for the services received, including tests. It is always a temptation for an equipment manufacturer to interpret the data in a manner that favors his own solution. It is the recommendation of the author that the plant engineer employ an independent consultant whose skills are adequate to establish the pertinent data and whose objectives are not influenced by subjective aberration. The establishment of such data will enable obtaining competitive bids from three or more competent suppliers. It is most probable that the total cost of this approach, including the cost of establishing the data, will be less than alternate approaches.

Having defined the gas-cleaning problem in terms of reliable and pertinent data, we can now calculate the collection efficiency required to meet the applicable air pollution code or process requirement. This second step will enable completion of the definition of the type of particulate control equipment we need including its overall collection efficiency.

In Chapter 3 the legislative trend was developed in depth, and it was emphasized that any collection efficiency selection should take into account the requirements of the stricter codes of the future. One solution is to install equipment that exceeds current requirements by a margin sufficient to meet the probable requirements of the forseeable future, or to allow additional space to upgrade the installed equipment to comply with future stricter codes.

Having defined the problem in meaningful engineering criteria including required collection efficiency, it is now possible for us to consider the four basic types of particulate control systems and to evaluate their operating characteristics within the frame of reference of each particular application.

GENERAL PRINCIPLES

The following fundamentals apply to all practical collectors of particulate matter.

(1) All collectors clean the gas by continuously removing the dust from a moving stream of gas. Typical order of process:

A. Source of dirty gas
B. Gas-cleaning device
C. Fan to move gas through total process
D. Stack to discharge cleaned gas to atmosphere

TABLE 8-1 FUNDAMENTAL OPERATING CHARACTERISTICS OF COMMERCIAL

Basic Type	*Specific Type*	*Basic Operating Force*	*Basic Measure and Unit of Capacity*	*Typical Capacity in cfm of Gas per Basic Measure of Capacity*
Mechanical Collectors	settling chamber	gravity	casing volume (cu ft)	20
	baffle	intertia	inlet area (sq ft)	1200–3600
	high-efficiency cyclones	centrifugal	inlet area (sq ft)	3000–3600
Fabric Filters	manual cleaning	filter cake filtration	fabric area (sq ft)	1–4
	automatic shaker cleaning	filter cake filtration	fabric area (sq ft)	1–4
	automatic reverse jet cleaning	felt fabric filtration	fabric area (sq ft)	3–8
Wet Scrubbers	impingement baffle	liquid capture	baffle cross-sectional area (sq ft)	500–600
	packed tower	liquid capture	bed cross-sectional area (sq ft)	500–700
	venturi	liquid capture	throat area (sq ft)	7,000 to 30,000
Electric Precipitators	single field	electric	collectrode area (sq ft)	5
	multiple field	electric	collectrode area (sq ft)	3

[a]Fob shop costs, installation and maintenance costs are based on 1970 mild steel construction costs, and do not include auxiliary equipment such as supporting structure, connecting flues, thermal insulation, foundations, stacks, etc. They are subject to substantial variation due to the specific requirements of each installation including geographic location. Use only for general comparison purposes.

You will note that the fan is usually located downstream of the gas-cleaning device to protect it from the erosive and possibly unbalancing effects of the entrained particulate.

(2) The size of the collector and its cost are directly proportional

PARTICULATE COLLECTORS

Typical Required Plan Area for Collector to Clean 100,000 cfm[e]	Typical Energy Required and % Collection Efficiency on Fine Dust[c]		COST IN $1000 FOR A COLLECTOR TO CLEAN 100,000 CFM: Cost to Purchase fob Shop and Install		COST IN $1000 FOR A COLLECTOR TO CLEAN 100,000 CFM: Annual Cost for Maintenance and Power		Effect On Collection of Operation at Over or Under Capacity[d]	
sq. ft.	inches w.g.	%	fob shop[a]	install	maintenance	power[b]	over	under
2600	0.2	25	10	3	1	0.5	down	up
300	0.5	40	13	4	2	1.3	up	down
125	4	80	20	5	2	10	up	down
1000	4	99	35	20	7	10	down	no change
1000	5	99	60	30	10	13	down	no change
600	8	99	80	35	15	20	no change	no change
300	4	80	30	15	7	10	up	down
250	7	90	40	20	9	17	down	up
100	30	99	50	50	11	75	up	down
270	0.5	95	75	55	4	1.3	down	up
500	0.5	99	100	70	5	1.3	down	up

[b] Energy costs are based on electric power at $0.015/kwh.
[c] Fine dust is considered to be 70% by weight minus 10 μ.
[d] Some collectors can be modified to maintain collection efficiency at over or under capacity.
[e] Plan areas do not include connecting flues.

to the gas quantity to be cleaned, the gas quantity being expressed volumetrically—usually in acfm (actual cubic feet per minute of dirty gas presented to the collector).

(3) The collection efficiency is calculated by the following for-

mula and is expressed as a percentage:

$$\% \text{ collection efficiency} = \frac{\text{weight of dust in} - \text{weight of dust out}}{\text{weight of dust in}} \times 100$$

(4) For all types of particulate collectors fine dust is harder to collect than coarse. Most dusts are a mixture of fine and coarse and the proportion of fines will have an important bearing on collector efficiency and therefore be a major factor in the selection of the type of collector needed to meet the collection efficiency requirement. Heavier concentrations of dust are usually easier to collect than low-grain loadings.

(5) The major types and subtypes of particulate collectors are summarized in Table 8-1.

MECHANICAL COLLECTORS

The unit weight of typical industrial dusts are one to two thousand times the unit weights of the gas in which they are entrained. Mechanical collectors take advantage of this difference in specific gravity to separate the heavier dust from the lighter gas. Basic types of mechanical collectors are the gravity settling chamber, baffle, and high-efficiency cyclones.

Gravity Settling Chamber

A typical gravity settling chamber is shown in Fig. 8-2. This collector slows the gas from conveying velocities to settling velocities for a sufficient period of time to enable the heavier dust to settle under the influence of gravity into the dust hoppers whence it is periodically removed. Settling velocities range from 60–600 ft/min.

Basic Gravity Settling Chamber Characteristics

Physical size	Very Large
Installed cost	Low
Energy cost	Very Low
Maintenance cost	Low
Collection Efficiency	Very Low
Reliability	Excellent
Efficiency at low loads	Increases
Efficiency at overloads	Decreases

The Unique disadvanage of this type is its very low collection efficiencies on fine and moderately fine dusts. This characteristic limits the use of the gravity settling chamber to that of a precleaner for a more efficient type when there are either expectionally high grain loadings or exceptionally coarse cinders which might damage or plug the downstream collector.

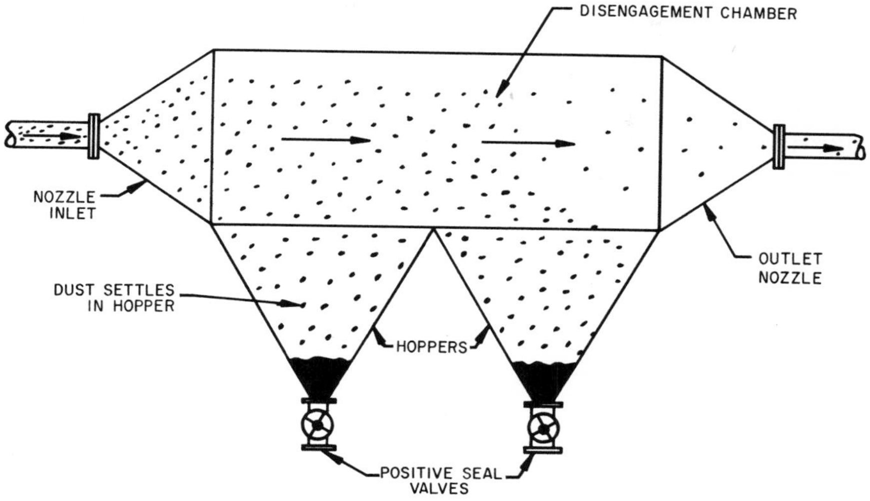

Fig. 8-2 Gravity settling chamber.

Recirculating Baffle Collector

Figure 8-3 shows a collector where the gas to be cleaned is introduced at high velocity under a horizontal baffle made up of rods spaced about a half-inch apart. To pass between the rods to reach the cleaned gas outlet chamber, the dirty gas must make a

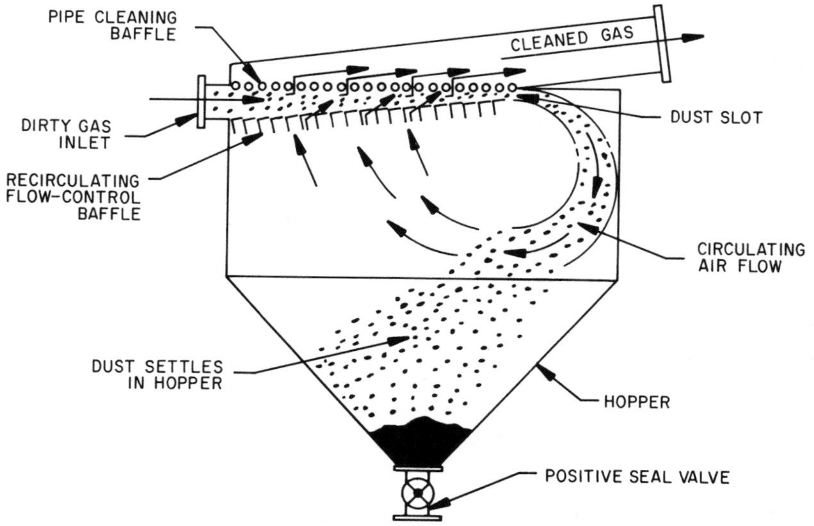

Fig. 8-3 Recirculating baffle collector.

sudden, high-velocity turn. The low specific gravity gas can easily make the sharp turn. The heavier dust, driven by inertia, is unable to make this sharp turn and is restrained below the baffle until it is captured by the dust slot. The dust is conveyed at a lowering velocity as the dust slot expands over the collector hopper into which it settles as in a settling chamber. The circulating flow is controlled at a nominal velocity by the expanding dust slot and the circulating flow control baffle.

Basic Characteristics of Recirculating Baffle Collector

Physical size	Small
Installed cost	Low
Energy cost	Low
Maintenance cost	Low
Collection efficiency	Low
Reliability	Excellent
Efficiency at low loads	Lowers slowly
Efficiency at overloads	Increases slightly

Unique disadvantages: while its collection efficiency is much better than that of the settling chamber, it is not sufficient to meet current emission control requirements except as a precleaner to a more efficient collector.

High-efficiency cyclones

Centrifugal collectors—commonly called cyclones—separate particulate matter from a carrier gas by transforming the velocity of an inlet stream to a descending outer vortex and an ascending inner vortex, both confined within the upper cylinder and lower cone of the cyclone. The rapidly rotating descending vortex holds the heavier dust against the walls of the cyclone by centrifugal force and throws it into the hopper from where it is periodically removed. The ascending inner vortex of cleaned gas, which is fed its entire length by the inner surface of the descending vortex, leaves the cyclone through the vortex finder at the top of the cylinder.

The flow pattern within a cyclone can be simple or complex depending upon many variables, including type of inlet, the dimensional proportions, and the inclusion or exclusion of a fines eductor or interposed vaning. Figure 8-4 shows the flow pattern of one common high efficiency type.

Basic Characteristics. Because of its simplicity, reliability, and high efficiency, the cyclone collector has been widely used through-

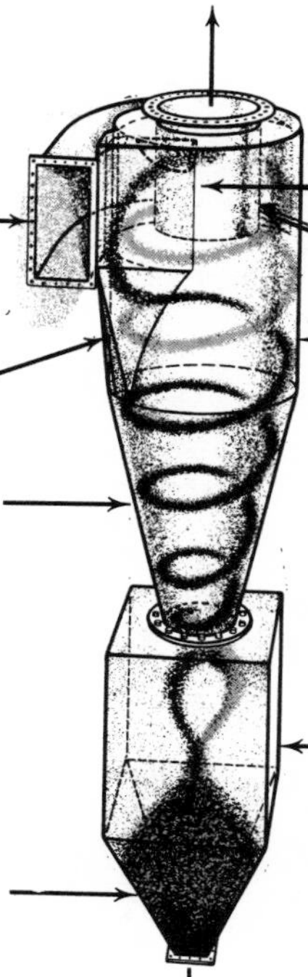

Fig. 8-4 Flow pattern in mechanical cyclone with vertical fines eductor. *(Courtesy Belco Pollution Control Corp.)*

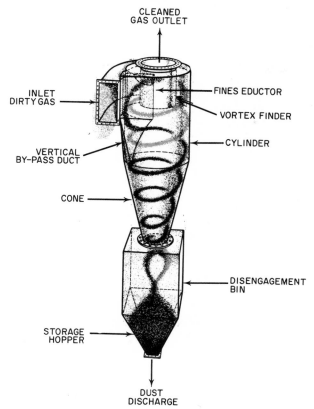

out the world for the past hundred years. To achieve its demonstrated versatility, there are hundreds of types of collectors using the cyclone principle to trap the dust. However, we shall consider only the three most commonly used high-efficiency types:

Small-diameter vane axial cyclone
Large-diameter cyclone having involute inlet
Large-diameter cyclone having involute inlet and fines eductor

Each of these widely used types has inherent advantages and disadvantages and is available in a wide range of dimensional variations which alter their operating characteristics. It is important to select carefully in order to maximize the advantages and minimize the disadvantages as they specifically relate to each intended application. The following characteristics are common to the three types which differ mainly in the manner in which the dirty gas is injected into the cylinder.

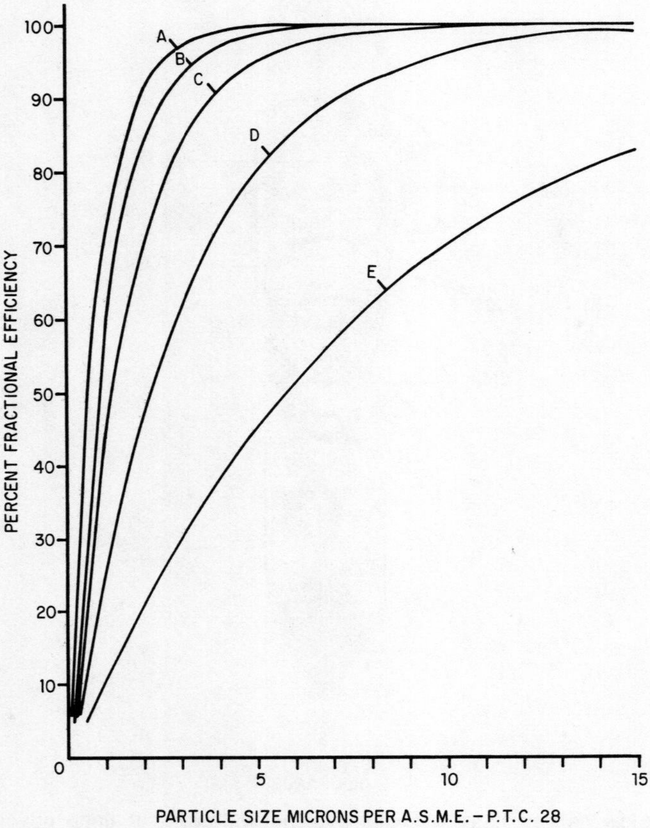

Fig. 8-5 Efficiency of cyclone collectors.

Fractional efficiency. All cyclone collectors have a distinct fractional efficiency curve such as shown in Fig. 8-5 wherein collection efficiency drops off rapidly below a certain micron size of dust. Other factors which affect collection efficiency are summarized below:

Increase In	*Effect On Collection Efficiency*
Dust specific gravity	Increase
Gas viscosity (temperature)	Decrease
Dust surface area	Decrease
Dust loading	Increase
Inlet velocity	Increase

Draft loss. Most cyclones operate at an inlet velocity of 3,000–4,000 ft/min. The energy requirement (draft loss) is proportional to the square of the inlet velocity. When handling atmospheric air the draft loss will be in the range of 4–6 in. water gauge when measured as a difference of static pressure between the inlet and the outlet. Lower draft losses are required for hotter gases of lower density.

The fundamental compromise. The design of each cyclone collector represents a compromise among three factors: collection efficiency, draft loss, and size. In general terms higher collection efficiencies require higher draft losses (expenditure of energy) or larger size (installed cost) or both. In order to meet better the great variety of industrial applications, some manufacturers of high-efficiency cyclones have devoted scores of years of research to develop families of cyclones representing a wide range of the fundamental compromise.

Collection efficiency criteria. While the design theory of high-efficiency cyclones is beyond the scope of this text, a change in cyclone dimension or ratio will affect performance characteristics as indicated below:

Increase of Cyclone Dimension or Ratio	*Collection Efficiency*	*Draft Loss*	*Capacity*
Inlet area to cylinder area	Down	Up	Up
Cylinder length	Up	Same	Same
Cone length	Up	Same	Same
Cylinder diameter	Down	Same	Up
Vortex finder diameter	Down	Down	Same
Vortex finder penetration	Up	Up	Same

It should be noted that the above factors cannot be varied without limit. The total performance characteristic of each cyclone is the result of the complex interrelation of all of its critical dimensions and ratios.

The paradox. There is a paradox of the relative collection efficiency of small-and large-diameter cyclones which results from over-simplification and which warrants clarification. Simply stated the paradox says. (A) Small-diameter cyclones are more efficient than large-diameter ones; (B) some large-diameter cyclones are more efficient than small-diameter ones.

The clarification states: (A) For any specific type of cyclone, that is, one whose ratio of dimensions are fixed, the collection efficiency will increase as the size (or cylinder diameter) is decreased; (B) as summarized above it is possible to design cyclones of increased collection efficiency by changing certain dimensions and ratios but at the cost of either increased draft loss or increased size, or both. Thus it is possible to have a 6-ft cyclone of a collection efficiency superior to that of a 6-in. model but only at the cost of other important operational or installed cost criteria.

The resolution of the paradox is simple: compare fractional efficiency guarantee curves. See Figure 8-5 and Table 8-2 for a comparison of fractional and overall collection efficiencies for different types of high-efficiency cyclones of diameters of from 10 to 59 in.

TABLE 8-2

RANGE OF OVERALL COLLECTION EFFICIENCY OF DIFFERENT TYPES OF HIGH EFFICIENCY CYCLONES COLLECTING PULVERIZED COAL FLY ASH[a]

Type Cyclone	*Diameter (in.)*	*Capacity (cfm)*	*Draft Loss (in. w.g.)*	*Fractional Efficiency Curve*	*% Overall Collection Efficiency*
Involute	50	12,000	2.7	E	67
Vane axial	10	700	2.4	D	84
Fines eductor	19	1,100	2.8	C	91
Fines eductor	50	6,500	4.5	B	94
Fines eductor	59	3,500	5.9	A	95.5

[a]Fly ash at true specific gravity 2.5; size 45% minus 10 μ; flue gas at 300°F.

Small-Diameter Vane Axial Cyclone Collector. Figure 8-6 shows a typical collector, consisting of twenty-five cyclones arranged in a common shell, having a common inlet, common cleaned gas outlet, and common dust storage hopper. In this collector the individual cyclones are arranged in five rows of five cyclones each. The collector shown is equipped with twenty-five cyclones of 10-in. diameter, and a total capacity of 20,000 cfm. This collector requires a plan area of about 25 sq ft, which emphasizes the compact size of the vane axial cyclone collector.

In this typical arrangement the dirty gas enters through the common dirty gas inlet into the dirty gas plenum from which it turns downward to feed the individual cyclones. Cyclonic spin is imparted to the descending gas by the interposed Vanes which cause the gas and dust to spiral downward in the descending outer vortex. The centrifugal forces generated by the high rate of spin

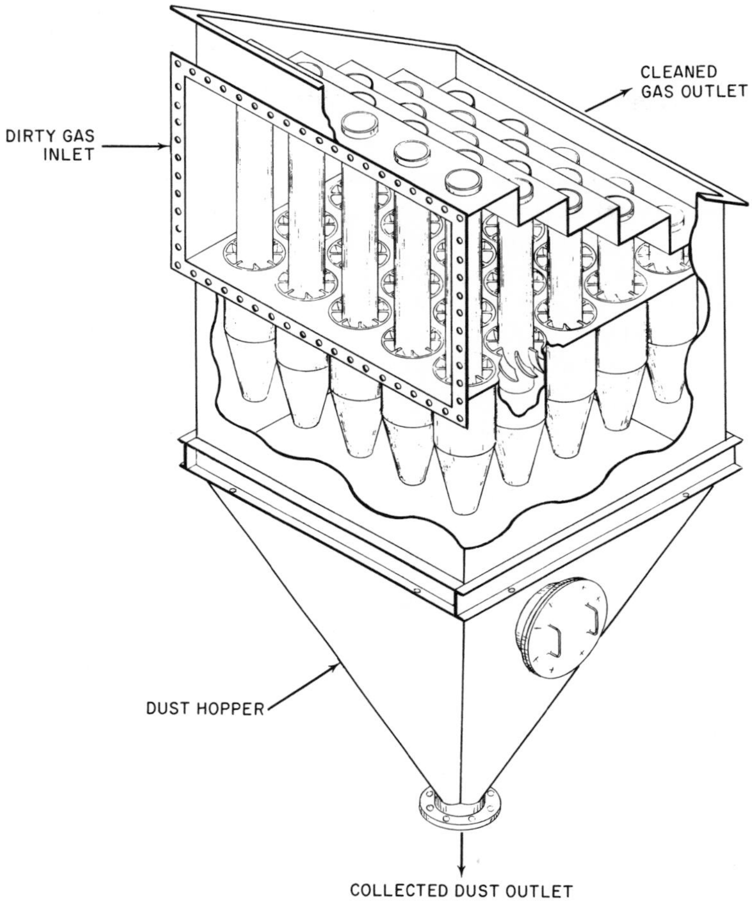

Fig. 8-6 Vane axial cyclone collector. *(Courtesy Belco Pollution Control Corp.)*

move the dust outward then hold it against the walls of the cylinder and cone until it is flung radially outward through the dust outlet into the storage hopper. The ascending inner vortex of cleaned gas starts slightly below the dust outlet and is fed its entire length by the cleaned inner surface of the descending vortex. The cleaned gas from each cyclone is picked up and piped through the dirty gas plenum into the cleaned gas plenum by the vortex finder tubes of varying lengths.

Various manufacturers have standarized on different cylinder diameters; the common commercial sizes are 6, 8, 9, and 10 in. In the past diameters as small as 1½ and 3 in. have been tried, however, they exhibited a strong tendency to plug on most applications and their usefulness today is limited to special applications such as natural gas pipelines and the cleaning of atmospheric air at compressor stations.

The capacity of individual tubes varies with diameter, vane design, and maximum acceptable gas velocities. Commercial capacities are generally within the range of 500 to 1000 cfm per tube. By selecting the correct number of tubes and arranging them in rows, it is possible to assemble a compact collector for any required capacity. It was said that the alteration of cyclone dimensions and other factors have a material effect on performance including collection efficiency and draft loss. In the vane axial collector we must add another important factor—the number and shape of the vanes which impart the spin to the incoming dirty gas. Over the years manufacturers have researched vane design

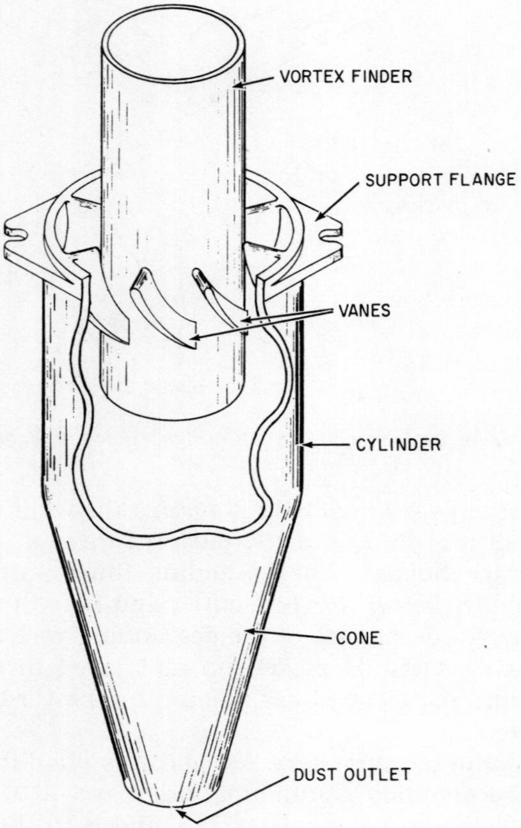

Fig. 8-7 Vane axial tube. *(Courtesy Belco Pollution Control Corp.)*

to maximize efficiency, and to minimize draft loss and vane erosion. Some manufacturers offer a choice of vanes in order to match the operating characteristic of the collector to the requirements of the application. Other manufacturers have concentrated on the devolpment of aerodynamically shaped vanes such as shown in Fig. 8-7 which take advantage of laminar controlled gas flow in order to induce rather than to force the gas to spin. Such vanes are true airfoils of streamline shape designed to minimize turbulence and erosive wear. It is also claimed that this longer vane provides improved flow control over a wide range of capacities.

Standard construction for heavy-duty industrial applications consists of a 3/16-in. mild steel shell, heavier tube sheets, and cast iron tubes and vanes, with the leading row of vortex finders protected by erosion shields. For lighter duty service welded tubes of 10-gauge mild steel are offered. For heavier duty service 1/4-in. shells with cast iron tubes of 500 Brinnel hardness are available.

The energy requirements of the vane axial cyclone are in the range of 4 to 6 in. W.G. on ambient air, which is typical of cyclone collectors. Draft losses are proportionately less on higher temperature gases having lower densities.

Unique advantages: The absence of moving parts makes for extreme simplicity and reliability. High capacity, compactness, and cost savings resulting from using a standardized tube combine to make this the least expensive of high efficiency cyclone collectors, and as such the least expensive of all particulate collectors. When its materials of collection are suitably selected, this collector is able to withstand rugged service conditions including moderately heavy grain loadings of abrasive dust, high temperatures, and long runs with a minimum of maintenance. Whatever maintenance is required can be performed by plant personnel without extensive training.

Unique disadvantages: The interposed vanes are subject to erosive wear which may become critical on coarse dust of abrasive characteristic including high grain loadings and high specific gravity. The relatively small size of the openings between the vanes are subject to product buildup and plugging on dusts having an inherent tendency to buildup or to adhere or cohere while dry or in the presence of moisture. This plugging usually starts in the rear row of cyclones which tend to bear the heaviest dust loading.

Because a great number of cyclones share a common hopper, any imbalance in gas flow or dust distribution will result in gas circulation within the hopper from the overloaded cyclones to the underloaded. For this reason a conservative manufacturer will discount the efficiency of the single tube when furnishing a multitube collector. To combat the disadvantages of erosion and plugging some manufacturers offer replaceable vanes and tubes, and for some services intermittent water sprays are used to clean the vanes and tubes.

Like all high-efficiency cyclone collectors, the vane axial loses collection efficiency as the dust gets finer. To evaluate a particular collector for a particular service, or to compare one collector to another, it is necessary to know both particle size distribution of the dust and fractional efficiency curve of the collectors.

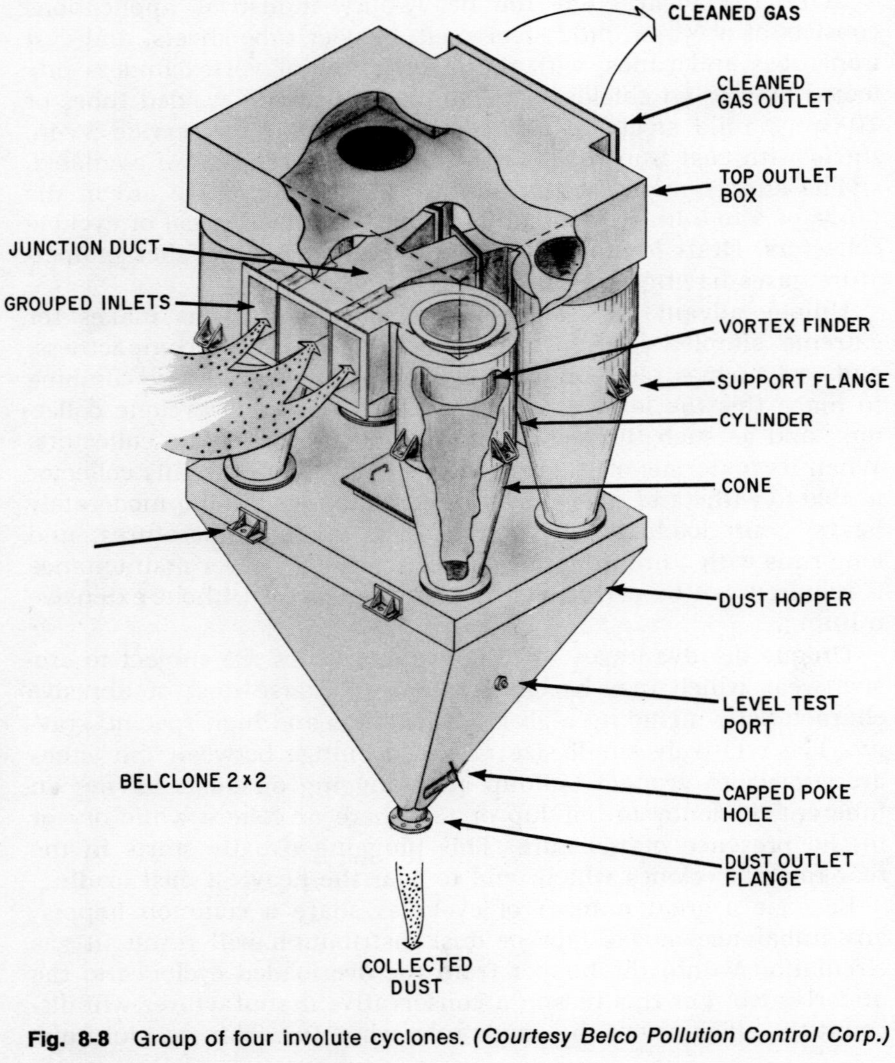

Fig. 8-8 Group of four involute cyclones. *(Courtesy Belco Pollution Control Corp.)*

Large-Diameter Cyclone Having Involute Inlet. Figure 8-8 shows a typical collector consisting of four cyclones arranged in parallel so that each cyclone handles 25 percent of the total gas flow. In order to simplify the connecting flues and to insure a balanced flow among the four cyclones, they are arranged in two rows of two with the four inlets grouped together with the assistance of a junction duct. The four cyclones discharge the dust into a common collected dust hopper and the cleaned gas is discharged into a common top outlet box whence it is discharged through a single cleaned gas outlet located to the rear, to either side, to the front, or through the top of the top outlet box, whichever best suits the layout.

Figure 8-9 shows a cut through several possible arrangements of involute cyclones at an elevation on the horizontal centerline of the cyclone inlets. It should be noted that some of the cyclones are left-handed, that is the inlet is on the left of the cylinder when facing the inlet. It should also be noted that the design of the junction ducts and the positions of the rear cyclones is such as to divide both gas and dust flow evenly among the grouped cyclones and in a manner that will insure proper flow into each cyclone. By using parallel groups of four or sixes it is possible to develop any required capacity. Some manufacturers will not furnish a collector of more than six cyclones per hopper because of the danger of circulation between overloaded and underloaded cyclones within the common hopper. Such circulation results in a reduction in collection efficiency.

Figure 8-10 shows a single involute cyclone with major parts labeled and illustrates the manner in which the dirty gas is injected into the cylinder. The rectangular involute inlet nozzle has its right-hand wall tangent to the cylinder, and it blends gradually with the cylinder through a fully open, 180° involute. This feature insures that the cylinder is of full effective diameter throughout its entire height including the portion opposite the inlet, thereby allowing the gas and dust to follow natural law in the development of the descending and ascending vortices without the need for vanes or a dipped roof. In this design, the other dimensions being correct, the descending vortex will be formed with a minimum of turbulence, thereby facilitating the precipitation of the dust under the influence of centrifugal force to the walls of the cylinder and cone.

In this cyclone, as in all high-efficiency types, the dirty gas induces the cyclonic spin in the cylinder. This energizes the descending outer vortex which extends downward into the vortex arrester where the rapidly rotating dust is flung radially outward, settling

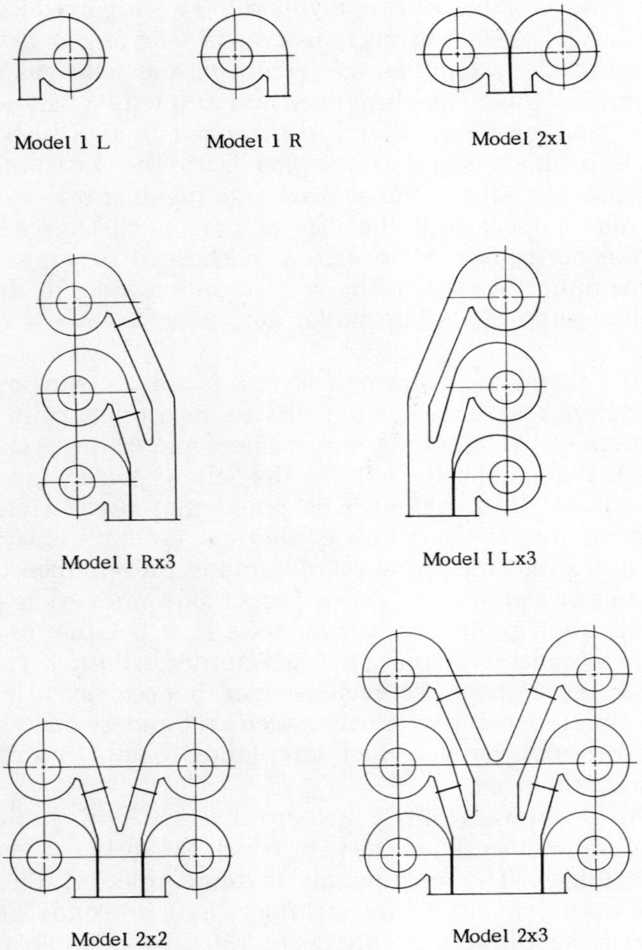

Fig. 8-9 Arrangements of involute cyclone collector. *(Courtesy Belco Pollution Control Corp.)*

into the storage hopper. The ascending inner vortex of cleaned gas starts within the vortex arrester and spirals upward, being fed and enlarged by the cleaned inner surface of the descending outer vortex. The cleaned gas is kept away from the inlet and conducted out of the cylinder by the vortex finder. The dimensional relations of inlet, inlet volute, cylinder, cone, vortex finder, dust outlet, vortex arrester, and storage hopper all have an important bearing on collector performance characteristics including collection efficiency and draft loss. One manufacturer offers the involute cyclone in

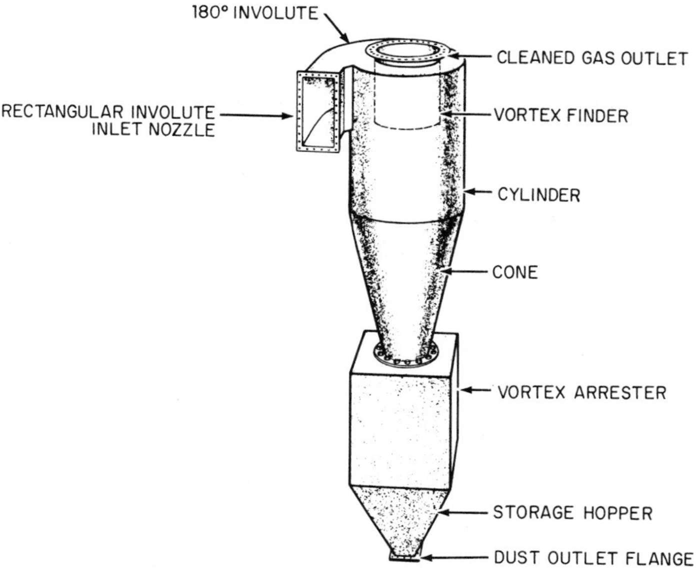

Fig. 8-10 Single high-efficiency involute cyclone. *(Courtesy Belco Pollution Control Corp.)*

seven efficiency series, single cyclone capacity ranging from 90 to 34,000 cfm. In most large-capacity applications it is common practice to use groups of cyclones each sized to handle between 5000 and 10,000 cfm rather than using a single very large cyclone.

In contrast to the vane axial cyclone collector, wherein the diameter and design of the individual cyclone is fixed, and increased capacity is accomplished by taking an increasing number of cyclones, the involute cyclone comes in the above-described variety of sizes and efficiency types. These variables provide the involute cyclone collector designer with great flexibility in matching collector performance to the requirements of each application.

Furthermore the involute cyclone collector can be manufactured from a wide range of construction materials including mild steel, firebox quality steel, low-alloy steel, stainless steel, lined steel, and lined alloy steel. For corrosive services cyclones can be lined with soft natural rubber, semihard, flexible ebonite, neoprene, butyl, or PVC. For foodstuff services cyclones are furnished with high-gloss-finished linings of teflon—polyethelene, polypropylene or phenolics. Material thickness can also be varied from 14 gauge to ½ in. In some instances it is economical to use different thicknesses in different parts such as the following thicknesses of mild steel plate:

Cylinder	1/4″
Cone	3/8″
Top outlet box	3/16″
Dust hopper	3/16″

The above selection reflects that the cylinder and cone are wearing parts, with the most severe erosion occurring near the bottom of the cone in the zone of highest velocity just before release into the vortex arrester.

The energy requirements of the involute cyclone collector are in the range of 4 to 6 in W.G. on ambient air and less on higher temperature gases having lower densities.

Unique Advantages: The absence of moving parts makes for extreme simplicity and reliability. Smooth interiors and the absence of vanes eliminate the wear of direct impingement and minimize turbulence, thereby minimizing the wear of sliding friction, as well as serving to defeat dust buildup and consequent plugging. Cyclones of lined construction are offered to resist one or more of the following: corrosion, erosion, or high temperatures. For corrosion cyclones can be lined with a great variety of material selected to resist the particular corrosive problem. For erosion the lining is accomplished by a replaceable castable lining held in place by hex mesh steel. For exceptionally high temperatures the castable lining is made of two layers—the layer next to the steel having a thermal insulation property, and the one next to the gas having erosive resistance. When its materials of construction are suitably selected, this collector excels at reliability under rugged operation including very heavy grain loadings, very abrasive dust, very high temperatures, and the strain of long runs without loss of collection efficiency or increase of draft loss.

Unique disadvantages: The cost and size of the involute cyclone increase as the efficiency requirement goes up. These economic penalties make the vane axial cyclone preferable whenever its operating characteristics can meet all requirements. When this is not possible, the involute cyclone collector is a logical choice. As previously developed under the "Paradox" the large-diameter cyclone can match or exceed the efficiency of the vane axial cyclone —however, at added cost.

Like all high-efficiency cyclone collectors, the involute cyclone collector loses collection efficiency as the dust gets finer. To evaluate a particular service or to compare one collector's suitability to another it is necessary to know both the size distribution of the dust and the fractional efficiency curves of the collectors.

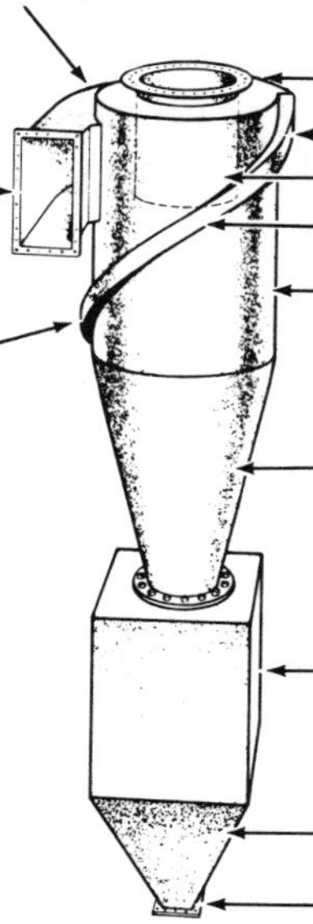

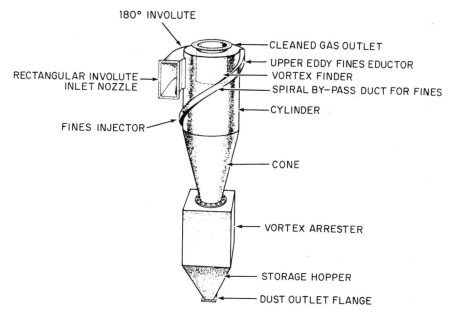

Fig. 8-11 Cyclone with involute inlet and fines eductor. *(Courtesy Belco Pollution Control Corp.)*

Large-Diameter Cyclone Having Involute Inlet and Fines Eductor. Figure 8-11 shows a single cyclone with major parts labeled. The flow pattern is similar to that of the standard involute inlet cyclone except that a fines eductor is added to the top of the cylinder to continuously remove the finer dust particles which tend to be trapped by the upper eddy. These fines are removed from the cylinder and transported by a spiral bypass duct to a fines injector where they are combined with the coarser dust particles at the level where the cylinder and cone meet.

This feature improves collector efficiency without increasing draft loss. It improves efficiency in two ways: (A) The addition of the upper eddy balances the lower eddy thereby further smoothing the internal flow pattern; (B) the removal of the fines from the top of the cylinder and their injection into the ribbon of coarser dust increases their capture and reduces reintrainment.

Unique advantages: This cyclone enjoys the same advantages as those above-described for the involute cyclone plus increased collection efficiency without increased draft loss.

Unique disadvantages: The addition of the fines eductor increases installed cost by a factor of about 10 percent. There is no increase in operating cost.

Miscellaneous Types of Mechanical Collectors

The following types, while of limited application, are mentioned because many remain in service.

Chip Trap or Cinder Trap. These combine gravity settling with intertial separation resulting from an interposed single or multiple baffle. Such traps are useful for collecting very large particles, or protecting a more efficient collector from large particles.

Louvre Separator. These collectors are a variation of the recirculating baffle collector and include the v-pocket and conical-pocket dust louvre. Some increase collection efficiency by using a cyclone collector in the recirculating cycle. Their collection efficiency falls between those of the baffle collector and the ordinary cyclone collector.

Dynamic Precipitator. This is a combination centrifugal fan and dust collector. Like many combination devices it excels as neither but has the advantages of compactness and low first cost.

Centrifugal Inertial Separator. This is a variation of the vane axial cyclone wherein both the inner and outer vortices flow in the same horizontal direction. While of lower collection efficiency it has the advantage of straight-through flow and can often be installed within the cross section of an existing duct.

Ordinary Cyclone. This cyclone is typically made of light-gauge material by the local tinsmith from handbook or rule-of-thumb dimension. It can be recognized by the relatively large cylinder diameter and small inlet. It operates at relatively low velocities and is suitable only for collecting coarse dusts such as sawdust. Its main advantage is low cost.

Wet Cyclone. This is a large-diameter type in which the capture and retention of the dust is increased by spraying a liquid such as water into the inlet. This spray is captured as is the dust and wets the walls of the cyclone. This technique is sometimes used to keep the temperature of the cyclone within structural design limits when cleaning very hot gases.

Cyclonic Droplet Collector. Because centrifugally precipitated droplets form a continuous wetted film, very high efficiencies at a moderate expenditure of energy are practical in specially developed droplet collectors. Unless there is a preponderence of liquid particulate to dry particulate, buildup may present a problem.

Ultrahigh-Efficiency Cyclone Collector. These cyclones come in two basic types: very small diameters or very tall with relatively small inlets. Both are capable of high efficiencies at the added costs of size, capital investment, and draft loss.

The development and testing of these is a highly sophisticated undertaking. Unfortunately only a few companies have the necessary equipment and trained personnel needed to generate and disperse ultrafine dust within a moving gas stream—an essential requirement for such a developmental program.

See Fig. 8-5 and Table 8-2 for a typical range of fractional efficiencies obtainable. Curves C and D represent high-efficiency cyclones and curves A and B represent ultrahigh-efficiency cyclones.

FABRIC FILTERS

Operating Principles

One of the most versatile collectors for the removal of dry, solid particulate matter from an air or gas stream is the fabric dust collector in which the dust-bearing gas is passed unidirectionally through a fabric filter medium of woven or felted cloth.

While fabric filters come in many generic types, all depend on trapping the dust on the dirty-gas side of the fabric while allowing the gas to pass through the interstices between the woven threads of the fabric. While these interstices are as large as 100 μ in a typical commercial bag filter having a woven fabric media, such a collector traps dust particles as small as 0.5 μ.

The capability of the woven fabric filter to collect fine dust is the direct result of the buildup of a fragile, porous layer of dust on the dirty-gas side of the cloth. This layer, called filter cake, blocks the larger interstices and captures the finer dust particles.

Fabric filters using thicker felted fabrics, a complex labyrinth mass of randomly oriented fine fibers, are less dependent upon filter cake for high efficiency. In general, felted bags cannot be cleaned effectively by shaking and must be cleaned frequently with high-pressure reverse jets.

The mechanisms which cause the filter cake layer to form include some or all of the following:

Agglomeration of upstream fine particles
Impaction of larger particles on fiber
Diffusion of submicron particles onto fiber
Electrostatic attraction and repulsion
Thermal effect
Coarse sieving by cloth
Fine sieving by filter cake layer

The complex interrelation of these factors are beyond the scope of this chapter; nevertheless a functional understanding of the process is essential to understanding the fabric filter.

By definition of the American Society of Testing Materials, the permeability of a filtering fabric is defined as the quantity of clean atmospheric air expressed in cubic feet per minute that will pass through 1 sq ft of cloth at a draft loss of 0.5 in. W.G.

Below are shown the typical ranges of permeability for the types of filter media used in common commercial fabric filters:

Fabric Type	*Yarn Type*	*ASTM Permeability*
Woven	Staple	20–100 cfm
Woven	Multifilament	10– 50 cfm
Felt	Fiber	15– 35 cfm

Because fabric filters operate at air-to-cloth ratios (CFM per 1 sq ft of cloth) of under 10 ft/min., upon start-up with new and clean cloth filter media, the draft loss of a fabric filter will be under ½ in. W.G. After operation for an indeterminate period, starting with low efficiency which gradually increases, the draft loss will approach 2–3 in. as the layer of filter cake builds up. On further operation the draft loss across the filter cake and cloth may approach approximately 6 in., at which time some form of cleaning must be employed to reduce the filter cake thickness to a point where the draft loss reverts to 2 or 3 in. In the automatic filter this cycle is repeated, usually compartment by compartment, to keep the draft loss within an acceptable range.

From the foregoing it can be seen that the draft loss through the filter is primarily due to the filter cake, rather than to the cloth upon which that cake is built.

Table 8-3 shows a typical operating cycle of one compartment of a compartmented, automatic fabric filter.

The following comments are keyed to cycle numbers:

(1) When new and clean, the draft loss across the fabric filter media is much below 0.5 in. W.G.

(2) During the precoat cycle the draft loss will gradually increase as the layer of filter cake is built up. The efficiency will in-

TABLE 8-3 OPERATING CYCLE OF ONE COMPARTMENT OF CONVENTIONAL COMPARTMENTED BAG HOUSE WITH WOVEN FABRIC

Cycle Number	*Filter Cycle*	*Cloth Status*	*Filter Cake Thickness*	*Air-to-Cloth Ratio or Filtering Velocity (ft/min)*	*Draft Loss (in./W.G.)*	*Collection Efficiency*
1	new	clean	zero	2–3	< 0.5	no dust
2	precoat	clean to coated	zero to minimum	2–3	< 0.5 to 2–3	low to high
3	filtering	coated	partial	2–3	3–5	high
4	filtering	full coated	maximum	2–3	4–6	high
5	cleaning	shaking	maximum to minimum	dampered off	Zero	not operating
6	filtering	coated	minimum	2–3	2–3	high
7	filtering	full coated	maximum	2–3	4–6	high

crease to operating levels of plus 99 percent; the gas discharge will become clear.

(3–4) During the filtering cycle the draft loss will increase to the maximum acceptable level as the layer of filter cake thickens.

(5) The conventional bag house will damper off the compartment being cleaned. Maximum filter cake is shaken off the cloth and drops into a hopper under the influence of gravity. The vigor and duration of shaking must be adequate to reduce filter cake thickness from maximum to minimum. Excessive shaking may reduce collector efficiency by reducing filter cake thickness below minimum.

(6) Compartment is restored to filtering cycle. Cycles (5), (6), and (7) are repeated for each compartment by a predetermined program so that no more than one compartment is removed from service at one time.

Fabric filters come in a great variety of sizes and types which can be distinguished by describing the various major components:

Casing. The casing may be single or it may be compartmented. It may have the gas drawn through it, or pushed through it, thus giving to general descriptions: open pressure casing, closed pressure casing, closed suction casing. The latter is the most common for commercial collectors, enabling the fan to operate on cleaned gas.

Filter Media. The filter media may be arranged in one or more of the following shapes: envelope, round tubular, or zigzag fabric. The tubular type is the most common with the vertical tube diameter varying between 5 and 36 in. and in height from 5 to 25 ft. Common bag sizes are summarized below.

Type Filter Cloth	*Bag Diameter*	*Bag Length*
Woven or felted	5″–8″	5′–15′
Felted	10″–20″	10′–20′
Fiberglass	12″	15′–25′

Filter Cloth. A great variety of cloths are available to a bag filter manufacturer: a major classification is into woven fabric or felted fabric.

The woven type comes in a great variety of weaves, such as taffeta, twill, or sateen, and has additional variations, such as type of fiber, including cotton, wool, synthetics, and lubricated fiber glass; yarn size; staple or multifilament yarn; napped or unnapped; high or low thread count; high or low yarn twist; etc.

Felted bags are more expensive than woven cloths and come in wool and needled synthetic fibers. Cotton and fiber glass do not make felted fabrics.

Fibers vs. Temperature. Natural fibers such as cotton or wool have been used for years for temperatures below 200°F. In recent years man-made fibers have extended the application of bag filters to temperatures in the 300 to 400°F range, and more recently lubricated fiberglass has extended application to temperatures as high as 550°F.

The selection of a particular filter medium is a complex one and must take into account the characteristics of the dust and of the gas, including the need for resistance to moist heat, abrasion, mineral acids or alkalies, and the need for tensile strength. Furthermore economic factors such as first cost, installation cost, power cost, probable life, and replacement cost must be evaluated.

Miscellaneous Features. Other distinguishing features of a bag filter will be the number and arrangement of bags, whether gas flow is vertically up or down, or horizontal, whether the flow is from the inside of the bag to the outside or from the outside of the bag in, and how the bag is cleaned.

Advantages and Disadvantages of Fabric Filters. A major advantage of the fabric filter is high efficiency at all loads from maximum down to zero gas flow.

Some disadvantages include large size and high maintenance due to the cost of bag replacement. Other problems include destruction of bags by too high a temperature, or blinding or plugging of fabric due to moisture resulting from operation below the dew point.

Basic Types of Fabric Filters

Of the infinite variety of fabric filters the following have the broadest commerical application:

Intermittent baghouse with manual or powered cleaning
Automatic conventional baghouse with mechanical shaking
Automatic reverse air flow cleaning
Travelling reverse ring jet cleaning
Automatic reverse pulsed jet cleaning

The fundamental construction and operational characteristics of these five types using tubes for filtering are summarized in Table 8-4.

TABLE 8-4 SUMMARY OF BASIC TYPES OF FABRIC FILTERS USING CLOTH TUBES

Basic Type of Baghouse	Type of Bag	NORMAL DIRECTION OF GAS FLOW DURING FILTERING CYCLE		CLEANING TUBES		Air to Cloth Ratios (ft/min)	Space Requirement	Cost per Square Foot of Filter Media
		Thru Collector	*Thru Tube*	*Method*	*Cycle*			
Intermittent	woven cloth tube	up	inside to outside	manual or powered shaking	periodically during shutdown	1-4	very large	low
Conventional	woven cloth tube	up	inside to outside	automatic mechanical or pneumatic shaking	intermittent by isolated compartment	1-4	very large	low
Reverse air flow cleaning	woven cloth tube	up	inside to outside	automatic tube collapse by low pressure reverse flow	intermittent by isolated compartment	1-4	very large	low
Reverse ring jet cleaning	felted cloth tube	down	inside to outside	travelling compressed air ring	continuous on heavy dust loadings or intermittent on light dust loadings while gas flows	3-8	large	high
Reverse pulse jet cleaning	felted cloth tube	up	outside to inside	high pressure pulsed jet	programmed by manifolds while gas flows	3-8	large	high

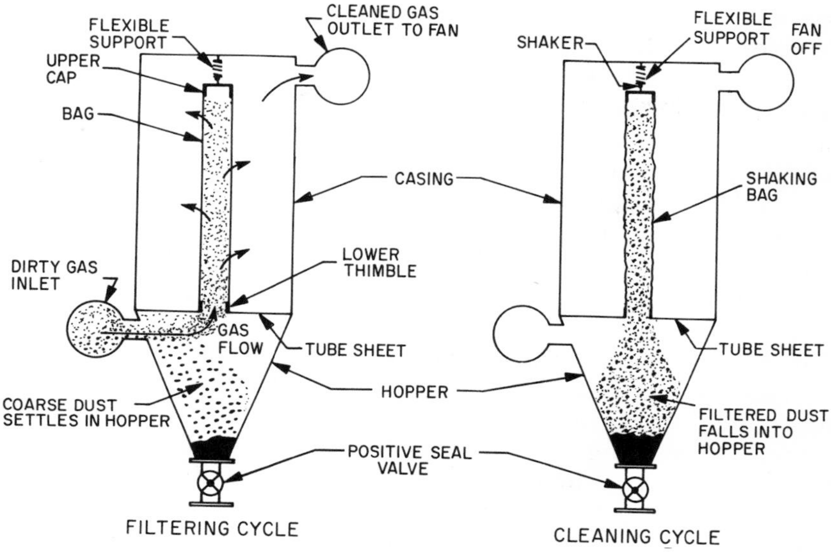

Fig. 8-12 Intermittent baghouse with manual or powered shaking.

Intermittent Baghouse with Manual or Powered Shaking. This is the oldest and simplest of all fabric filters and is still widely used in small sizes up to several thousand cubic feet per minute.

Major components. As can be seen in Fig. 8-12, the intermittent baghouse consists of a casing divided into an upper and lower part by a tube sheet. The upper portion contains the woven fabric filter which may be in the shape of one or more tubes, or in a great variety of other shapes all designed to squeeze sufficient filtering area into the available space. The lower portion may be in the form of one or more pyramidal hoppers, or if the casing is long there may be a single trough-type hopper running the full length of the casing. The latter is usually fabricated of light-gauge, mild steel plate. Small units are shop assembled, larger units are shipped knocked down and are equipped with punched flanges to facilitate field assembly. While only one bag is shown in Fig. 8-12, the collector can be furnished of any number to provide needed capacity. Each bag tube is stretched between a flexibly mounted upper cap and a lower thimble which is part of the tube sheet.

Gas flow during filtering cycle. Under normal operation the dirty gas is introduced below the tube sheet and into the upper portion of the hopper. Since the gas velocity is reduced upon entry, the coarse particles of dust will settle directly into the hopper, while the finer particles and all of the gas will pass upward into the fabric tubes. The gas will pass outward through the fabric; the dust will be trapped within the cloth tube. The cleaned gas is collected by the casing and discharged into a duct leading to the fan. Because the collected dust increases the thickness of the filter cake on the inner walls of the tubular bags, the draft loss—that is the difference of static pressure measured between the dirty and clean side of the bags—will increase. Periodically the collector must be taken out of service so that the bags can be cleaned.

Cleaning cycle. During the cleaning cycle the fan is stopped so that gas flow through the collector ceases. The upper ends of the bags which are flexibly mounted, are now shaken by mechanical means. The shaking must be vigorous enough to dislodge most of the filter cake deposited on the inner surface of the bag so that it will drop into the hopper. It is important that a sufficient layer of filter cake remain locked in the embrace of the cloth filter media so that collection efficiency will not be reduced. For some services, where exceptionally high collection efficiencies are required, precoating and recoating after shaking is accomplished by introducing coarse dust or other coating material. In sizing such a collector it is desirable to take into account the operating cycle of equipment in which the dirty gas originates. Thus for a noncontinuous industrial process a 5-hr continuous run capacity will allow shut down and cleaning between shifts and during lunch hour.

Special or alternate features. Particularly in the smaller sizes, the intermittent baghouse comes in a great variety of forms wherein the hopper is replaced with a flat bottom, perhaps with a drawer, or a door through which the dust can be removed. Such manual removal of dust can be practical whenever loadings are light so that dust removal can be accomplished at infrequent intervals.

If the intermittent baghouse is to be operated under pressure—that is where the dirty gas is blown into the collector—the upper casing becomes unnecessary. Indoors the casing may be replaced with an expanded metal screen or eliminated; outdoors it can be replaced with weather protection such as a roof and louvered sides.

Automatic Conventional Baghouse with Mechanical Shaking. This is one of the oldest and most widely used fabric filters.

Major Components. As can be seen in Fig. 8-13 the automatic conventional baghouse with mechanical shaking consists of a cas-

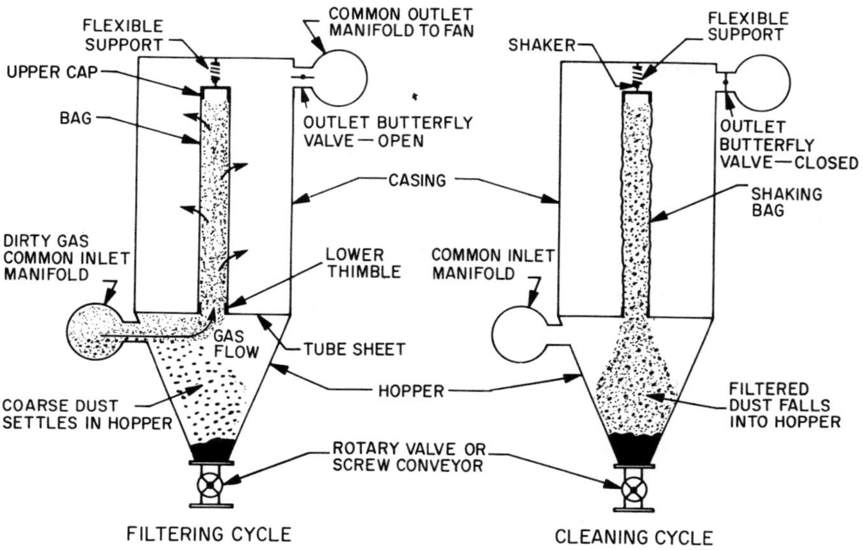

Fig. 8-13 Automatic conventional baghouse with mechanical shaking.

ing divided into an upper and lower portion by a tube sheet. The upper portion contains the woven fabric tubular bags, and the lower portion consists of pyramidal hoppers, each with its own valve, or a long trough-type hopper emptied by a sealed screw conveyor. The casing is usually fabricated of light-gauge mild steel plate. While only one bag is shown in Fig. 8-13, each compartment usually contains many. Each bag is stretched between a flexibly mounted cap and a fixed lower thimble on the tube sheet. To facilitate installation and replacement the bags are usually fastened to the caps and thimbles by quick-closing, spring-loaded, circular clamps. On large industrial filters intended for continuous, automatic service the casing is divided into several compartments in order that one may be removed from service while the bags are cleaned; the other compartments continue to operate without excessive draft loss or gas velocity. In a multicompartment collector, wherein each compartment has its own hopper, the gas is introduced into the individual hoppers through a common inlet manifold and the cleaned gas will be discharged through a common outlet manifold.

Gas Flow During Filtering Cycle. During normal operation the dirty gas is introduced below the tube sheet and into the upper portion of the hopper. Since gas velocity is reduced upon entering, the coarse particles will settle directly into the hopper. The finer particles and all of the gas will then pass upward into the fabric tubes, and then from the dirty inside of a tube through the fabric to the clean outside. The cleaned gas is collected by the casing and discharged into the common outlet manifold. As the collected dust increases the thickness of the filter cake on the inner walls of the tubular bags, the draft loss—that is the difference of pressure between the inlet and the outlet of the collector—will increase. Periodically, either with a predetermined program, or by monitoring the draft loss, the collector will go into its cleaning cycle.

Cleaning Cycle. During the cleaning cycle the normally open outlet butterfly valve is closed. The upper ends of the bags, which are flexibly mounted, often from a common support structure, are now shaken either by mechanical or pneumatic means. The shaking dislodges the filter cake from the inner surface of the bag causing the cake to fall into the hopper. In some cleaning cycles a small amount of atmospheric or secondary air is bled into the casing to facilitate the cleaning of the bags and speeding the dust fall. Upon conclusion of the cleaning cycle the outlet butterfly valve is opened, thereby returning the compartment to the filtering cycle. Individual compartments are thus cleaned one at a time to keep the draft loss across the collector at or near a nominal value.

Special or Alternate Features. In some designs a common trough-type hopper runs the full length of several compartments. In this design the dirty gas can be introduced into one end of the hopper which also serves as a common inlet manifold. In such a design the dust is removed by a sealed screw conveyor running the full full length of the hopper, the discharge of the screw conveyor being through a rotary valve. While the above-described mounting of the bags is typical, there are as many ways of mounting bags as there are manufacturers. The method of shaking can vary from manual to a wide range of mechanical and pneumatic methods either with or without the assistance of low-pressure reverse airflow.

Automatic Baghouse with Reverse Airflow Cleaning

Major components. The major components of a baghouse with reverse airflow cleaning is shown in Fig. 8-14. The casing is divided into two parts by a tube sheet. The upper portion contains the woven fabric tubular bags and the lower portion contains the hopper. The hopper may be a continuous trough-type running under many compartments, or a series of individual pyramidal hoppers

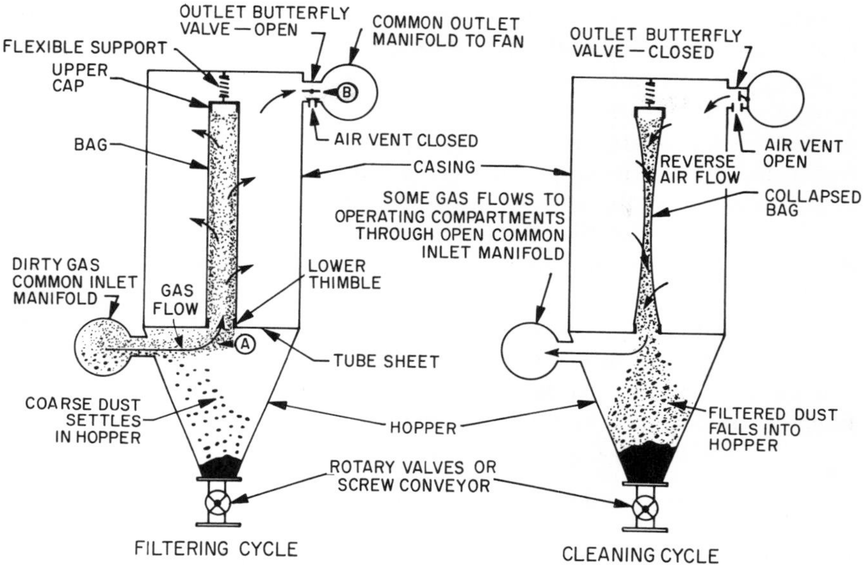

Fig. 8-14 Automatic baghouse with reverse airflow cleaning.

each serving one compartment. The dirty gas can be fed through a single inlet into the trough-type hopper or through a common inlet manifold serving many pyramidal hoppers. Each tubular bag is clamped to a thimble at its lower end and clamped to a flexibly mounted cap at the upper end.

Gas Flow During Filtering Cycle. During the filtering cycle the gas flow is through the common inlet manifold into the individual hoppers, upward into the tubes, through the fabric from the inside to the outside, and then out the common outlet manifold. The hopper serves as a settling chamber for the coarse dust.

Cleaning cycle. During the cleaning cycle the normally open outlet butterfly valve is closed. At the same time a small air vent valve is opened, allowing atmospheric air to rush into the casing and collapse the bags. This will break the filter cake and allow it to drop into the hopper aided by the reverse airflow. This secondary reverse flow is maintained by the suction through the still open common inlet manifold. Upon conclusion of the cleaning cycle the outlet butterfly valve is opened and the air vent is closed. Individual compartments are cleaned one at a time on a predetermined schedule activated either by a programmer or by monitoring the draft loss.

Special or Alternate Features. Some filter tubes are equipped with rings spaced periodically along their vertical height, thereby improving the cleaning action of the collapsed bag. Some cleaning programs open and close the valves several times, thereby repeatedly collapsing and snapping the bags open. Some manufacturers shake the bags during the cleaning cycle.

In some designs cloth tubes are grouped into compartments which are periodically cleaned by reverse flow high-pressure jets directed either on the "dirty side" or on the "clean side."

If on the "dirty side" the jet is directed into the dirty gas inlet at point A in Fig. 8-14, thereby collapsing the bags by interfering with the inflating effect of the incoming gas. If on the "clean side" the jet is directed into the casing at point B, thereby collapsing the bags by increasing the pressure within the casing.

Baghouse with Traveling Reverse Ring Jet Cleaning.

Major Components. As shown in Fig. 8-15 the baghouse with reverse jet cleaning consists of a casing divided into three parts by an upper tube sheet and a lower one. The space above the upper sheet serves as a common inlet manifold to several compartments. The central portion of the casing houses the felted filter tubes and the lower portion consists of either individual pyramidal hoppers, each served by its own valve, or a common trough-type hopper

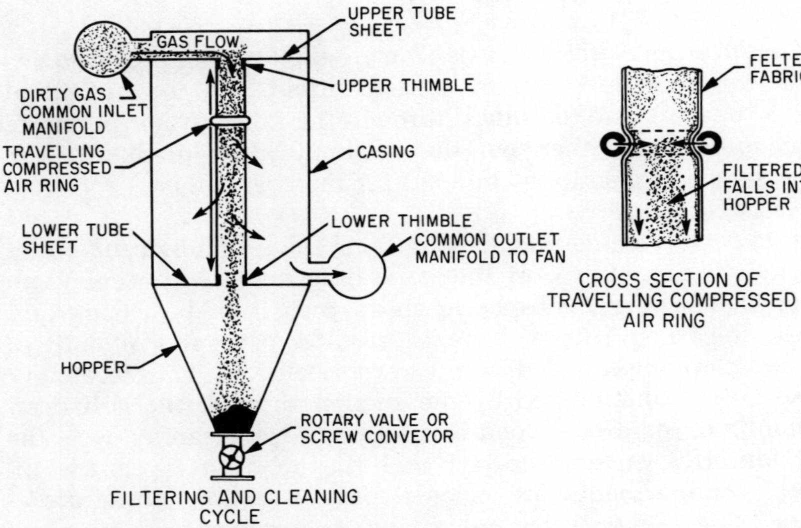

Fig. 8-15 Automatic baghouse with reverse ring jet cleaning.

served by a sealed screw conveyor. The filter bags are stretched between and clamped to thimbles on the upper and lower tube sheets.

Gas Flow During Filtering Cycles. The dirty gas enters the space above the upper tube sheet passes downward into the filtering tubes, and outward through the filtering media wall. This deposits the dust on the inner wall of the filter tube. The cleaned gas is collected by the casing and is discharged through a common outlet manifold.

Cleaning Cycle. Each bag is cleaned by a traveling compression air ring whose inside diameter is slightly smaller than the outside diameter of the tube. These rings are alternately raised and lowered in groups along the full length of the filter tube. There is an annular slot in the ring along its inner diameter which directs the compressed air through the fabric and into the tube. Velocity is sufficient to clean the tube of its filter cake. Compressed air is fed into the traveling rings through flexible tubes. Note that the mechanism which raises and lowers the groups of compressed air rings operates on the clean side of the bag filters. Note also that the filtering and cleaning cycles take place simultaneously. In this type of filter compartmentation is not necessary inasmuch as all tubes are operating continuously.

Special or alternate features. In heavy dust loadings the cleaning cycle will be continuous. On lower dust loadings the cleaning cycle will be programmed so as to operate periodically.

Automatic Baghouse with Pulse Jet Cleaning.

Major Components. The casing is divided into an upper and lower portion by a tube sheet as shown in Fig. 8-16. The upper, smaller portion serves as a common cleaned gas discharge manifold. For each felted filter tube there is a venturi-shaped thimble. Mounted above each tube is a compressed air jet. The fabric filter medium is mounted over an internal frame with a closed bottom. The lower portion of the casing is open to either a single trough-type hopper or to multipyramidal types.

Gas flow during filtering cycle. The gas is introduced to the lower part of the casing just above the hopper. The latter serves as a settling chamber and settles out coarse dust without the need for filtering. The gas flow is then directed upward into the casing around the filtering tubes, and then through the filtering tubes

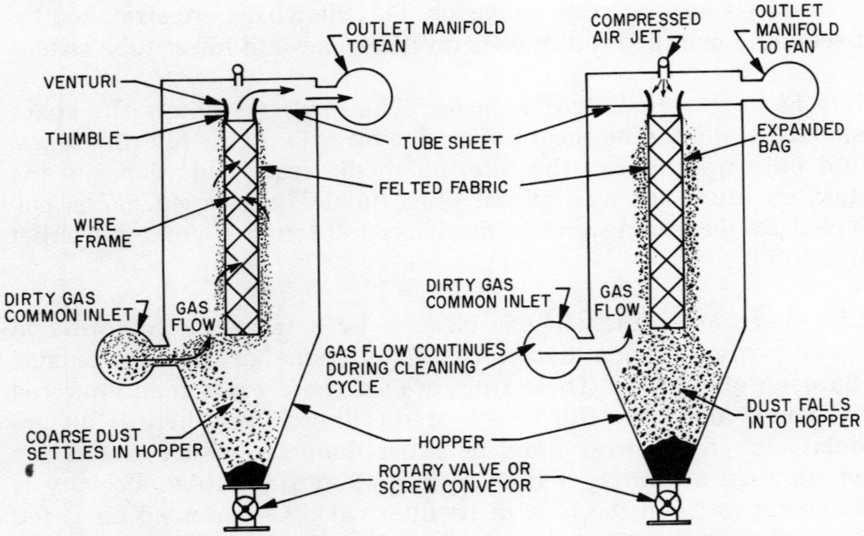

Fig. 8-16 Automatic baghouse with pulse jet cleaning.

from the outside in. Note that in this instance the filter cake is built up on the outside of the tube, the tube being held circular by an inner frame. The cleaned gas passes upward through the tube sheet and out through the outlet manifold.

Cleaning cycle: Periodically sets of compressed air jets are activated thereby blowing air at high pressure through the venturi into individual tubes. The force of the pulse jet snaps the tube outward, thereby dislodging the accumulated filter cake. The cleaning cycle usually cleans several tubes at a time through sets of manifolds. The frequency of cleaning is limited only by considerations of bag filter life.

Special or alternate features. Each manufacturer features proprietary designs having distinctive nozzle, venturi, and tube arrangements to optimize cleaning and bag life while minimizing draft loss and compressed air consumption. A careful evaluation of competitive offerings is warranted.

Miscellaneous Types of Fabric Filters

Variations retain most of the operating characteristics of the basic type. Selection should be based on an evaluation of installed and operating costs including bag life, cost of bag replacement, and labor cost of such replacement.

Envelope-Shaped Filter Media. There are several designs of the automatic conventional baghouse which use envelope-shaped bags rather than tubes. In some designs the envelope has seams which form the envelope into a series of joined tubes.

Alternate designs of the automatic baghouse with reverse airflow cleaning include those equipped with envelope-shaped bags mounted horizontally on frames. Cleaning is accomplished by a traveling blow-back device which runs across the width of the collector.

Pressure Jet Reverse Flow. In this variation of the automatic reverse flow cleaning collector a pulse jet is added which directs a blast down the tubes. The reverse flow and jet are programmed to alternate during the cleaning cycle.

Disposable Filter Cartridge. This type of collector, usually of small capacity, is used where the character of dust is such that it cannot be removed effectively from the filter media.

WET SCRUBBERS

Operating Principles

Wet scrubbers use a liquid, usually water, to capture then remove particulate matter from a moving stream of gas.

Spray Chamber. To illustrate one principle of wet-scrubbing, Fig. 8-17 shows one of the simplest scrubbers—a gravity settling chamber scrubber with sprays, in which fine water sprays wash the gas with water and settle the dust as sludge in the sludge tank.

The dirty gas is slowed in the inlet nozzle, fights its way through the turbulence caused by the force of the fine sprays, then passes through the demist section, and is speeded up in the outlet nozzle to duct velocity.

As in all settling chambers, some coarser dust will settle onto the surface of the water pool under the influence of gravity. Major collection is accomplished by collision between a dust particle and a water droplet, resulting in the capture of the former by the latter. Such collosions are caused by the following mechanisms:

Impingement of larger dust particles on droplets
Interception from diffusion of fine particles
Electrostatic forces

Thermal gradients
Condensation of moisture on dust particles

Spray Section. The above mechanisms result in the dust particle becoming larger and heavier because it has been captured by a water droplet. This increase in size and weight helps the collection process by strengthening the forces which help separate the heavier dust from the lighter gas. These forces are gravity, inertia, and centrifugal force.

Once captured the wetted dust particle acts as if it were a droplet of water. These readily coalesce into still larger droplets, then into wet films on surfaces, and finally into pools of liquid.

As with all collectors, fine particles are harder to capture than coarse particles. The probability of capture of fine particles is largely dependent upon the amount of energy causing the mixing between gas, dust, and droplets.

In the spray chamber shown in Fig. 8-17 almost all mixing energy is supplied by the pressure of the sprays. Higher pressures can be used to produce finer droplets, more droplets, and more turbulence. All of these tend to improve collection efficiency, but this device is basically a low-efficiency scrubber operating at low energy levels and suitable only for the collection of the coarser dusts.

Dust Cycle. Captured, wetted dust particles settle as droplets under the influence of gravity to the surface of the water pool. The droplet

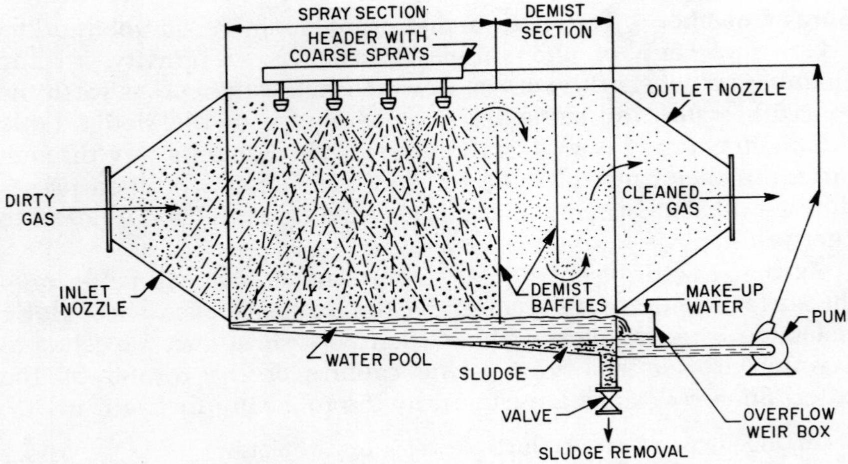

Fig. 8-17 Gravity settling chamber scrubber.

readily joins the pool and releases the dust which, having a higher specific gravity than the water, settles to the bottom to form sludge, which is periodically drawn off.

Demist Section. The demist baffles encourage settling of the droplets by reversing the direction of gas flow over the water pool. The forces of inertia and gravity combine to impinge the droplets on to the surface of the pool.

Water Cycle. In Fig. 8-17, water is recycled at a rate of 2 to 10 gal/min for every 1,000 cu ft of gas cleaned. Makeup water is proportional to the amount of water evaporated by the gas. Once-through systems are practical only in locations where there are unlimited quantities of river, lake, or ocean water. In such cases the water effluent is usually discharged to a large settling pond. The recirculation of water tends to plug and erode spray nozzles. Practical solutions include the use of special or coarser nozzles, and the filtration of the liquid.

Introduction of Scrubbing Liquid. The scrubbing liquid can be introduced into the scrubber in many ways, such as by the following:

Fine spray nozzles
Coarse spray nozzles
Very coarse, spray nozzles, i.e.: ¾-in. pipe
Overflow weir
Impingement of gas onto liquid pool
Introduction of gas under surface of liquid pool

The liquid sprays can be of low or high pressure; direction of flow can be cross-flow, concurrent flow—that is, in the same direction as gas flow, or counterflow.

All of the above methods are used in commercial scrubbers.

Wetted Impingement Baffle Scrubber. To illustrate an additional important principle, Fig. 8-18 shows a spray chamber to which has been added three wetted impingement baffles. These baffles will improve efficiency of collection by capturing dust particles which impinge upon and are captured by the wetted surface, and are then flushed away in the moving liquid film.

The collection efficiency of this collector nevertheless remains low because the energy level is low. In commercial wet scrubbers, collection efficiency is increased by increasing inertial and cen-

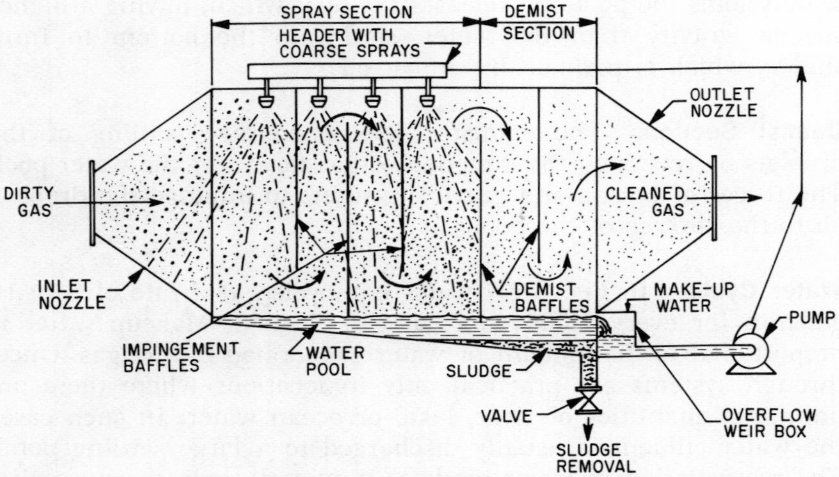

Fig. 8-18 Wetted impingement baffle scrubber.

trifugal forces due to the application of increased energy resulting mainly from high gas velocities.

Collection Efficiency. The collection efficiency of all well-designed wet scrubbers can be related to the total energy expended. Low-energy collectors have low efficiency, high-energy types have high efficiency. The energy can be introduced either in the water cycle or the gas cycle. In most commercial collectors almost all of the energy is introduced in the gas cycle and can be measured as draft loss in inches of water.

Advantages and Disadvantages of Wet Scrubbers. A major advantage of the wet scrubber is the great variety of types, which allows the selection of a collector suitable for almost any collection problem including collection efficiencies as high as 99 percent.

Some disadvantages of the wet scrubber include the disposal of a wet sludge, the high energy cost of the high-efficiency scrubber, the high material cost related to services where there is chemical corrosion, and the potential problems of plugged nozzles, unavailability of scrubbing liquid of sufficient clarity, and the treatment of corrosive scrubbing liquids.

A unique disadvantage of the wet scrubber is the very visible white plume which is the inherent characteristic of all wet-scrubber stacks discharging to atmosphere without downstream gas conditioning. Adverse public criticism is sometimes mistakenly directed

TABLE 8-5 SUMMARY OF BASIC TYPES OF SCRUBBERS

Basic Type	Specific Type	WATER VS. GAS FLOW (Concurrent or counter or cross)	WATER CIRCULATION (gpm per 1,000 cfm)	DRAFT LOSS (Inches water gauge)	PERCENT COLLECTION EFFICIENCY ON FINE DUST: Low	Moderate	High
Impingement baffle	tangential inlet wet cyclone	concurrent or cross	3–5	1–4	X	X	
	spiral baffle wet cyclone	concurrent	1–2	4–6	X	X	
	single plate	concurrent	2–4	1–8	X	X	
	multiple plate	concurrent	3–5	6–12		X	
Packed tower	fixed bed	concurrent or counter	10–20	2–4	X		
	fluidized bed	counter	15–30	4–12	X	X	
	flooded bed	concurrent	2–4	4–8		X	
	multiple bed	counter	20–40	4–12		X	
Submerged Orifice	wide slot	concurrent	15–30	2–15	X	X	
	circular slot	concurrent	15–30	2–15	X	X	
	multiple slot	concurrent	15–30	2–15	X	X	
Venturi	high-pressure	cross or concurrent	5–7	30–100			X
	medium-pressure		3–5	10–30		X	X
	low-pressure		2–4	3–10	X	X	
	flooded disc		5–6	30–70		X	X
Miscellaneous and combination scrubbers	cross-flow packed	cross	1–4	2–4	X	X	
	centrifugal fan	concurrent	1–2	[a]	X	X	
	multiple venturi	concurrent	4–6	20–80		X	X
	combination venturi	concurrent	5–7	15–60		X	X
	combination fan type	concurrent	2–3	[a]		X	

[a] ½ hp per 1,000 cfm.

against the striking plume. See Table 8-5 for a summary of the basic types of wet scrubbers and their operating characteristics.

Impingement Baffle Scrubbers

The collection efficiency of this type is dependent upon a cycle which starts by impinging dry or prewetted dust onto a wetter surface, capture by the draining liquid film of the wetted surface, and ultimately flushing the dust out of the collector as sludge along with the scrubbing liquid.

Wet Cyclone Scrubber. In Fig. 8-19 dirty gas at high velocity is introduced tangentially into the lower portion of a vertical cylinder. Centrifugal force throws the dust onto the wetted wall. Water is introduced radially outward across the gas flow through coarse nozzles. The upward angle of spray is caused by the upward gas flow.

The upper portion of the cylinder acts as demister, the demisted

Fig. 8-19 Cyclone scrubber. *(Courtesy Chemical Construction Corp.)*

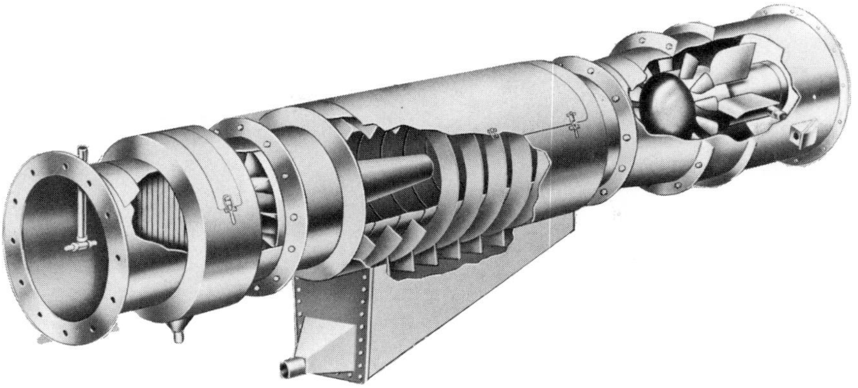

Fig. 8-20 Multiple-action scrubber. *(Courtesy Western Precipitation Div. of Joy Mfg. Co.)*

and cleaned gas exiting through the cleaned gas outlet. Water and entrapped dust drain from the bottom of the cylindrical tank.

Special or Alternate Features. Note that the scrubber shown Fig. 8-19 has a swinging inlet damper which can be closed partially at low gas loads to keep inlet velocities high, thereby keeping centrifugal force and collection efficiency high. This feature is essential for applications where there are variations in gas flow.

Almost any dry cyclone can be converted to a scrubber by spraying water either into the inlet or into the top of the cylinder.

Wet Cyclone with Spiral Baffles. In Fig. 8-20 dirty gas at high velocity is wetted by a concurrent spray in the inlet, then impinged on a series of angled wetted baffles, and then caused to spin by a set of wetted spiral baffles. The spray and dust are flung outward by centrifugal force, and then drain from all surfaces into the sludge tank.

Alternate or Special Features. This collector can be furnished with an axial fan, making an external fan unnecessary.

Impingement Baffle-Plate Scrubber

As shown in Fig. 8-21, this scrubber consists of a tower equipped with one or more wetted impingement baffles, each consisting of a perforated plate having from many hundred to a few thousand small holes per square foot.

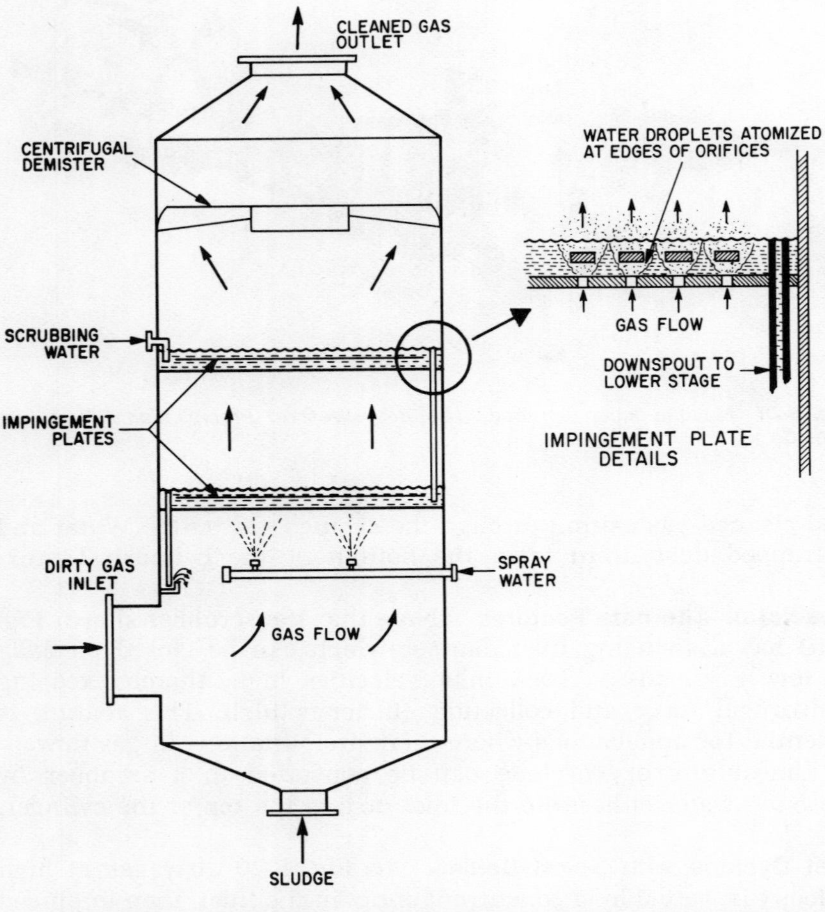

Fig. 8-21 Impingement plate scrubber. *(Courtesy U.S. Dept. of Health, Education, and Welfare)*

In the design shown, the gas and dust is first conditioned by a coarse spray, then passed up through the flooded impingement baffles. The small holes divide the total gas stream into many small ones. Because of the high velocity, perhaps 100 ft/sec, the edge of the holes atomize the liquid into 10-μ droplets. Fine droplets increase the collection of fine dust. Each hole is equipped with a small target plate upon which the dust and droplets impinge at high velocity.

Liquid flow control is effected by weirs and interstage overflow piping. The upper section of this scrubber is equipped with a vane axial demister. The dirty water drains out of the bottom.

Special or Alternate Features. Impingement baffle scrubbers may be of single or multiple stage, and may have a tangential inlet, thereby facilitating collection and wetting of dust by using centrifugal force as in a tangential inlet cyclone.

Packed-Bed Scrubbers

In this type of scrubber, the dirty gas is typically passed upward through the tortuous instertices formed by a thin or thick bed of small packing wetted by the counterflowing wash liquid. Since the paths of gas and dust flow are opposed and tortuous, and since the area of wetted surface is great, there is repeated opportunity for the dust to be captured by a wetted surface, or by the descending wash liquid.

A great variety of packings are used, including glass or plastic spheres, raschig rings, berl saddles, tellerettes, partitioned rings, etc.

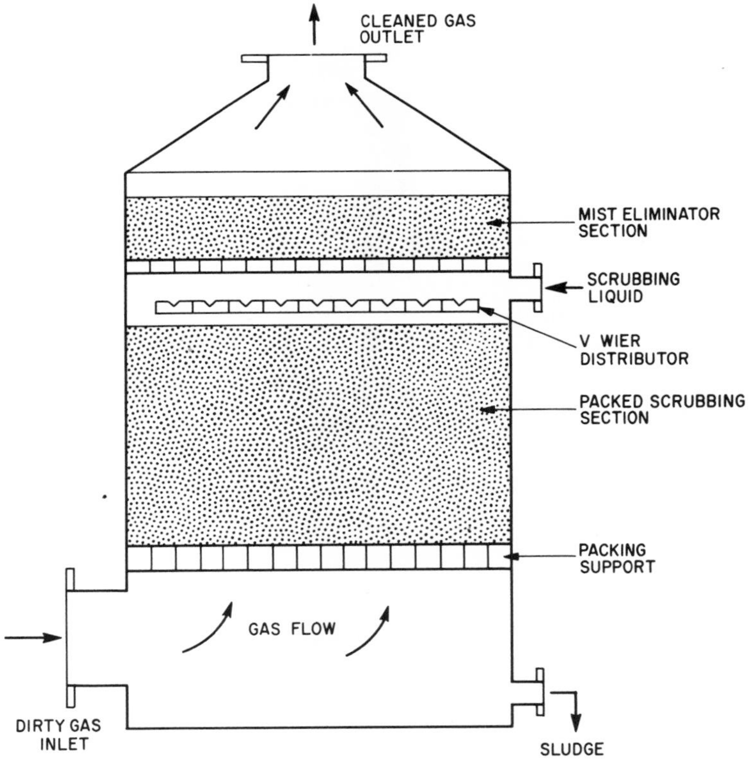

Fig. 8-22 Countercurrent flow scrubber. *(Courtesy U.S. Dept. of Health, Education, and Welfare)*

Fixed-Bed Scrubber. Figure 8-22 shows a typical single-bed scrubber. The packed scrubbing sections can be thin, thick, or multiple. Multiple beds, with redistribution of water between them, precludes gas channeling and the tendency to dry out a portion of the bed.

Water distribution can be accomplished by means of sprays, or a series of notched weirs.

Alternate or Special Construction. Because of the large liquid surface exposed to the gas, deep-bed packed towers are often used to remove gaseous fumes and entrained droplets.

A packed bed is sometimes used as a demister in the upper section of the tower.

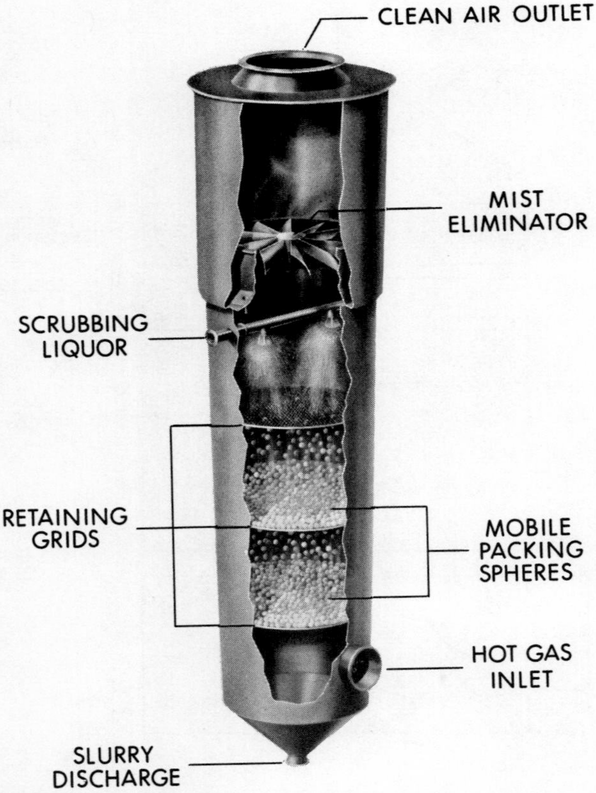

Fig. 8-23 Fluidized bed scrubber. *(Courtesy U.O.P. Air Correction Div.)*

Fluidized Bed Scrubber. Figure 8-23 shows a typical scrubber in which the gas flow rate is sufficiently high to keep the packed bed of spheres in constant motion. This motion tends to defeat fouling, plugging, and channeling. The escape of spheres is prevented by retaining grids.

As in the fixed bed, liquid flow is counter to gas flow. Initial wash water distribution is accomplished by multiple spray heads wetting the entire cross section of the bed. A vane axial mist eliminator spins the escaping droplets against the cylindrical wetted wall before the demisted gas is discharged through the cleaned gas discharge. Sludge is discharged out of the bottom.

Alternate or Special Construction. A tangential inlet can improve performance by using the lower portion of the scrubber as a wet cyclone. Wetting can be enhanced, erosion can be minimized, and gas temperature and volume can be reduced by installing inlet sprays.

Flooded-Bed Scrubber. Figure 8-24 shows a concurrent-flow tower packed with glass marbles operating as a flooded-bed scrubber. The gas flow is insufficient to raise the marbles from the bed, but sufficient to keep them moving in a gentle self-cleaning rubbing action.

The scrubbing liquid flow is concurrent and is introduced by sprays beneath the packed bed. The rate of gas flow through the bed is sufficiently high to support a turbulent layer of liquid above the bed. A centerline overflow limits the depth of the turbulent layer and drains the liquid into the lower sludge tank. A packed or vane axial demister demists the cleaned gas.

Alternate or special construction. Multiple beds can be used for higher collection efficiency. As with all multiple bed-packed scrubbers, redistribution of scrubbing liquid between beds is important.

Because of the large interface between gas and liquid, the flooded-bed scrubber may be used to control emissions of noxious gaseous fumes. Because of the cleaning action of the moving flooded bed, high grain loadings can be handled without plugging.

Submerged-Orifice Scrubber

With this scrubber the dirty gas, at high velocity, is directed through a restricted passage onto or into a pool of liquid. This causes extreme turbulence which insures a thorough mixing of gas and scrubbing liquid. The turbulence and mixing wets the dust particles and facilitates their capture.

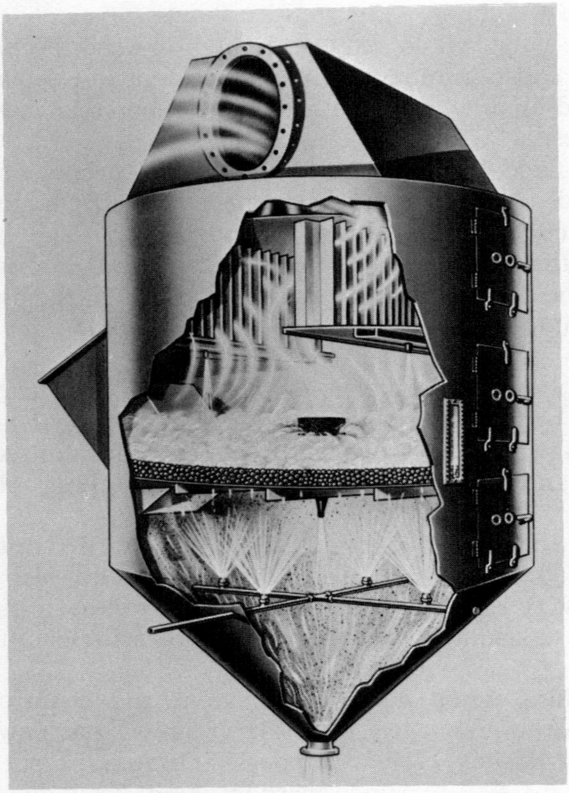

Fig. 8-24 Flooded bed scrubber. *(Courtesy Environeering, Inc.)*

Figure 8-25 shows a submerged-orifice scrubber having a long narrow slot partially submerged in the liquid pool. The dirty gas is introduced into a plenum chamber above the liquid. Its only escape route is in forcing its way through the partially submerged slot at high velocity, causing extreme turbulence and mixing.

The slot outlet is S-shaped thereby adding centrifugal forces to enhance turbulence and help separate the heavier wetted dust and liquid from the lighter cleaned gas. The gas is further demisted by angled baffles.

In the submerged-orifice scrubber, large quantities of scrubbing liquid are circulated—as much as 20 gpm per 1,000 cfm of gas—all without nozzles, piping, or pumps. All energy is supplied by the gas flow, which energy can be expressed as loss of static pressure when passing through the collector.

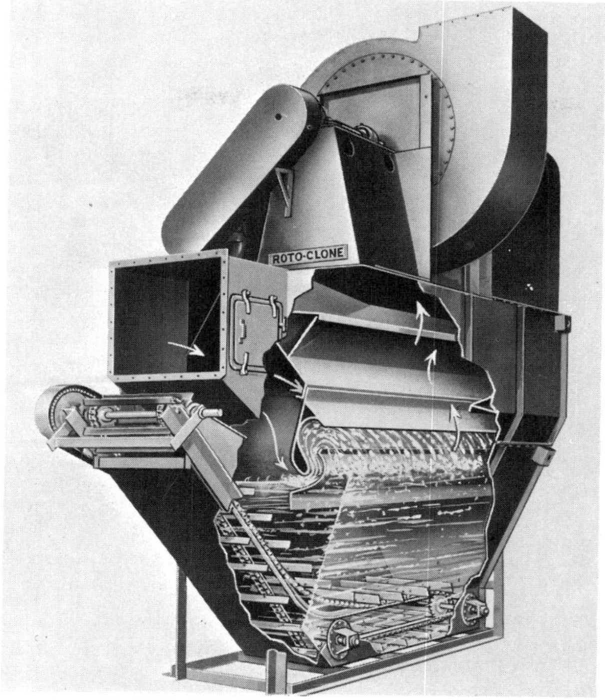

Fig. 8-25 Submerged-orifice scrubber. *(Courtesy American Air Filter Co., Inc.)*

Special or Alternate Construction. For effective collection the velocity of gas through the slot must be sufficiently high to cause violent turbulence. The slot can be long and narrow, circular and narrow, single slot or multiple slot. Figure 8-26 shows an orifice scrubber employing multiple, circular slots penetrating the liquid pool, followed by angled demisting baffles. The narrow slots limit the capacity of this collector to small and moderate sizes.

Venturi Scrubbers

In the venturi scrubber, the dirty gas is passed through a rectangular or round venturi at high to very high velocities. The cleaning liquid is introduced into the throat (vena contracta) of the venturi through coarse nozzles.

Gas velocities in the range of 10,000 to 40,000 ft/min (114–455 mph) through the venturi throat impact the liquid, causing very small droplets to form. This results in extreme turbulence and mixing.

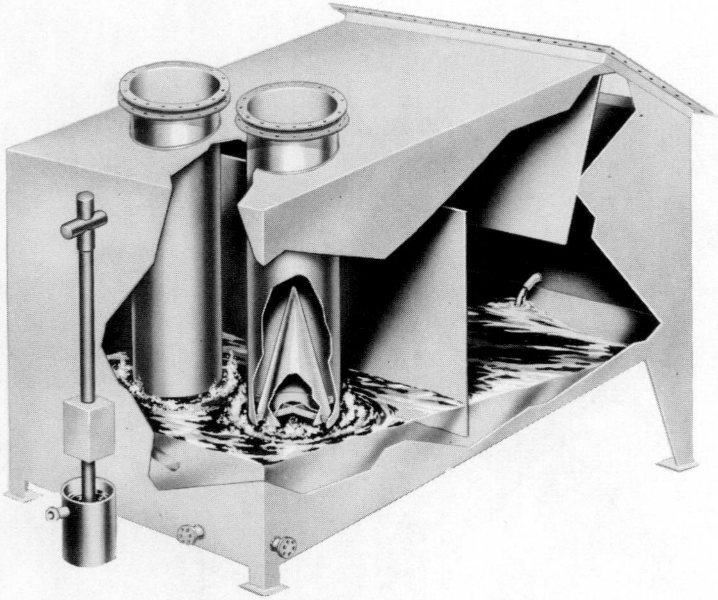

Fig. 8-26 Multiple submerged-orifice scrubber. *(Courtesy Western Precipitation Div. of Joy Mfg. Co.)*

Because of high gas velocities, the venturi is the most compact of all scrubbers.

Collection Efficiency. The collection efficiency of the venturi scrubber is directly related to the energy expended. At expenditures of energy of 3 to 10 in. W.G., collection efficiency is comparable to a wet cyclone; at pressure drops of 10 to 30 in. W.G., collection efficiency is comparable to that of a multiple impingement baffle and at pressure drops of 30 to 100 in. W.G., efficiency exceeds all other wet scrubbers and can match that of the bag filter and the electric precipitator.

Rectangular, Horizontal Gas Flow Venturi Scrubber. Figure 8-27 shows a rectangular, horizontal gas flow venturi scrubber into which the scrubbing liquid is introduced by coarse nozzles into two sides of a narrow, rectangular throat.

Rapid changes in gas velocity, gas pressure, and gas temperature all encourage condensation of moisture on the dust. The dust is accelerated as the cross section of the venturi reduces, and is flung into the maelstrom of the throat at or near gas velocity. This

Fig. 8-27 Rectangular, horizontal gas flow venturi scrubber. *(Courtesy Chemical Construction Corp.)*

high velocity enhances capture of the dust by impingement upon the water droplets. As droplets accelerate from zero to gas velocity in the expanding diffusing section of the venturi, the noncaptured dust overtakes the droplets at decreasing relative velocity. Capture of fine dust by fine droplets through the mechanism of diffusion is increased as relative velocities of dust and droplet approach zero.

Demisting of the horizontal venturi scrubber is usually effected by a tangential inlet, centrifugal demister.

Alternate or Special Construction. To keep efficiency high at lowered gas velocity, venturis having a variable throat are offered. Control of throat width can be manual or automatic.

Vertical Downward Gas Flow Venturi Scrubber. Figure 8-28 shows a venturi scrubber in which scrubbing liquid is introduced into the tapered inlet section of the venturi by overflowing a circular weir. Liquid is fed tangentially into the weir through a pipe, thereby eliminating nozzle plugging.

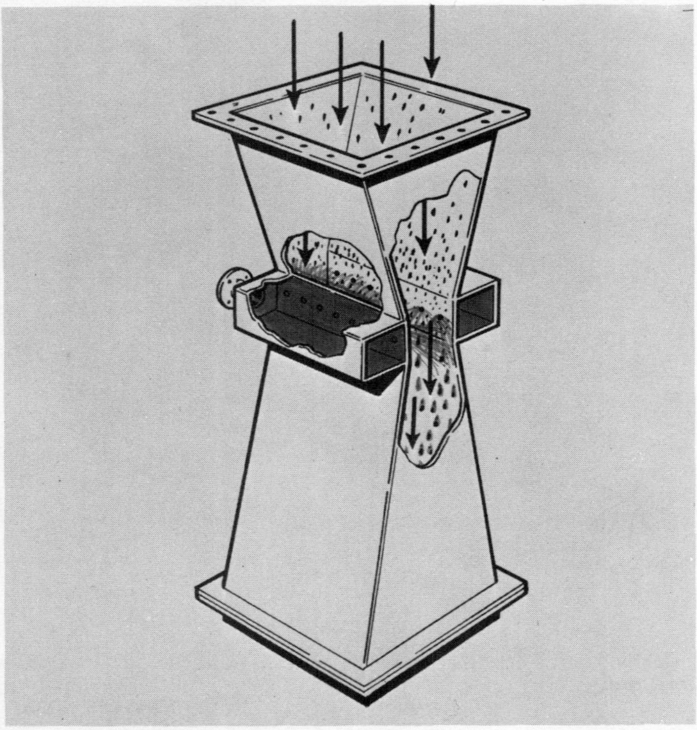

Fig. 8-28 Vertical downward gas flow venturi scrubber. *(Courtesy U.O.P. Air Correction Div.)*

This design can use water containing large quantities of solids. It also protects and cools the inlet section of the venturi.

Alternate or Special Construction. Figure 8-29 shows a similar venturi which introduces scrubbing liquid by means of two straight weirs. Figure 8-30 shows a vertical upward gas flow venturi scrubber. In this design scrubbing liquid is introduced into the throat through pipes under the gravity head of liquid from the feed tank, which in turn is fed by gravity from the mist eliminator, thereby eliminating circulating pumps. Gravity increases turbulence within the diffuser section. The design of the feed tank accepts makeup liquid and discharges slurry.

Flooded-Disc Venturi Scrubber. Figure 8-31 shows a downward flow venturi scrubber whose throat is blocked by an adjustable

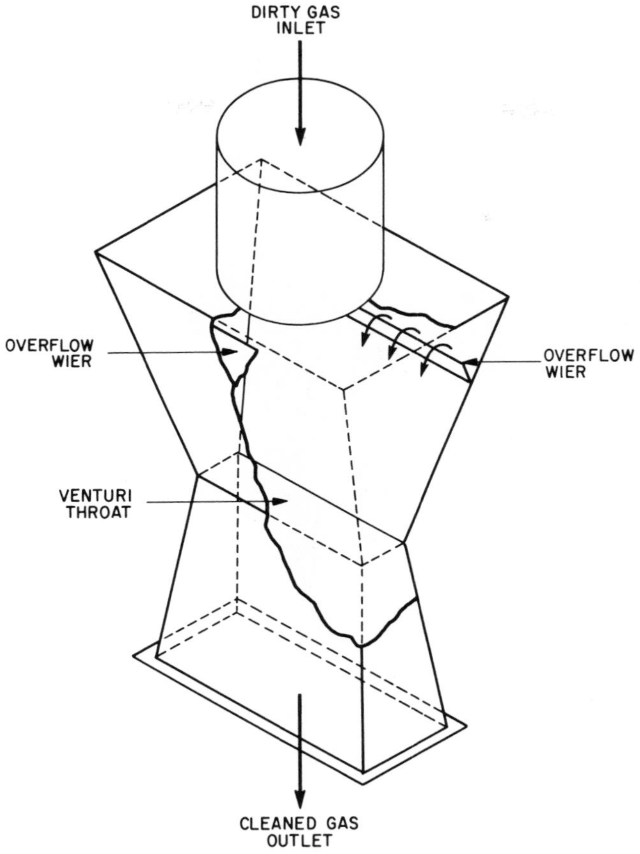

Fig. 8-29 Vertical downward venturi with weirs. *(Courtesy U.O.P. Air Correction Div.)*

disc. Raising the disc increases gas capacity by widening the circular slot between disc and throat. At low loads, gas velocity and collection efficiency can be kept high by lowering the disc.

Scrubbing liquid is introduced via the disc-positioning pipe, pipe cap, and flooded disc.

Alternate or special construction. Disc position can be manually or automatically controlled.

Scrubber Demisters

Because collection efficiency of the fine dust is improved as the number of droplets is increased and as the size of droplets

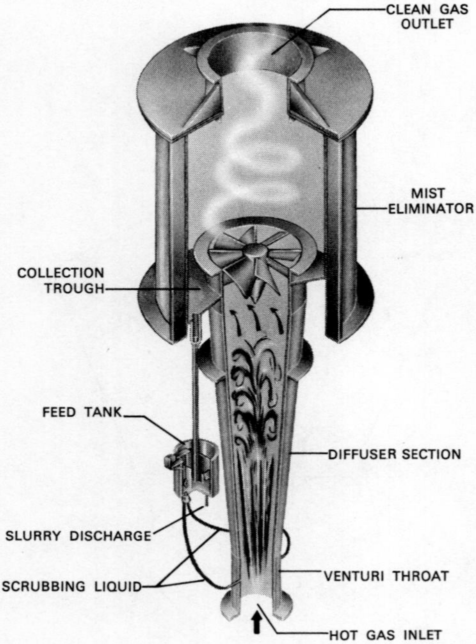

Fig. 8-30 Vertical upward gas flow venturi scrubber. *(Courtesy U.O.P. Air Correction Div.)*

is decreased, entrained moisture in the form of billions of fine droplets would be carried out of the scrubber unless captured before discharge. The loss of particulate moisture increases makeup water substantially, and may result in droplet fallout.

Fortunately water droplets are easily captured by impingement upon a wetted surface. Many scrubbers have a built-in demister of one of the following types as previously illustrated.

Tangential inlet centrifugal, Fig. 8-27
Vane axial centrifugal, Fig. 8-30
Zigzag baffle, Fig. 8-24
Dry-packed bed, Fig. 8-22
Reverse flow inertial, Fig. 8-17

Miscellaneous and Combination Scrubbers

Cross-Flow Packed Scrubber. Figure 8-32 shows such a scrubber. This low-efficiency scrubber has the advantages of low pressure drop and low recirculation water.

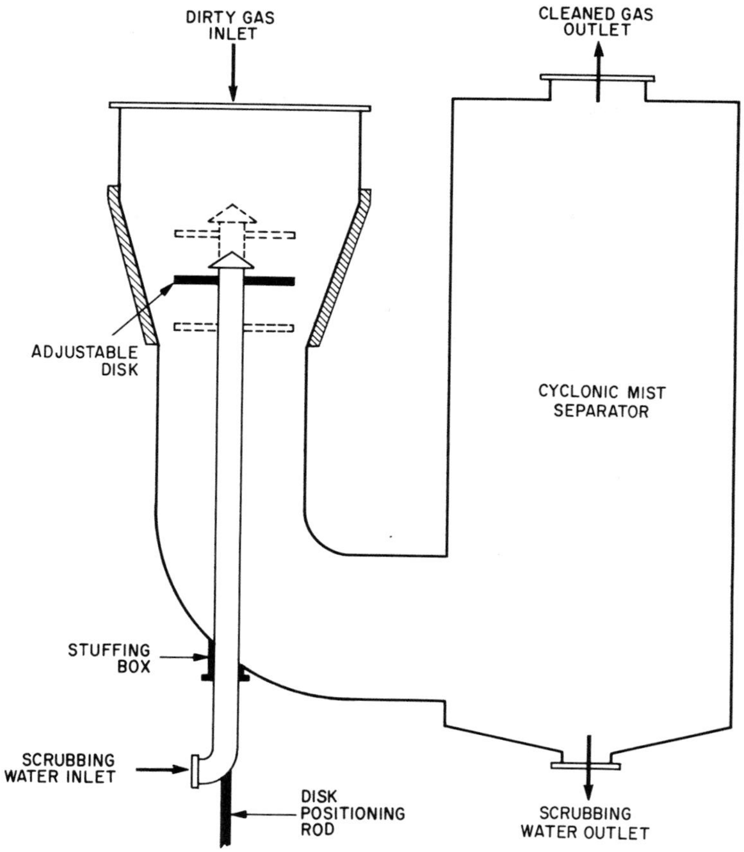

Fig. 8-31 Flooded disc venturi. *(Courtesy Research-Cottrell)*

Centrifugal Fan-Type Wet Scrubber with Water and Sludge Outlet

Figure 8-33 shows such a wet scrubber. Advantages include compactness, low scrubbing liquid consumption, elimination of external fan, and moderate collection efficiency. Disadvantages include accelerated erosion and reduced fan efficiency. Other designs use flat blades and depend on a downstream demister for final collection.

Multiple Venturi. In this design, many small venturis are grouped in parallel, each handling a fraction of the total gas quantity. Figure 8-34 shows that gas flow is vertically downward, and scrubbing liquid is introduced by means of a fine, cone-shaped spray directed into the inlet of each venturi tube.

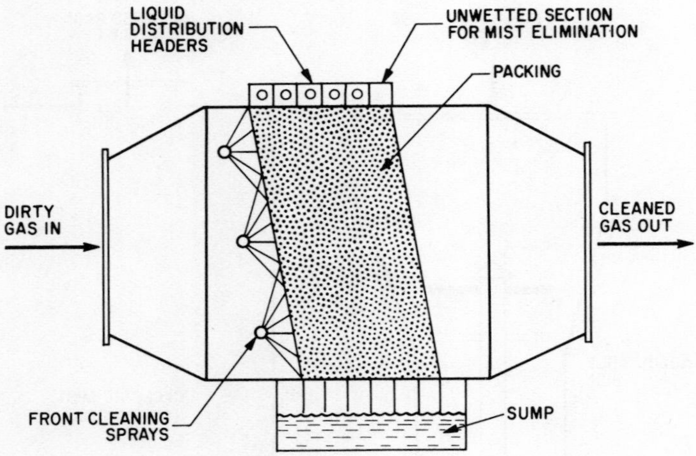

Fig. 8-32 Cross-flow packed scrubber. *(Courtesy U.S. Dept. of Health, Education, and Welfare)*

Fig. 8-33 Centrifugal fan-type wet scrubber. *(Courtesy American Air Filter Co., Inc.)*

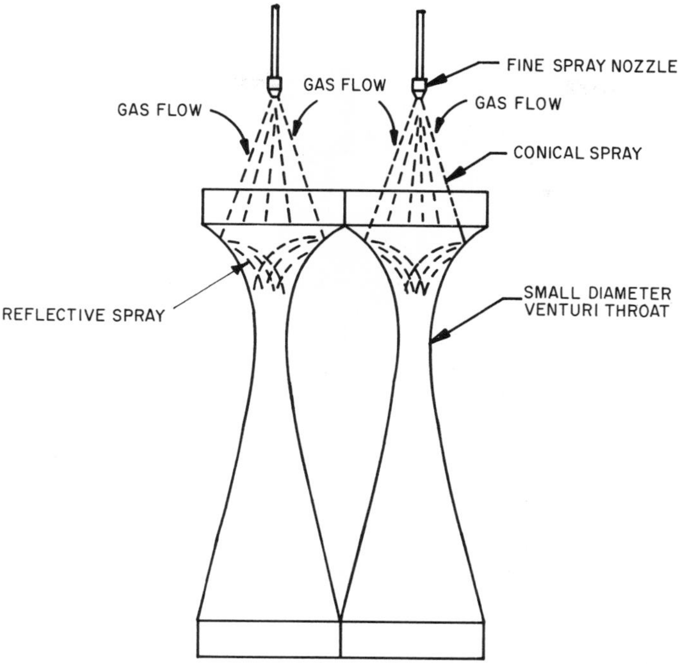

Fig. 8-34 Multiple venturi.

Advantages include having the dirty gas pass through the fine conical spray before entering the small venturi throat. Disadvantages include requiring spray water of low solids content, plus the difficulty of replacing internally mounted nozzles.

Combination Venturi Scrubber. Figure 8-35 shows a combination wet scrubber consisting of the following stages of scrubbing:

Round, overflow-weir venturi scrubber
Reverse flow inertial demister
Fluidized, packed-bed scrubber
Vane axial demister

Combination Fan Scrubber. Figure 8-36 shows a combination wet scrubber consisting of the following stages of scrubbing:

Tangential inlet wet cyclone
Vane axial demister
Centrifugal fan-type wet scrubber
Tangential inlet wet cyclone
Vane axial demister

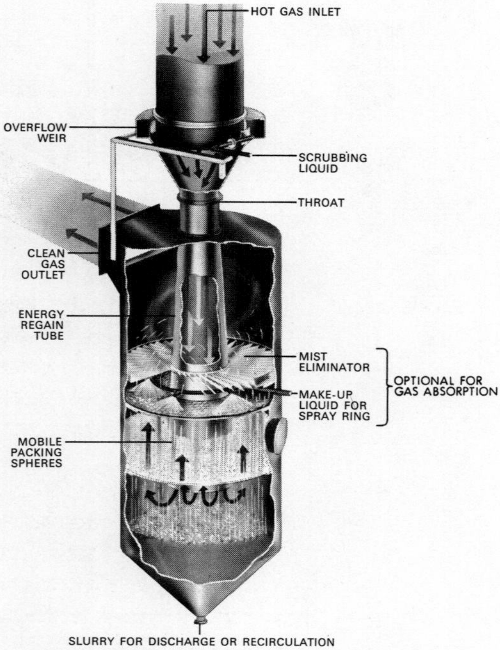

Fig. 8-35 Combination wet scrubber—I. *(Courtesy U.O.P. Air Correction Div.)*

Wastewater Disposal Systems

One disadvantage of the wet scrubber is the inherent problem of sludge disposal. A wastewater treatment problem has been substituted for an air pollution problem.

The following common types of waste disposal should be evaluated:

Settling tank	Continuous filtration
Settling pond	Liquid cyclones
Clarator	Continuous centrifugal
Clarifier	Chemical treatment
Thickener	

The relative merits of the above systems are beyond the scope of this chapter, nevertheless the prospective purchaser of a wet scrubber must assure himself of the acceptability of the wastewater disposal system as well as the collection efficiency and installed and operating cost of the wet scrubber.

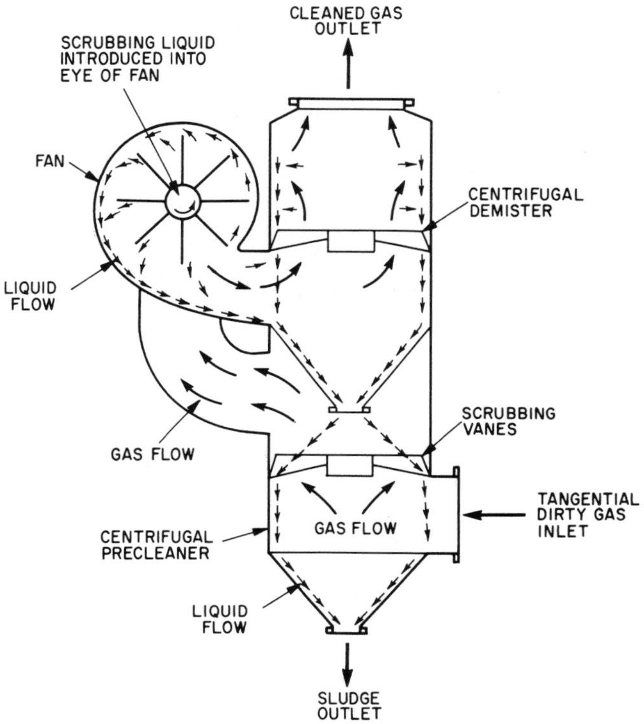

Fig. 8-36 Combination wet scrubber—II. *(Courtesy The Ducon Co.)*

Chemical Corrosion

Most wet scrubbers are available in a wide variety of construction materials including mild steel, stainless steel, other corrosion-resistant metals, as well as in rubber-lined and plastic constructions. If gas or liquid contains or accumulates corrosive elements, the final water treatment system must resist or correct the chemically corrosive condition.

ELECTRIC PRECIPITATORS

Operating Principles

The industrial electric precipitator was invented in 1910 by Frederic Gardner Cottrell while he was an instructor at the University of California. Electric precipitators have been applied to a great variety of gas-cleaning problems at collection efficiencies

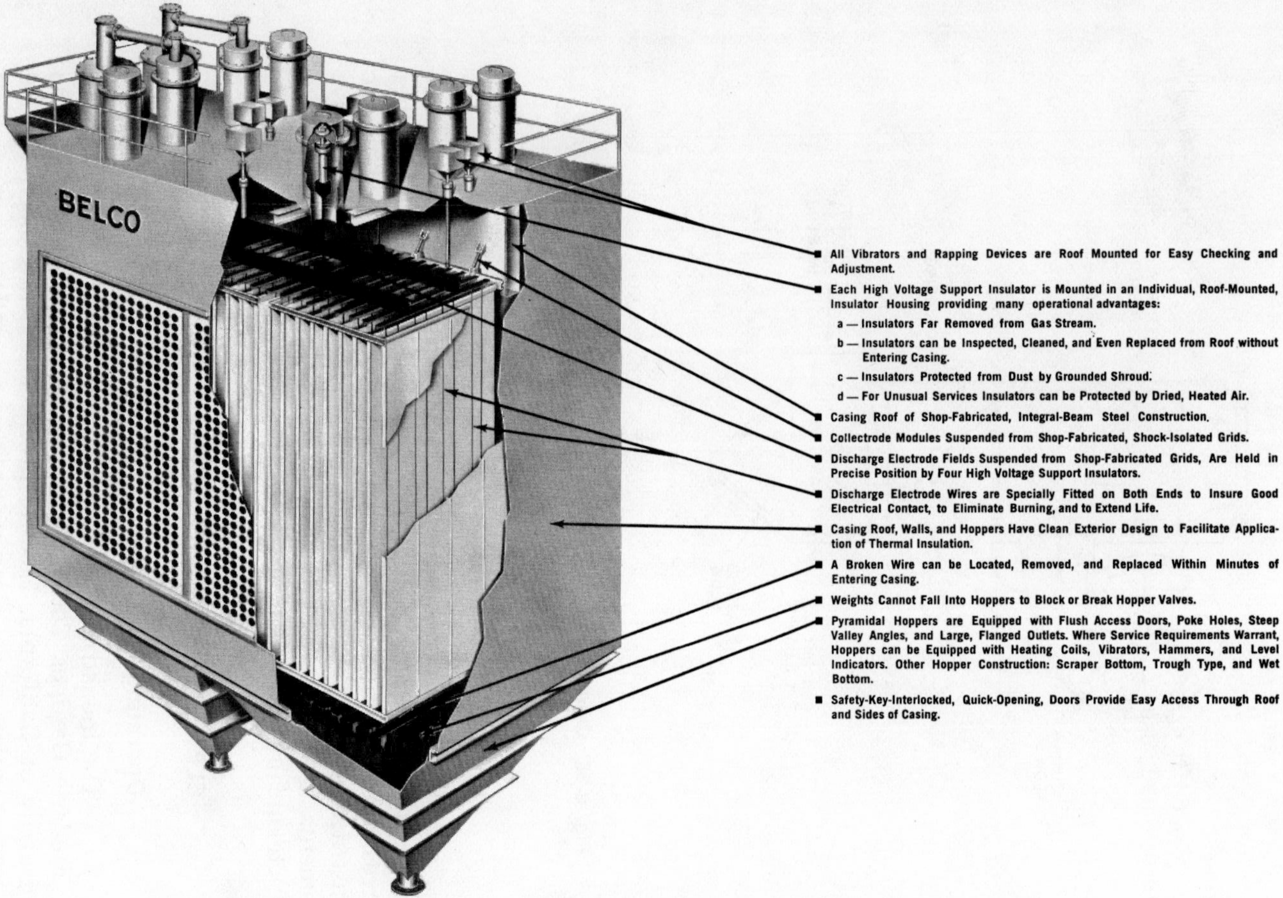
BELCO
All Vibrators and Rapping Devices are Roof Mounted for Easy Checking and Adjustment.
Each High Voltage Support Insulator is Mounted in an Individual, Roof-Mounted, Insulator Housing providing many operational advantages:
a — Insulators Far Removed from Gas Stream.
b — Insulators can be Inspected, Cleaned, and Even Replaced from Roof without Entering Casing.
c — Insulators Protected from Dust by Grounded Shroud.
d — For Unusual Services Insulators can be Protected by Dried, Heated Air.
Casing Roof of Shop-Fabricated, Integral-Beam Steel Construction.
Collectrode Modules Suspended from Shop-Fabricated, Shock-Isolated Grids.
Discharge Electrode Fields Suspended from Shop-Fabricated Grids, Are Held in Precise Position by Four High Voltage Support Insulators.
Discharge Electrode Wires are Specially Fitted on Both Ends to Insure Good Electrical Contact, to Eliminate Burning, and to Extend Life.
Casing Roof, Walls, and Hoppers Have Clean Exterior Design to Facilitate Application of Thermal Insulation.
A Broken Wire can be Located, Removed, and Replaced Within Minutes of Entering Casing.
Weights Cannot Fall Into Hoppers to Block or Break Hopper Valves.
Pyramidal Hoppers are Equipped with Flush Access Doors, Poke Holes, Steep Valley Angles, and Large, Flanged Outlets. Where Service Requirements Warrant, Hoppers can be Equipped with Heating Coils, Vibrators, Hammers, and Level Indicators. Other Hopper Construction: Scraper Bottom, Trough Type, and Wet Bottom.
Safety-Key-Interlocked, Quick-Opening, Doors Provide Easy Access Through Roof and Sides of Casing.

Fig. 8-37 Modern dry-plate-type electric precipitator. *(Courtesy Belco Pollution Control Corp.)*

as high as 99.9 percent, at capacities to 4,000,000 cfm, and at gas temperatures to 1,000°F. In 1970 there were over 5,000 installations in the United States cleaning 600 million cu ft of gas per minute originating in thermal power stations, incinerators, and industry—an investment in clean air of over $1 billion.

An electric precipitator separates entrained particulate matter from a gas stream by first charging the dust to a negative voltage of about 50,000 V, precipitating it onto grounded collecting electrodes, then dropping the agglomerated dust into a hopper. Despite the high voltages used. energy consumption is low and its draft loss is 0.5 in. W.G., the lowest of all high-efficiency collectors.

In full-sized, commercial units the dirty gas is passed horizontally through narrow, vertical gas passages formed by parallel rows of grounded collecting electrodes. Electrically insulated, high-voltage wires are spaced precisely on the center lines of each gas passage, thereby causing the dirty gas to pass between the high-voltage wires and the grounded plates.

A three-dimensional drawing of a modern precipitator is shown in Fig. 8-37. Typical gas-passage dimensions on a large commercial precipitator are: 9-in. wide, 30-ft high, and 27-ft long. The electrical distance—that is the distance between the high-voltage discharge electrode wires and the grounded collecting plates is one-half the gas-passage width, or 4½ in.

A typical cross section of a four-gas-passage collectrode module, one of many in a modern electric precipitator, is shown in Fig. 8-38. The 6-ft collectrode panels are shop assembled from 18-in.,

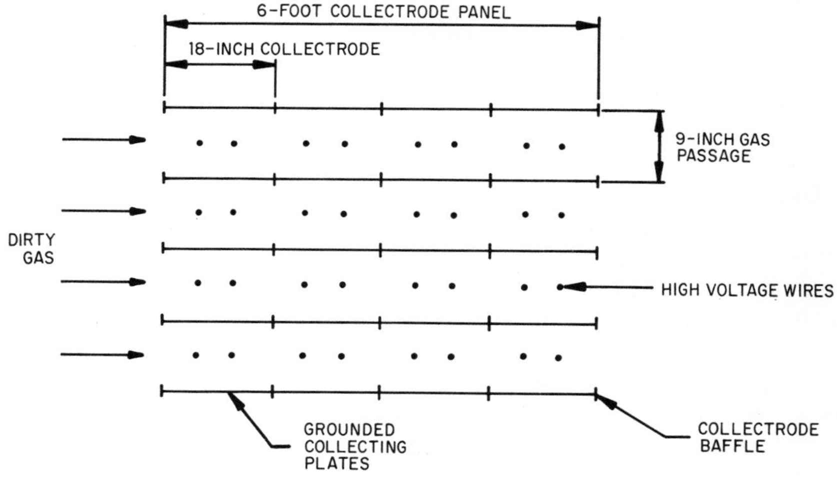

Fig. 8-38 Cross section of four-gas-passage collectrode module.

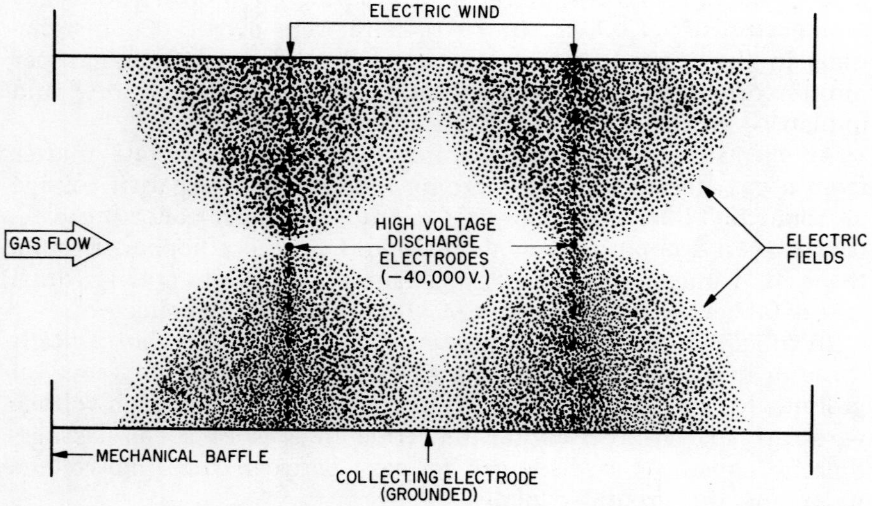

Fig. 8-39 High-voltage electric field.

roll-formed steel sheet. They are hung on 9-in. centers to form the gas passages. High-voltage discharge electrode wires are hung precisely on the center line of each gas passage.

Figure 8-39 shows the pattern of current flow between wires and plates. This flow, also called "electric wind," is the result of a massive migration of negatively charged gas ions away from the negatively charged wire toward the relatively positively charged grounded collecting plate.

Figure 8-40 shows the six essential steps of the Cottrell Process. Steps 1, 2, 3, and 4 are continuous. Steps 5 and 6 are at programmed intervals ranging from a minute to several hours depending on many operational factors.

1. Gas Ionization. Electric precipitators operate at 40,000 to 50,000 V below ground potential. This potential is sufficient to cause billions of electrons to boil off the wires and to bombard the gas molecules in the region close to the wires. The force of this bombardment forms both positive and negative gas ions whose presence is evidenced by a visible blue corona.

The ions respond to the powerful electrostatic force by moving toward the oppositely charged electrodes. The positive gas ions return to the negative wire and regain their lost electrons; negative gas ions move toward the grounded (relatively positive) collectrode,

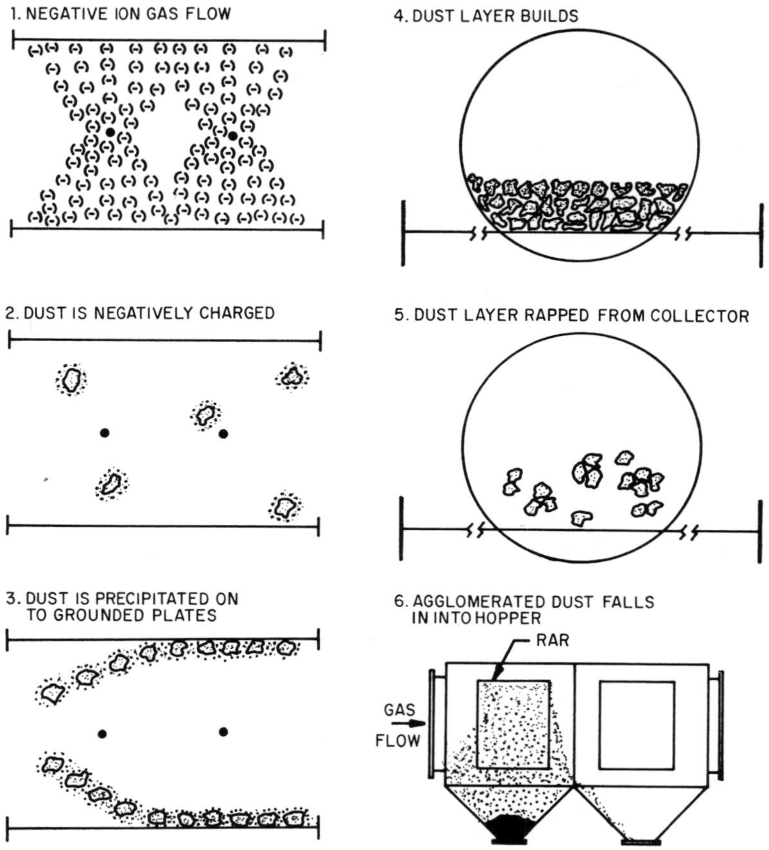

Fig. 8-40 Six-step Cottrell process.

thereby flooding the gas passage with billions of negative gas ions. Most current flow within an electric precipitator is due to the flow of negative gas ions from wires to plates.

2. Dust Charging. As the dust particles are carried through the gas passages by the entraining gas, they collide with the negative gas ions and are thus charged negatively. Because the gas ions are thousands of times smaller than even submicron dust and because of their great number, there are ample charging forces for almost any collecting problem.

3. Dust Precipitation. The negatively charged dust particles are moved rapidly toward the grounded (positive) plates to which they attach themselves, being held by powerful electric forces.

4. Dust Layer Builds. The dust particles build a thickening layer on the collectrodes, and the negative charge gradually bleeds from the dust into the grounded collectrode. As the layer thickens, the charges on the recently precipitated dust must be conducted through the layer of that previously precipitated. The resistance of the dust layer to this current flow is termed "dust resistivity." Precipitators are successfully operating on dusts whose resistivity is in the range of 10^7–10^{11} Ω/cm.

5. Collectrode Rapping. After a $\frac{1}{16}$–$\frac{1}{4}$-in. layer of dust has been precipitated, the dust particles next to the collectrode have lost much of their charge to the grounded plate. The electrical attraction has been weakened. The recently arrived dust particles still hold much of their charge, being electrically insulated from the plate by the dust layer, thereby holding the entire layer against the collectrode. A sharp rap causes the dust layer to shear away from the collectrode. The force of rapping is restrained so as to allow the layer to be shattered into relatively large agglomerations of dust particles, but not back into the original submicron particles.

6. Dust Fall into Hopper. The relatively large agglomerates fall into the hoppers under the influence of gravity just as in the gravity settling chamber. Most electric precipitators are operated at gas velocities in the range of 3 to 6 ft/sec to allow ample settling time. The high voltage remains on to recharge and reprecipitate any fine particles which may become re-entrained during the rapping cycle.

Advantages and Disadvantages of the electric Precipitator

The major advantage of the electric precipitator is high collection efficiency with a minimum of operating cost. It requires the least energy of all high-efficiency collectors, and its maintenance cost is low because it is of steel construction, operated dry, above the dew point. Additional advantages include high reliability at any required collection efficiency from maximum to zero gas flow.

Unique disadvantages include high installed cost, particularly in sizes below 50,000 cfm, unpredictable collection efficiency on certain high-resistivity dusts, and loss of collection efficiency at gas flows above designed rating.

Dry-Plate-Type

The most common industrial collector is known as the high-voltage, single-stage electric precipitator because charging and precipitation of dust are accomplished within the same section.

Construction. Referring again to Fig. 8-37, the major components consist of:

Gas-tight casing including hoppers
High-voltage discharge electrode system
Grounded collectrode system
High-voltage supply

Sizable precipitators are divided into many bus sections as shown in Fig. 8-41.

Precipitator casing. The precipitator casing is of weatherproof, gas-tight construction, suitable for outdoor or indoor installation. It is shop fabricated in modular weldments of the largest practical size, giving due consideration to transportation and erection.

Major casing parts include shell, hoppers, inlet and outlet nozzle connections, inspection doors, and insulator housings. The standard shell (wall sections, end sections, and roof sections) is fabri-

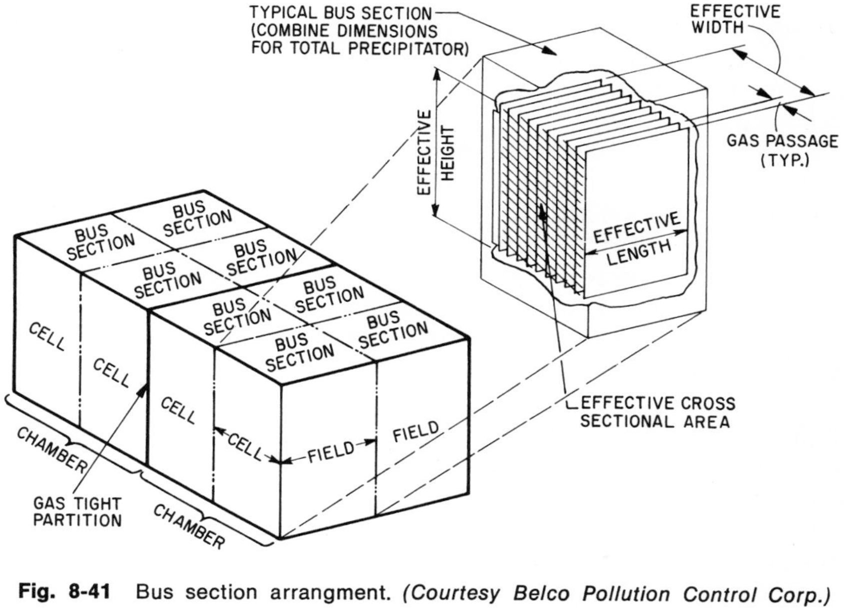

Fig. 8-41 Bus section arrangment. *(Courtesy Belco Pollution Control Corp.)*

cated of welded, mild steel plate suitably reinforced by columns, beams, and stiffeners to control the maximum positive or negative design pressure of the gas being handled and to withstand the maximum design temperature.

The shell roof sections are designed to support the weights of the internal and roof-mounted parts, and thermal insulation, and to meet specified environmental requirements such as earthquake, wind load, snow load, and live load, etc.

Gas-tight integrity is preserved by gasketed inspection doors and sealed insulator housings.

The casing and insulator housings of the precipitator form a continuous, grounded steel chamber which completely encloses all high-voltage elements, thereby insuring safety of personnel and complete electrical shielding. Since the casing is gas tight and completely enclosed, access to the high-voltage elements can be accomplished only through the access openings. All access openings are equipped with warning signs and safety grounding hooks. Quick-opening access doors are also equipped with safety-key interlocks tied in with the overall safety-key interlock system of the precipitator and its related high-voltage equipment.

The grounded bus duct provides electrical shielding between the transformer-rectifier unit and the high-voltage feed-through insulator housing.

Grounded collectrode system. Standard planar collectrodes are precision roll-formed from cold-rolled steel sheet to insure their being flat and true, and are then locked into precise collectrode panels on a full-sized jig.

For special operating conditions, collectrode panels are available in low-alloy, alloy, and other corrosion-resistant metals. The panels are furnished in lengths ranging between 3 and 9 ft, and in heights ranging from 10 and 36 ft.

Collectrode panels are grouped within the precipitator casing to form independently suspended and independently rapped collectrode modules. The modules are rapped periodically by electromechanical means.

High-Voltage Discharge Electrode System. Each precipitator contains one or more high-voltage discharge electrode fields which are individually energized. These fields are electrically isolated from the remainder of the precipitator, arranged in modules to match the collectrode modules, and positioned so that the discharge wires hang precisely along the center lines of the gas passages, and consist of the following major parts:

H-V bus duct
1 Feed-through support insulator
1 Vibrator insulator
3 Tension support insulators
Upper support grid
Discharge electrode wires
Lower alignment grid
Weight tensors

Discharge electrode wires are usually of round, 12-gauge steel spring wire and are reinforced top and bottom to insure good electrical contact and to resist mechanical and electrical erosion.

An independent rapping system keeps the H-V discharge electrode system continuously clean.

High-voltage supply. High-voltage silicon diode power packs have been specifically developed for the severe load conditions encountered in supplying H-V direct current to electric precipitators.

Because maximum precipitator collection efficiency is predominantly dependent upon maximum voltage, power packs should have substantial overcapacity both as to voltage and current. In practice, maximum precipator voltage is the highest level that can be maintained without having a continuing electric arc (arc-over) form between the wire and the collectrode at one or more points within the precipitator.

Because arc-over voltages change from second to second as operating conditions vary, modern power packs are equipped with a system of automatic voltage regulation which continuously regulates the high voltage just short of arc-over.

The modern power pack consists of three self-contained components:

Remote control cabinet
Saturable reactor
Transformer-rectifier unit

Installation consists of setting in place, bolting down, and the running of external conductors.

The H-V circuit, as shown in Fig. 8-42 consists of the following major elements:

1. Power supply (ac, 1Ph, 440V, 60 Hz)
2. Saturable reactor to regulate power (ac)
3. H-V transformer to increase voltage (ac)
4. Rectifier to change ac to dc
5. Precipitator electrode systems complete circuit (dc)

The remote control cabinet contains all required control devices

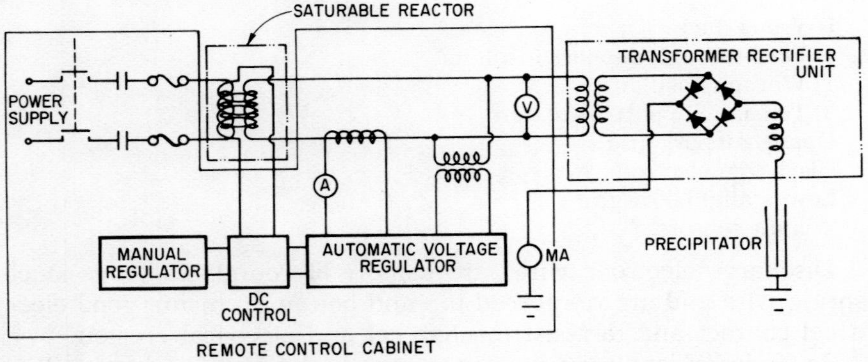

Fig. 8-42 Simplified schematic diagram. *(Courtesy Belco Pollution Control Corp.)*

(except for the separately mounted saturable reactor) and the instruments needed to monitor and control the operation of an electric precipitator. It is usually located at a convenient control point and may be located far from the precipitator.

Modern silicon diode transformer-rectifier units are specially designed for the severe load conditions and corrosive atmospheres encountered in supplying H-V dc power to electric precipitators. The hermetically sealed transformer-rectifier package protects and cools the immersed internals either with oil or a noninflammable insulating liquid such as askarel.

There are no moving parts to wear out. Units require no lubrication or expensive replacement tubes, and the sealed package requires virtually no maintenance, except the routine checking of the coolant's dielectric strength.

The major components of a transformer-rectifier unit include the tank, the H-V transformer, silicon rectifier stacks arranged in full-wave or double half-wave bridge circuit, the H-V switchgear, and the H-V output bushing(s).

Powering arrangements. As shown in Fig. 8-43 there are many possible powering arrangements of the fields and cells of an electric precipitator.

Automatic Voltage Regulation. The major objective of automatic voltage regulation is to hold the highest voltage short of flashover on each precipitator bus section even under widely varying operating conditions in order to assure continuous maximum collection efficiency. Years of operating experience have demonstrated that plant personnel are unable to optimize the voltage within a precipitator by manual means.

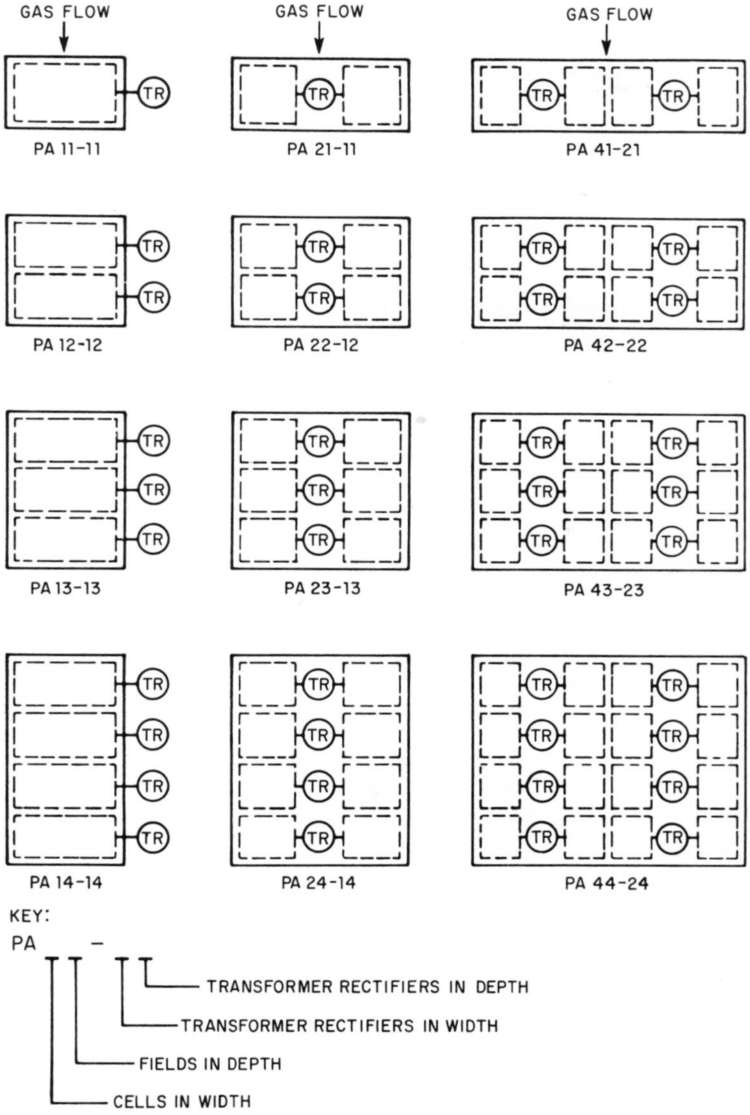

Fig. 8-43 Powering arrangements. *(Courtesy Belco Pollution Control Corp.)*

Location of Power Packs. Modern power packs are designed to meet the installation preferences of experienced operators of electric precipitators. While the flexibility of the three separate units affords many arrangements, the following locations are usually followed in modern practice.

The remote control cabinet is incorporated in the main control center or in a secondary control center along with the controls for the dust disposal system. The saturable reactor is located near, but not next to, the remote control cabinet in an area of good ventilation to dissipate heat. The transformer-rectifier unit is mounted on the roof of the precipitator in a specific location selected to simplify and shorten the runs of the H-V bus duct, and to facilitate structural support, easy access, and physical replacement.

Essential Design Criteria of a Successful Electric Precipitator Installation

The following criteria are essential for the reasons summarized:

Positive Gas-Flow Control. Electric precipitators operate at gas velocities of 3 to 8 ft/sec. Conveying velocities within flue systems are in the order of 50 ft/sec or about 10 times those suitable for precipitation.

A typical electric precipitator, sized to clean 1,000,000 cfm will have 140 parallel gas passages 9 in. wide by 30 ft high—each handling 7,150 cfm.

The twofold problem is to reduce gas velocity from 50 to 5 ft/sec, and to divide the 1,000,000 cfm equally into 140 gas passages each capable of handling 7150 cfm. Failure to control gas flow results in harmful eddies and points of high gas velocity. These materially reduce precipitator collection efficiency.

This velocity reduction and division of gas flow is primarily accomplished by the careful design of the connecting flue systems and by perforated plates mounted in the precipitator inlet. Controlling devices within the flue systems include splitters, turning vanes, distribution baffles, perforated plates, and settling hoppers. An important tool of the flow control engineer where complex flue systems must be designed is the use of scale-model flow studies.

Uniform Electrode Emission. A large electric precipitator may contain up to 250,000 sq ft of collecting electrodes. The discharge electrode wires, 32 miles of them at an average negative polarity of 45,000 V (dc), are positioned precisely on the center lines of each passage at an electrical distance of only 4½ in. from the grounded collectrodes.

Because even a single reduction in electrical distance limits voltage for an entire bus section, and thereby reduces efficiency, good precipitator design should provide for definite structural relationships between the two electrode systems.

Effective Electrode Cleaning Without Puffing. The high-efficiency requirements of the new air pollution codes do not allow for reduced efficiencies due to ineffective electrode cleaning, nor do the codes permit visible puffing caused by excessive reentrainment losses.

Modern precipitators are equipped with two interdependent electromechanical rapping systems: one for keeping the H-V discharge electrode fields continuously clean, a second for the sequential rapping of the collectrode modules. Both are designed for automatic electrode cleaning while the high voltage remains full on.

Modern rapping eliminates puffing and minimizes reentrainment by using:

Baffled collecting electrodes
Controlled, low, gas velocities
Sequential rapping of small, shock-isolated, collectrode modules
Vertical, shear-action, parallel rapping
Rapping intensity matched to dust adhesion
Rapping interval matched to dust loading

Even in those cases where the operating conditions are relatively stable, it is important that the rapping program take into account the great difference between the operating conditions in the inlet of the precipitator and those near the outlet. By dividing the total collecting surface of the precipitator into small, shock-isolated collectrode modules, it is practical to approach the ideal wherein each square foot of collectrode is rapped no more often and no more powerfully than necessary to clean that particular square foot of collectrode.

Safe, Reliable, Automatic H-V Supply. Because precipitator collection efficiency is preponderantly dependent upon the high voltage carried on the discharge electrode wires, it is essential to hold these voltages as close as possible to flashover amounts. Since optimal voltage cannot be maintained by manual control, modern H-V units are equipped with automatic voltage regulation which includes sensing the H-V circuitry.

Solid state, automatic voltage regulation holds each bus section of the precipitator at maximum voltage short of flashover, despite varying operating conditions, including rapping cycles, and thereby continuously provides maximum efficiency without attention from plant personel.

Multifield Electric Precipitators for High Efficiency. The number of fields (electrical stages in the direction of gas flow) to achieve any required collection efficiency is a compromise between the high cost of many short fields and the lower cost of a single long one. Table 8-6 represents typical design practice.

TABLE 8-6 MULTI-FIELD PRECIPITATORS

% Collection Efficiency	NUMBER OF FIELDS *U.S.*	*Europe*
95	1 or 2	1
96–98	2 or 3	2
99	3 or 4	2 or 3
99+	4 or 5	3

Classifying Effect. Table 8-7 illustrates how the collection efficiency of an identical field is reduced due to the "classifying effect" as its position within the casing is located further downstream. The "classifying effect" is due to the first field collecting more of the easily collected dust and less of the hard-to-collect. Thus, as the dust is borne through a multistage collector, it gets more and more difficult to collect.

Custom Application with Economy. The modern electric precipitator, field assembled of shop-fabricated modules, provides mass-produced economy with custom versatility. One such design is offered in the following standard ranges of effective dimensions:

	Minimum	*Maximum*
Effective length	6 ft	45 ft
Effective height	10 ft	36 ft
Effective width	7½ ft	150 ft
Number of fields	1	15
Number of cells	1	8
Gas passages per cell	10	50

Selection with such modular flexibility enables the precipitator engineer to fit into tight layouts without compromising key design criteria such as:

TABLE 8-7 PERCENT COLLECTION EFFICIENCY OF MULTISTAGE ELECTRIC PRECIPITATOR

Total Number of Fields	FIELD ONE		FIELD TWO		FIELD THREE		FIELD FOUR		*Total % Collection Efficiency*
	%	% *Accu-mulated Loss*	%	% *Accu-mulated Loss*	%	% *Accu-mulated Loss*	%	% *Accu-mulated Loss*	
1	80	20							80
2	80	20	75	5					95
3	80	20	75	5	70	1.5			98.5
4	80	20	75	5	70	1.5	60	0.6	99.4

Low gas velocity
Length to height ratio
Inlet aspect ratio
Optimal H-V energization
Systematized cyclic rapping
Simplified field erection
Provision for future operating conditions and to meet tightening codes

Easily Maintained, Rugged Construction. It is poor economy to shut down a plant or major producing unit to repair a poorly designed or undersized electric precipitator. The modern electric precipitator has many features designed to lenthen runs between shut-downs and to make for easy inspection and maintenance. The following are some major features of modern design that contribute directly to operational reliability:

Vibrators and rapping devices roof-mounted for easy checking and adjustment
High-voltage support insulators mounted in individual, roof-mounted, insulator housings provide operational advantages:

A. Insulators far removed from gas stream
B. Insulators can be inspected, cleaned, or replaced from roof without entering casing
C. Insulators protected from dust by grounded shroud
D. For unusual services, insulators can be protected by continuously washing with dried heated air

Casing roof of shop-fabricated, integral-beam steel construction
Collectrode modules suspended from shop-fabricated, shock-isolated grids
Discharge electrode fields suspended from shop-fabricated grids, held in precise position by not less than four H-V support insulators
Discharge electrode wires specially fitted on both ends to insure good electrical contact, to eliminate burning and to extend life
Casing roof, walls, and hoppers of clean exterior design to facilitate application and continuity of thermal insulation
Automatic location of broken wire by tilted weight visible from hopper and prevented from falling by the lower alignment grid into hoppers to block or break hopper valves
Pyramidal hoppers equipped with flush access doors, poke holes, steep valley angles, and large, flanged outlets. Hoppers can be equipped with heating coils, vibrators, hammers, and dust level indicators. Other hopper construction: scraper bottom, trough type, and wet bottom
Safety-key-interlocked, quick-opening, doors provide easy access with safety through roof and sides of casing into suitably sized access

passages, above, below, and between every H-V discharge electric field

Miscellaneous Types

In response to special operating requirements, manufacturers have developed several special types of electric precipitators.

Acid Mist Precipitators. Figure 8-44 shows an acid mist precipitator designed to collect sulphuric acid mist. All parts exposed to the gas stream are fabricated of lead. The basic design includes an upper cylindrical chamber and a lower one, the two chambers being connected by 10-in. diameter lead tubes. Star-shaped, weight-tensioned H-V lead wires with steel cores hang on the center lines of the tubes whose inner surfaces constitute the collectrodes.

Gas flow is horizontally into the upper portion of the lower chamber, downward then upward into the belled inlets of the

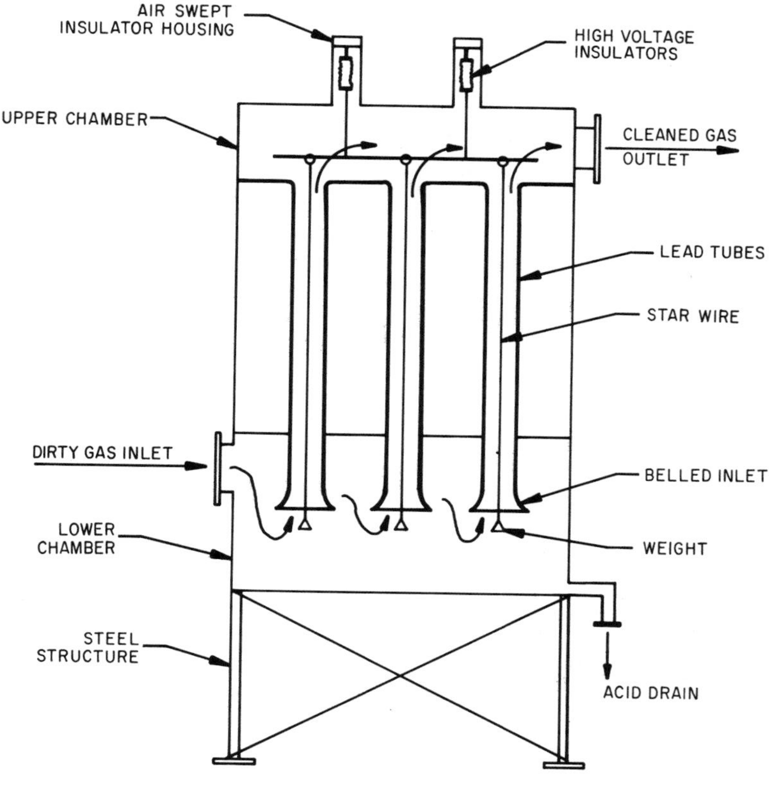

Fig. 8-44 Acid mist precipitator.

TABLE 8-8 TYPICAL USE OF PARTICULATE COLLECTORS

INDUSTRY AND SPECIFIC APPLICATION	TYPES OF COLLECTORS[a]		
	Most used	*Also used*	*Seldom used*
ROCK PRODUCTS			
Wet process cement kiln	MC, EP	CC, WS	FF
Dry process cement kiln	MC, EP, FF	CC	WS
Cement mills	FF	EP, MC	WS, CC
Cement silos	FF		
Raw material preparation	MC, FF	EP	WS, CC
Asphalt plant dryer	MC, WS	EP, FF	CC
Coal drying	MC, WS		FF, EP
Coal mills	FF	MC, WS	EP, CC
Alumina kiln	MC, FF, EP	WS	CC
Bauxite kiln	MC, EP	FF, WS	CC
Gypsum kiln	MC, FF, EP	WS	CC
Magnesium oxide	MC	FF, WS, EP	
Lime kiln	MC, EP	FF, WS	
Phosphate kiln	MC, WS, EP	FF	
Finishing	FF	MC, EP	WS, CC
IRON AND STEEL			
Blast furnance	WS, EP	MC	FF
Open hearth	WS, EP	FF	MC, CC
Basic oxygen furnance	WS, EP		

Electric furnance	FF, WS	EP	MC, CC
Sintering	MC, FF, EP	WS, CC	
Cupolas	WS	MC, FF, EP	CC
Ore beneficiation	MC	FF, EP, WS	
Ore pellitizing	MC	WS, EP	
Ore roasting	MC, WS, EP		FF, CC
Taconite	MC	FF, WS, EP	CC
Pyrites roasting	MC, WS, EP		FF, CC
Hot scarfing	WS, EP		MC, FF, CC
Coke ovens	EP	WS	MC, FF, CC
Ferro-manganese blast furnance	MC, EP	CC, WS	FF
Scrap preheating	WS		MC, FF, EP
MINING AND METALLURGY			
Ore preparation and beneficiation	MC, FF, WS	EP	CC
Aluminum pots	FF, WS, EP	MC	CC
Elemental phosphorous	EP	WS	FF, WS
Copper converter	EP	FF, WS, CC	MC
Copper reveratory furnace	EP	WS	MC, FF, CC
Copper roaster	MC, EP	WS, CC	FF
Ilmenite ore dryer	MC, EP	CC	FF, WS
Molybdenum	WS	EP	FF, MC, CC
Precious metals	FF, EP	MC, WS	CC
Lithium kiln	EP		
Lead furnace	FF, WS		MC, EP
Titanium dioxide	FF, WS	EP	MC, CC

TABLE 8-8 (Continued)

INDUSTRY AND SPECIFIC APPLICATION	TYPES OF COLLECTORS[a]		
	Most used	*Also used*	*Seldom used*
Zinc cupola	EP	WS	
Zinc roaster	MC, EP	FF, WS	CC
Zinc smelter	FF, WS, EP		MC, CC
FOUNDRY AND INDUSTRIAL			
Ferrous cupola	WS	MC, EP	FF, CC
Non-ferrous cupola	FF, EP	MC, WS	CC
Foundry cleaning room	FF, WS	EP	MC, CC
Sand preparation and handling	FF	WS, EP	MC, CC
Ventilation	FF		MC, FF, EP
Machining	MC, FF	WS	EP
Feed and flour milling	MC, FF		WS, EP, CC
Lumber mills	MC	FF	WS, EP
Wood working	MC, FF	WS	EP
PULP AND PAPER			
Black liquor recovery furnace	EP	WS	MC, FF
Lime kiln	MC, WS	EP, CC	FF
Chemical dissolver tank	WS	MC, FF, EP	
ELECTRIC POWER—TYPE OF FUEL			
Stoker fired coal	MC	EP, CC, WS	FF
Spreader stoker fired coal	MC	EP, CC, WS	FF
Puverized coal fired	EP, CC	WS	FF, MC

Oil fired	MC	EP	FF, WS, CC
Natural gas fired	none	none	
Lignite fired	MC, EP, CC	WS	FF
Wood and bark fired	MC	WS, EP	FF, CC
Bagasse fired	MC	WS	FF, EP, CC
Fluid coke	MC, EP	WS	FF, CC
WASTE DISPOSAL INCINERATORS			
Apartment house	WS	EP	MC, FF, CC
Industrial (thermal oxidizer)	EP	WS	MC, FF, CC
Municipal	EP, WS	MC, CC	FF
CHEMICAL AND OIL REFINING			
Refinery cat cracker—reactor	MC	none	
—regenerator	MC	none	
—CO boiler	EP	MC, WS	FF, CC
Sulphuric acid mist	EP, WS	MC	FF, CC
Phosphoric acid mist	WS		
Nitric acid mist		WS	
Carbon black	FF	MC, EP, WS	CC
Oil shale distillation	MC, CC	EP	FF, WS
Raw material preparation	MC	WS, EP, FF	CC

[a]MC = mechanical collector, FF = fabric filter, WS = wet scrubber, EP = electric precipitator, CC = combined collector = MC + EP.

lead tubes, upward through the tubes into the upper chamber, then out the cleaned gas outlet.

Figure 8-44 shows just three tubes. Acid mist precipitators are available in sizes approaching five hundred tubes. Parallel and staged groupings can handle any gas quantity. The acid mist is charged negatively and then precipitated as it passes upward between the wires and tubes. Being a mist it forms a liquid film which drips into the lower chamber and out the acid drain. Water sprays are often installed in the upper chamber to flush out precipitated particulate periodically.

Acid mist precipitators are an essential part of many processes producing acid.

Wet Bottom Precipitator for Paper Industry. This design was developed to recover soda ash after a recovery furnace fired by black liquor. Because dry hopper bottoms proved troublesome, these designs drop the dust rapped from the collectrodes into an agitated pool of black liquor. Some older designs feature casings built of hollow, glazed tile.

Wetted Wall Tube-Type Precipitators. The functional design of these is similar to that of the acid mist type; they are fabricated of steel, however. The upper tube sheet is continuously flooded with water to flush the walls of the tubular collectrodes. The effluent drains as a sludge.

Low-Voltage, Two-Stage Electrostatic Precipitator. This device was originally designed to purify air, and is used in conjunction with air conditioning systems including those for purification of incoming air at hospitals, industrial and commercial installations, as well as the compact units supplied for home use.

The first (ionizing) stage consists of fine (0.007-in.) positively charged wires and parallel grounded rods. Spacing is 1–2 in., and direct current voltage is from 12 or 13 kV. Positive polarity is used to minimize ozone formation. The second (collecting) stage consists of oppositely charged parallel metal plates spaced an inch or less apart. In some designs alternate plates are charged positively and negatively at 6 or 6½ kV. Other designs use positively charged plates to precipitate the dust on grounded ones.

These collectors are suitable only for low dust loadings and are not equipped with automatic electrode cleaning devices. Periodic, manual washing of the plates is required to remove precipitated particulate. One exception is the oil mist precipitator whose precipitate drains as a liquid.

SELECTION OF PARTICULATE COLLECTORS

Because of the great number of variables the selection of a particulate collector for a specific application is very complicated. For many applications the choice of the type of collector is well established by operational precedent (See Table 8-8). Consult your trade association or one of the following:

Air Pollution Control Association
Industrial Gas Cleaning Institute
Your trade publications
National Air Pollution Control Administration
Local Air Pollution Control Authority
Consulting engineer

If precedent fails to help, the following approach may prove helpful.

1. Determine whether a mechanical collector can meet the required collection efficiency. If not, consider the following three high-collection-efficiency alternatives:
2. If the temperature is under 500°F and the gas volume under 100,000, cfm, a fabric filter may justify its high maintenance cost.
3. If corrosion and liquid waste disposal are not serious considerations, a wet scrubber may justify its high operating cost.
4. If none of the above satisfies your requirements, you may be able to justify the capital cost of an electric precipitator.

See Table 8-1 for the basic economic factors of the four basic types of particulate collectors.

Air Pollution Aspects of the Iron and Steel Industry, USDHEW, PHS, Division of Air Pollution, Cincinnati, Ohio (PHS Pub. No. 999-AP-1)

Air Pollution Engineering Manual, USDHEW, Superintendent of Documents, U.S. Government Printing Office, Washington, D.C. 20402

Air Pollution and the Kraft Pulping Industry, An Annotated Bibliography, USDHEW, PHS, Division of Air Pollution, Cincinnati, Ohio (PHS Pub. No. 999-AP-4)

Air Quality Criteria for Particulate Matter, Summary & Conclusions, USDHEW, PHS, Consumer Protection & Environmental Health Service, NAPCA, 801 N. Randolph Street, Arlington, V. 22203

Air Quality Control for Sulphur Oxides, Summary and Conclusion, USDHEW, PHS, Consumer Protection & Environmental Health Service, NAPCA, 801 N. Randolph Street Arlington, V. 22203

The following three are publications of The American Society of Mechanical Engineers, United Engineering Center, 345 East 47 Street, New York, N.Y. 10017: *Dust Separating Apparatus* (PTC 21-1941) *Determining Dust Concentration in a Gas Stream* (PTC 27-1957)

Determining the Properties of Fine Particulate Matter (PTC 28-1965)
Methods for Determination of Velocity, Volume, Dust and Mist Content of Gases, Western Precipitation Division, Joy Manufacturing Company, 1000 West Ninth Street, Los Angeles, Calif. 90015
Atmospheric Emissions from Petroleum Refineries, A Guide for Measurement and Control, USDHEW, PHS, Division of Air Pollution, Cincinnati, Ohio (PHS Pub. No. 763)
Atmospheric Emissions from Wet-Process Phosphoric Acid Manufacture, USDHEW, PHS, Environmental Health Service, NAPCA, Raleigh, N. C. (NAPCA Pub. No. AP-57)
APCD Source Testing Manual, County of Los Angeles, Los Angeles County Air Pollution District, 434 South San Pedro Street, Los Angeles 13, Calif.
The Chemical Industry Technical Manual No. 3, Air Pollution Control Assembly, 4400 Fifth Avenue, Pittsburgh, Pa. 15213
Control Techniques for Particulate Air Pollutants, Jan. 1969, USDHEW. PHS, Consumer Protection and Environmental Health Service, NAPCA, Washington, D.C.
A Digest of State Air Pollution Laws, USDHEW, PHS, Division of Air Pollution Washington, D.C. 20201 (PHS Pub. No. 711)
The Electric Utility Industry, Technical Manual No. 4, Air Pollution Control Assembly, 4400 Fifth Avenue, Pittsburgh, Pa. 15213
An Electrostatic Precipitator Systems Study, 1st Quarterly Progress Report to NAPCA, Cincinnati, Ohio (Contract CPA 22-69-73) Southern Research Institute, 2000 9th Avenue S., Birmingham, Ala. 35205
Emissions from Fuel Oil Combustion, An Inventory Guide, USDHEW, PHS, Division of Air Pollution, Cincinnati, Ohio (PHS Pub. No. 999-AP-2)
All the following are publications of Industrial Gas Cleaning Institute, Inc., Box 448, Rye, N.Y. 10580:

Terminology for Electrostatic Precipitators (E-P 1)
Procedure for Determination of Velocity and Gas Flow Rate (E-P 2)
Criteria for Performance Guarantee Determinations (E-P 3)
Evaluation Bid Form (E-P 4)
Information Required for the Preparation of Bidding Specifications for Electrostatic Precipitators (E-P 5)
Pilot Electrostatic Precipitators (E-P 6)
Gas Flow Model Studies (E-P 7)
Cyclonic Mechanical Dust Collector Criteria (M-2)
Gravity, Louver and Dynamic Mechanical Collector Criteria (M-3)
Information Required for the Preparation of Bidding Specifications (M-4)
Gaseous Emissions Equipment: Product Definitions and Illustrations (G-1)
Fundamentals of Fabric Collectors and Glossary of Terms (F-2)
Operation and Maintenance of Fabric Collectors (F-3)

Mineral Facts and Problems, Bulletin 630, Bureau of Mines, 1965 Ed.,

Superintendent of Documents, U.S. Government Printing Office, Washington, D.C. 20402

1968 National Incinerator Conference, American Society of Mechanical Engineers, United Engineering Center, 345 East 47th Street, New York, N. Y. 10017

Optical Properties and Visual Effects of Smoke-Stack Plumes, 5555 Ridge Avenue, Cincinnati, Ohio 45213 (PHS Pub. No. 999-AP-30)

"Plant Engineering," Richard A. Young, Special Report Sept. 4, 1969

Watkins Cyclopedia of the Steel Industry 12th Ed. (1965), Steel Publications, Inc., 624 Grant Building, Pittsburgh, Pa. 15230

9 Selection of Equipment for Gaseous Waste Disposal

INTRODUCTION

Chapter 8 discussed the selection of equipment for the removal of particulate matter from a waste gas stream. Assuming that the waste gas stream does not have any particulates but still constitutes an air pollution problem, then ways must be devised to make it acceptable to the atmosphere. Examples of such waste gases might be those which contain high concentrations of sulfur dioxide, oxides of nitrogen, hydrocarbons, or other atmospheric pollutants coming from chemical processes, natural phenomena, or manufacturing operations.

It is true that waste gases also contain particulates which, if removed, still do not necessarily make the waste gas suitable for discharge into the atmosphere. For example, a power boiler could generate a stack gas containing high quantities of sulfur dioxide along with fly ash. With this type of waste it would be necessary not only to remove the fly ash but also to reduce the sulfur dioxide level. In this particular case it would probably be desirable to remove the fly ash by mechanical means and then use an absorption process to remove the sulfur dioxide. If, however, one had a combustible organic fume containing particulate matter, then it would probably be wise to incinerate the organic material before removing the particulate because of the possibility of an explosion in the particulate collection equipment if the organic fume was not removed first.

With fumes containing both gaseous pollutants and particulates it is wise to evaluate the logical sequence of action, depending upon what they contain, and then make plans appropriately.

The purpose of this chapter is to talk about the removal, reduction, dispersion, or destruction of the gaseous contaminants in any fume. In the field of waste gas disposal the problem is usually one of quantity. By this we mean that the contaminants in any waste effluent are usually quite low in concentration but the total quantity of the effluent is very high. This means that removal or destruction of the contaminant is very expensive, because all of the effluent must be handled as if it were a contaminant, even though the greatest portion of the effluent may be air or inert gas. For example, let us assume that we have a fume containing 100 ppm hydrogen chloride in 5,000 ft^3/min of air. 100 ppm is a very small amount of contaminant but is considerably above the 5 ppm specified as the acceptable threshold limit for HCl, and this means that the level must be reduced by at least a factor of 20 before it can be released into the atmosphere. In order to do this we must not only treat the 100 ppm of HCl but also the 5,000 ft^3/min of air. It is the treatment of this large volume which is expensive, not the treatment of the contaminant. Wet-scrubbing probably would be selected as the best method for removal because HCl is soluble in water, but both the size and the cost of the scrubbing apparatus would be predicated on the 5,000 cu ft of air rather than on the 100 ppm concentration of HCl.

The same type of statement can be made for almost any waste gas effluent and the purpose here is to discuss a few of the possible methods by which waste gases may be treated before release to the atmosphere.

The few alternatives which we have at our disposal for the treatment of a waste gas are (1) dispersion, (2) absorption in either a dry or wet medium, (3) adsorption, (4) incineration or other oxidation methods, and (5) treatment with odor counteractants. It is impossible to give all of the design details of each type of waste gas processing; however, adequate references are given so that further research into any of these areas can be continued by the reader.

DISPERSION WITH STACKS

Perhaps the oldest method of handling a waste gas was to put it up a chimney or stack and vent it directly into the atmosphere. Chimneys were initially conceived as a means of removing smoke from indoor fireplaces and then later from the blacksmith's forge and eventually from the electric power generation station burning fossil fuels. In this context the stack acted as a draft generator which pulled in sufficient combustion air to maintain combustion of the fuel whether it was wood, coal, oil, or gas. The higher the stack the greater the draft and usually the better the combustion.

In this sense the stack certainly was not an air pollution control device nor is it today but it is being passed off as one in a good many areas.

The tall stack, as differentiated from a low chimney or a vent stack from a process plant, is a tool which can be used to reduce ground concentrations of air pollutants by discharging them at a height where sufficient dispersion will take place so that the ground concentration at any point from the base of the stack will not exceed the allowable level. The tall stack, therefore, might better be called a ground level pollution control device rather than an air pollution control device since it does not reduce by a gram the amount of effluent which is being poured into the atmosphere.

Today, tall stacks or chimneys are used as an ultimate disposal method for wastes which would be exceptionally difficult or expensive to treat. The primary effluent which is emitted from tall stacks or chimneys is sulfur dioxide. Sulfur dioxide comes from two major sources: the first is smelting operations, and the second is electric power generation plants using a fossil fuel to generate power. In the first case the concentration of SO_2 is relatively high; in the second case it is considerably lower. In either case the object of the stack is to disperse the SO_2 at an altitude high enough to prohibit ground concentrations at any point or at any distance from the stack which will be above the tolerable level. Stacks of this type are usually at least 150 ft high and in many cases are over 500 ft high. Tall stacks are designed to reach above the inversion level so that they will disperse smoke, particulates, or waste gas at a point where it will not be trapped underneath the inversion layer, which would result in almost immediate fallout.

The inversion layer has a significant effect on the stack operation. Normal atmosphere should decrease 1°C in temperature for every 100 m in height. This is known as the adiabatic lapse rate. On a clear night the land is cooled by radiation and the lapse rate may become zero or even produce a higher temperature at a height increase, and this is known as a temperature inversion. As the sun begins to heat the land the following day, the air temperature also rises and the stable air in the inversion layer will cause a stack plume that does not extend through the inversion layer to descend to the ground rather than to rise, producing ground level concentrations 10 to 20 times the predicted level.

In investigating the use of a stack for disposal of a waste gas it is necessary to determine the acceptable ground level concentration of the toxic materials or the particulate matter in the waste gas. We also should have some idea of the topography of the area so we can locate the stack with respect to buildings and terrain which might introduce a factor of air turbulence into the operation of the stack, and we also should be aware of the meteorological conditions prevalent in the area, such as prevailing winds, humidity, rain fall, etc. We should also have an accurate knowledge of the constituents of the waste gas and their physical properties.

The permissible ground level concentration is usually available from literature but it certainly will be available from state and community air pollution regulations. While stacks and chimneys are used today for a variety of waste gases, the future probably will outlaw this method of disposal unless the concentration of the effluent gas is near or below the allowable threshold value for the toxic constituent. Therefore it would be difficult to encourage any operator of an industrial plant to count on the fact that a stack built today will be satisfactory and will meet air pollution regulations 6 months from today. This author would do everything possible to discourage a plant operator from installing an expensive stack for the dispersion of a toxic effluent even though such dispersion methods may be presently acceptable and safe from a practical point of view. The method does not eliminate any waste material but merely spreads it over a wider area.

In the design of any stack for the dispersion of a waste there are three nonmeteorological factors; (1) area of the source, (2) the effluent velocity, and (3) the effective height of the source.

There are also two important meteorological factors; (1) the wind speed and (2) the lapse rate, which was discussed earlier. If this lapse rate is positive and large enough there is a tendency for convection currents to be set up as the warm air below attempts to rise, and in this state the air is considered unstable. In unstable air, stack effluents theoretically rise indefinitely, although in practice, stable air is eventually reached. This would contrast with a negative lapse rate which would occur when an inversion layer was present and the end of the stack did not protrude through that layer.

At the present time there are at least 20 formulas or equations relating to rate of rise of a plume from a stack based upon both meteorological and nonmeteorological variables, and most of these formulas contain dimensional constants suggesting that all relevant variables have not been properly included. Each expert or student of the tall stack has his own favorite formula. Each of them appears under close scrutiny to have some definite shortcomings. For this reason no specific formula for the calculation of stack height or rate of plume rise will be given in this chapter. However, it is suggested that those brave souls who wish to design their own tall stack or chimney carefully review the literature shown in the bibliography for this chapter.

Considerable work is being done at the present time by the National Air Pollution Control Administration, using computers to determine the acceptable height of stacks under various conditions and the direction and degree of fallout of pollutants under specific meteorological situations.

Similarly, there has been a significant amount of work done using large models to attempt to determine the possible effects of a tall stack if constructed in a particular area. Models emit smoke from the top of the stack and change wind direction and velocity in an attempt to predict the direction of fallout, but even these have been of questionable value in determining an overall approach to the problem.

The dispersion of a contaminant in the atmosphere depends on two primary factors, average wind speed and characteristics of atmospheric turbulence. Obviously, as the wind speed increases the dispersion is greater and the smaller the concentration of the contaminant. The atmospheric turbulence factor consists of both horizontal and vertical eddies which mix the contaminated air with clean air surrounding it. Turbulence decreases the concentration of contaminants in the plume and increases the concentration outside the plume. Therefore the stronger the turbulence and the stronger the wind the more the pollutants are dispersed (see Fig. 9-1).

Turbulence is increased with wind speed and with convection currents. Convection occurs when the temperature decreases rapidly with the height; that is, whenever the lapse rate exceeds 1°C per hundred meters. Mechanical turbulence occurs often when the wind speed changes and the direction changes with height. For example, if there is no wind at ground level and there is wind at 100 ft, mechanical turbulence will contribute to the dispersal of a pollutant.

Topography also has an effect on the amount of turbulence which is generated. Turbulence is usually greater over rough terrain than it is over smooth, as can be evidenced when flying an airplane over mountains as compared with the desert.

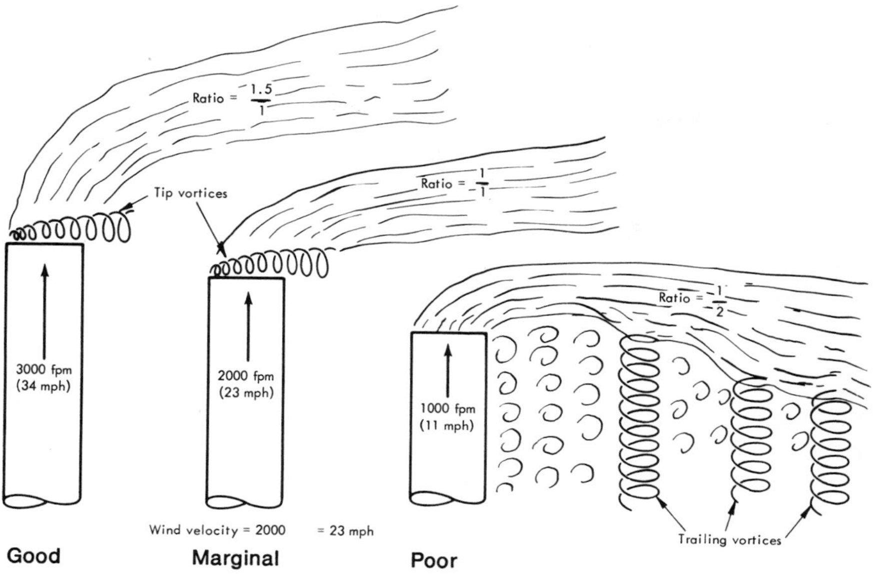

Fig. 9-1 Relationship of stack velocity to wind velocity. *(Courtesy Heating, Piping & Air Conditioning)*

There are several rules of thumb in the design of a stack which may be used by the average plant engineer without getting him into serious trouble. These are as follows: (1) The stack should be 2½ times the highest surrounding buildings or surrounding topography so that significant turbulence is not introduced by these factors. (2) The gas injection velocities from the stack to the atmosphere should be greater than 60 ft/sec so that the gases will escape the turbulent wake of the stack. In many cases it is desirable to have a gas velocity considerably higher, up to 100 ft/sec if possible. However this is not always possible because of pressure drop in the stack. There is a critical wind velocity for every stack exit velocity. Above this critical velocity the wind shears off the gas as it leaves the chimney and there is no corresponding rise in the waste gas due to the exit velocity. Then gas temperature and flow rate no longer affect the ground level concentration for which the exit velocity was initially designed. (3) Gases from stacks with diameters less than 5 ft and less than 200 ft in height hit the ground part of the time and ground concentrations may be excessive. In effect, we are saying that for these conditions stack design is unpredictable. (4) The maximum ground concentration of stack gases subjected to atmospheric diffusion occurs about 5 to 10 stack heights from the stack in a horizontal direction under a windless condition. (5) When stack gases are subjected to atmospheric diffusion and building turbulence is not a factor, ground level concentrations on the order of 0.001–1 percent of the stack concentration are possible for a properly designed stack. As you can see, this is a wide range and gives testimony to the lack of science involved in the design of stacks. (6) Ground concentrations can be reduced through the use of higher stacks. The ground concentration varies inversely as the square of the effective stack height.

In industrial areas we often find multiple stacks a short distance from one another. This can cause problems in the ground concentrations of certain contaminants, especially if they are emitting the same pollutant. If two such stacks blend their plumes this will undoubtedly create a greater concentration of the pollutant at some point after the blending occurs.

In the average industrial plant the design of a stack is a complicated procedure and unless the organization considering the stack has a large engineering department with competent designers in this particular field, a consultant should be used. However, this is only after every other possible consideration has been given to the destruction or disposal of the waste effluent by other means than the dispersion technique.

Materials of Construction

Small-diameter stacks or chimneys are often constructed entirely of metal, either steel or stainless steel, and are either self-supporting or held in position by guy wires. Small diameter stacks several feet in diameter have been installed up to 200 ft high by proper guying, and by using forced draft such stacks can be made acceptable and will discharge their effluent at a height well above process equipment. Often such stacks are used to vent incinerators, boilers or other heat-releasing process equipment and the effluent may not be a pollutant to the atmosphere. It is merely necessary to discharge the high-temperature steam or products of combustion away from the work area.

Normally, when we talk about a tall stack for dispersion, such as those used for a power generating station, we must consider a lined chamber usually constructed of steel and lined with firebrick or suitable castable material having a high enough heat resistance to handle the exit gases. This type of construction is suitable for stacks of 200–250 ft, but when one gets to large-diameter and extremely high chimneys then we must consider designing in two major sections, the liner and the column or outer shell. The liner is designed to withstand the temperature and the corrosive condition of the gas while the outside column, which may be either steel or masonry, protects and supports the liner. Large stacks are always freestanding and consequently are much larger in diameter at the base than they are at the top.

When handling fumes containing sulfur dioxide one of the considerations, using an inner liner or, for that matter, no liner at all, is the condensation of sulfuric acid at the top of the stack. If the inside stack temperature drops below the condensation temperature of sulfuric acid then the stack must be designed so that it will not corrode. Glass linings and special metals have been used in such cases but, naturally, this adds to the cost of the stack. Handling any acid gas is something that requires the consideration of corrosion.

A phenomenon that has been noticed in some very tall steel chimneys is wind-excited oscillations. While hydraulic dampers could be put in the wires of guyed chimneys, nothing could be done with self-supporting chimneys until work at the British National Physical Laboratory showed that helical fins welded on the upper one-third of the chimney would destroy the periodic vortex that caused the trouble.

Many industrial plants may come to the easy conclusion that a tall chimney or stack is the answer to its problem and this may be based erroneously on the idea that the cost of such a system is much less than any other type that can be installed. Often, nothing is farther from the truth. For example, a 400-ft stack with an 8-ft-diameter inside liner will cost from $300,000 to $350,000 and this price increases very rapidly with height and with the inside diameter of the stack. Every possible method should be considered before resorting to a dispersion method which may not be satisfactory in the long run.

ABSORPTION AND CHEMICAL REACTION

Another method of treating industrial gaseous wastes is absorption. Absorption can take place either with or without a chemical reaction. For example: waste gas containing hydrogen chloride can be scrubbed with water to form hydrochloric acid. In such a system hydrogen chloride gas is absorbed by water to create a weak hydrochloric acid solution. On the other hand, if we wish to effect both absorption and neutralization at the same time we can scrub with a weak solution of sodium hydroxide or a weak solution of calcium hydroxide which will result in the ultimate product being sodium chloride or calcium chloride. In one case we have a true absorption system and in the other we have a chemical reaction.

While there is a definite difference between the two systems, the apparatus used to carry out the intended task can be essentially the same. It could be a packed or plate type column, or a variety of other types of contacting equipment which would produce the same end result.

It is desirable for the plant engineer to examine his gaseous effluent to decide the best approach for treatment of the waste before employing any method. Obviously if a waste gas has a constituent which has economic value and is recoverable, then the recovery of this material should be considered first. If, however, the waste gas does not have a material in sufficient quantity to justify recovery then the most expedient method of final disposal should be paramount.

Absorption or reaction of gaseous waste constituents is not limited to liquid-gas contactors. Gases can be trapped or reacted by passing them through solid chemical beds in which the reaction will occur. So in essence we have two considerations: gas to liquid contacting and gas to solid contacting, both by absorption and chemical reaction.

In the treatment of waste gases which commonly emanate from manufacturing processes there are literally thousands of possibilities and it is not the intention of this book to discuss these in any detail. Many constitute an integral part of a chemical process and therefore are not considered waste gas treatment. The others, however, can be used to trap, absorb, or change various effluents by chemical reaction and are certainly worth mentioning. In order to do this it is desirable to divide the effluents into two major categories: (1) those gaseous wastes which can be absorbed by water-scrubbing in some type of conventional scrubbing apparatus with essentially complete removal from the waste gas stream, and (2) those materials which cannot be scrubbed out by water.

Water Soluble Gaseous Effluents

In order for a waste gas or vapor to be suitable for water absorption or scrubbing it should have a high degree of solubility in water. Obviously, unless the waste gas is nearly totally soluble in water, it cannot be absorbed by water-scrubbing. Water-scrubbing in various types of contacting apparatus is therefore practical for materials such as hydrogen chloride, dimethylformamide, acetone, hydrogen fluoride, silicon tetrafluoride, and ammonia. Such waste gases with high solubility are easily absorbed in water, however, the resulting mixture will be an acid or a base or a mixture of an organic in water which ultimately must be neutralized or reclaimed by distillation. Absorption in water alone does not solve the pollution problem because the pollutant, being soluble in water, must ultimately be reclaimed or disposed of in some other manner. At low concentrations it may be possible to put this mixture into the normal water treatment plant associated with the manufacturing facility, if one exists, or it may be possible to send it to a municipal sewage plant, but in almost every case the solutions must be very dilute. Therefore caution should be used in selecting water absorption as a means of gas purification since it usually requires another step in a waste disposal process.

Any solution of a gas in a liquid exerts a definite partial pressure at a specific temperature and concentration. The lower the partial pressure of the dissolved gas in a solution of any given concentration the more soluble it is said to be. Therefore a gas like ammonia exerts a lower partial pressure at a given concentration in water than a gas like sulfur dioxide which is moderately soluble and much less than a gas such as oxygen which is only slightly soluble in water.

The solubility of a gas in an absorbing medium can be approximated by Henry's law.

$$p = Hc$$

where p = partial pressure of the dissolved gas
c = concentration of the dissolved gas
H = a constant

Henry's law is accurate where solubility curves follow a straight line, i.e., partial pressure versus concentration, and this generally holds true for gases which do not change molecular form in solutions such as H_2, O_2, N_2, CH_4 in water. Gases such as NH_3, HCl, and SO_2, which dissociate or combine with water and are more soluble, tend to deviate from Henry's law.

As the concentration of the solute increases, the partial vapor pressure of this material also increases up to the equilibrium point. This will be the point where the partial pressure of the solute in the gas phase is equal to its partial pressure exerted by the liquid phase and this is the strongest solution that possibly can be produced in any gas absorption process. At this point some of the gas which is to be absorbed will begin to pass through this system into the atmosphere.

In the case of organic solvents which are to be absorbed in water, it is usually intended and desirable to recover these organic materials by subsequent distillation. Therefore the maximum concentration which can be obtained in the absorber is desirable. This can be accomplished by recycling the water or other liquid absorbant until the equilibrium concentration is reached, and then a side stream of this equilibrium concentration taken off to a distillation tower where subsequent recovery can be achieved. Organic solvents are seldom recovered by absorption in water unless subsequent distillation is intended because such materials can create problems in water treatment systems.

Other Absorption Processes

It is impossible to discuss all of the liquid absorption and gas purification processes. One process which might be considered in certain instances is the use of ethanolamines for the absorption of hydrogen sulfide and carbon dioxide, both of which are very soluble in various amines. Monoethanolamine is used to remove CO_2 from flue gas either for its recovery or for the purification of flue gas to nitrogen. The use of ethanolamines for the removal of H_2S is usually limited to purifying a sour gas, since H_2S is ultimately released for some other type of disposal. This is not normally a waste gas treatment system. CO_2 and H_2S may also be removed by using aqueous ammonia with the ultimate production of ammonium sulphate. This type of waste gas treatment is regularly used for coke oven by-product gas before release to the atmosphere. There is an economic advantage here as well, since ammonium sulphate is recovered for sale as a fertilizer. Alkaline salt solutions can also be used for absorbing H_2S and CO_2, noting that sodium and potassium carbonate solutions will remove carbon dioxide in much the same way as monoethanolamine. There is poor CO_2 recovery efficiency with these systems and usually they are not considered practical.

Another means of purifying sour gas streams or air containing H_2S is to contact it with sodium carbonate, but the acid gas must be removed from the solution in a later step by heating and must be disposed of in some other manner. Potassium phosphate has been used in a similar manner for the cleaning of acid gases.

Sulfur dioxide can be removed by liquid absorption and purified for later collection, compression, and resale. A number of processes have been developed using liquid absorption techniques for this waste gas but most are not practical for the casual requirement. The sulphidine process uses a mixture of xylidine and water in approximately 50–50 percent mixtures and the by-products are sodium sulfate and pure SO_2.

The ASARCO process developed by American Smelting & Refining Company for the removal of SO_2 from smelter gases uses dimethylaniline as the absorbant to remove the gas from the stream. Either dilute sulfuric acid or liquid SO_2 is the by-product.

SO_2 is often absorbed by aqueous ammonia to produce ammonium sulfite, with subsequent oxidation to the sulfate. Another system which absorbs SO_2 uses an aqueous solution of ammonium sulfite.

Several processes which were used experimentally after World War II for the absorption of SO_2 from power plant flue gases utilized a chalk or calcium carbonate slurry in one case, and a calcium sulfate slurry in another to absorb the waste gas. The zinc oxide process contacts the gas with a solution of sodium sulfite and sodium bisulfate which results in an increase in the bisulfite content. This solution is then mixed with zinc oxide and zinc sulfite is precipitated. The zinc sulfite is filtered and calcined, releasing concentrated CO_2 and regenerating zinc oxide.

All these systems are of questionable use to the plant engineer trying to solve a relatively small problem.

Absorption Equipment

The equipment used for the absorption or chemical reaction of gaseous wastes is very similar and in many cases identical to the equipment used for particulate removal described in Chapter 8 under Wet-Scrubbing Apparatus. Gas absorbers for most industrial waste processes consist of one of five types of apparatus: (1) packed columns, (2) plate columns, (3) sparging systems, (4) spray towers, and (5) miscellaneous high energy scrubbers.

Packed Columns. Packed columns are usually designed of material resistant to the corrosion of the absorbed gas-liquid mixture. They consist of a vertical column with one or more packed sections containing a packing made of ceramic, metal, or plastic which provides a large amount of surface area per unit volume. A distributor is used to distribute the liquid phase over the packing (see Fig. 9-2). Recycling in one or more packed sections is employed to increase the concentration of the absorbed vapor or gas. The gaseous effluent, normally air, containing contaminants which are now below the acceptable minimum will exit from the top of the tower and the solution will be taken from the bottom of the tower. Packed towers are normally used where high liquid flow rates are permissible. They are also suitable for the absorption and dissipation of heat from waste gas. Packed columns are gen-

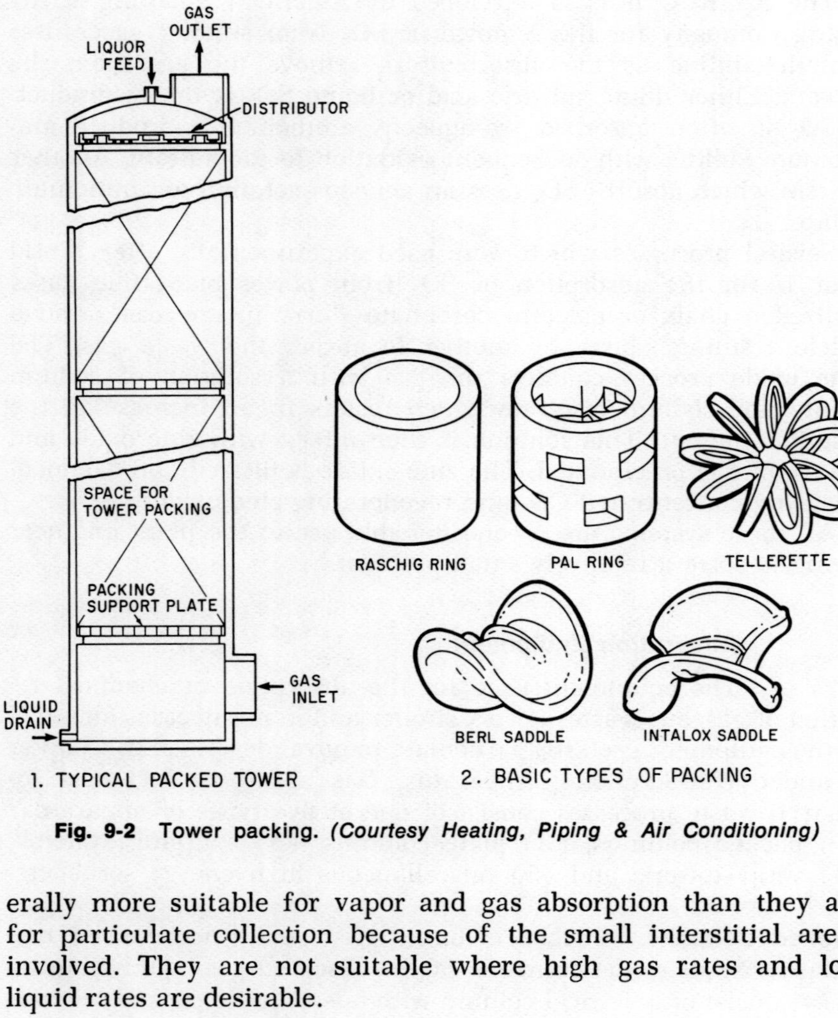

Fig. 9-2 Tower packing. *(Courtesy Heating, Piping & Air Conditioning)*

erally more suitable for vapor and gas absorption than they are for particulate collection because of the small interstitial areas involved. They are not suitable where high gas rates and low liquid rates are desirable.

Plate Columns. Plate columns are used for the absorption of gases and vapors when low liquid rates are desirable (see Fig. 9-3). This is particularly true when the gas contains small quantities of highly soluble constituents. In the plate column, because of the distribution mechanism, smaller quantities of liquid solvent can be used for the absorption of small quantities of vapor or gas solute. Plate columns may be used in a variety of types, the standard being the bubble cap, but similar bubble trays, sieve trays, etc. may be used in a similar manner.

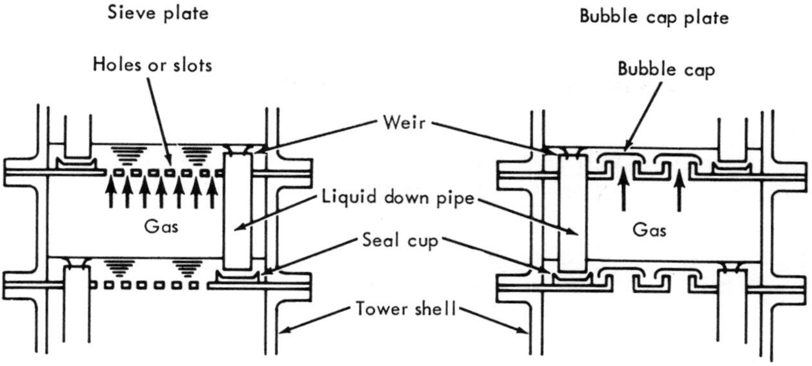

Fig. 9-3 Types of plates. *(Courtesy Heating, Piping & Air Conditioning)*

Sparging Systems. Sparging systems are generally used for a high gas rate which must be distributed through a large liquid volume, such as the carbonation of municipal water supplies with carbon dioxide. Generally spargers are not applicable to waste gas absorption systems but they can be used where the waste gas or vapor is highly soluble. A submerged sparger discharging the gas under several feet of water in small bubbles will provide a large surface area for the absorption of the waste gas or vapor.

Spray Towers. Spray towers are normally used to reduce the temperature of a waste gas in the most economical manner (see Fig. 8-19). Large quantities of water or other absorbent are sprayed in fine droplets into a tower or chamber and contacted with the gas and absorption of the soluble vapor or gas occurs within the chamber. Spray towers generally are less efficient than packed or tray type columns and are not used extensively for gas absorption.

High-Energy Scrubbers. High-energy scrubbers generally are utilized for removal of particulates as described in Chapter 8. They also can be utilized for gas absorption in the case of materials which are highly soluble in water or some other scrubbing liquid (see Fig. 8-27). The high energy has no particular advantage for gas absorption except size and first cost. The contact time is less and the scrubbing action is no better than with the packed or plate type column, but the requirements in terms of the pressure required for both the scrubbing medium and the waste gas is usually quite high.

Conclusions

In general, the absorption mechanism is applicable for the removal of soluble waste gas components in accordance with Henry's law. The least expensive and most effective systems are usually those which provide the greatest degree of cleanup at the lowest operating expense and generally result in contacting devices such as packed or plate columns which have been used for years for this purpose. The theories of gas absorption are readily found in a wide number of chemical engineering textbooks and are therefore not discussed in detail here.

Gas-Solid Absorption Systems

In the past few years much work has been done on solid phase absorption or reaction systems, especially with respect to the difficult effluent of sulfur dioxide. Such systems are not usually compatible with small emission requirements and therefore should not be considered for streams of relatively low flow. They offer, however, a possibility when dealing with high gas flow rates containing substantial quantities of SO_2 which, if discharged directly to the atmosphere, would cause ground level concentrations of the

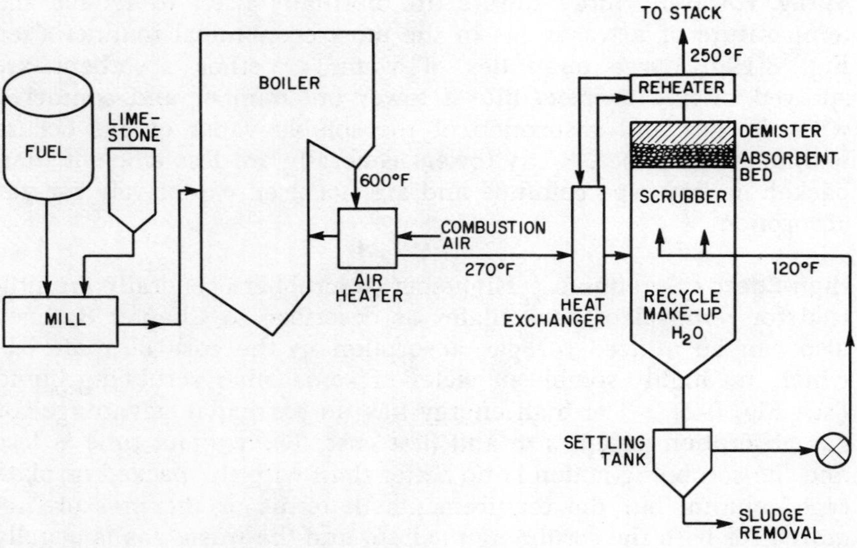

Fig. 9-4 Dolomite injection process—wet-scrubbing flow diagram. *(Courtesy Chemical Engineering Progress)*

gas well above the normal standards. Prior to the introduction of some of these systems a tall stack had been used to disperse SO_2 effluents but with the trend to eliminate the pollutant from the atmosphere, other methods, although not economical in themselves, have been developed for the ultimate recovery of the gas in one of several forms. The four primary processes which have been used to date for SO_2 removal are (1) the dry Dolomite injection process with wet-scrubbing, (2) catalytic oxidation, (3) the alkalized alumina process, and (4) the dry absorption or Reinluft process.

The first process developed by Combustion Engineering utilizes dry dolomite and limestone injected into a boiler burning a fossil fuel (Fig. 9-4). The limestone reacts with sulfur dioxide in the flue gases directly in the boiler forming calcium sulfate, or gypsum, which is removed by subsequent wet-scrubbing.

The second or catalytic oxidation process (Fig. 9-5) developed by the Monsanto Company passes the flue gases, after they have gone through a particulate removal step, through a vanadium pentoxide catalyst bed, producing about 77 percent sulfuric acid solution.

The alkalized alumina process (Fig. 9-6), developed by the U.S. Bureau of Mines, passes the flue gas through a bed of alkalized alumina, which is a physically active form of sodium aluminate, in the form of spheres about 1/16 in. in diameter. The absorber is actually a fluidized bed and the alkalized alumina reacts with the SO_2 in the flue gases at a temperature between 300 and 350°C.

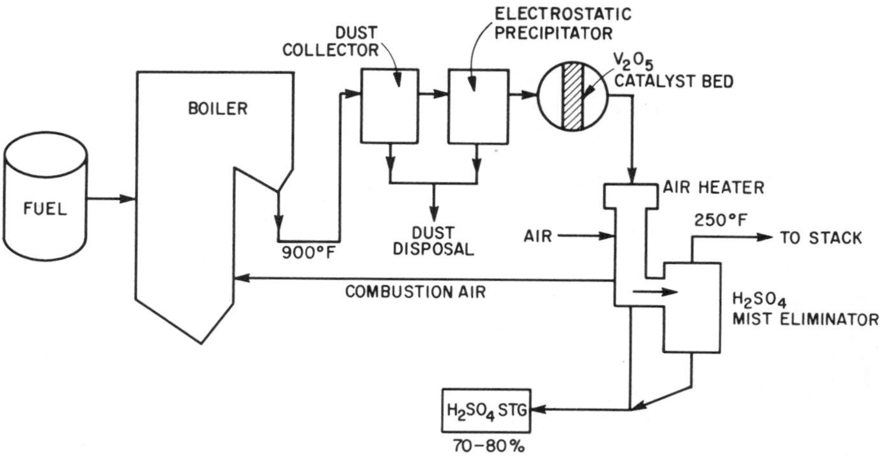

Fig. 9-5 Catalytic oxidation flow diagram. *(Courtesy Chemical Engineering Progress)*

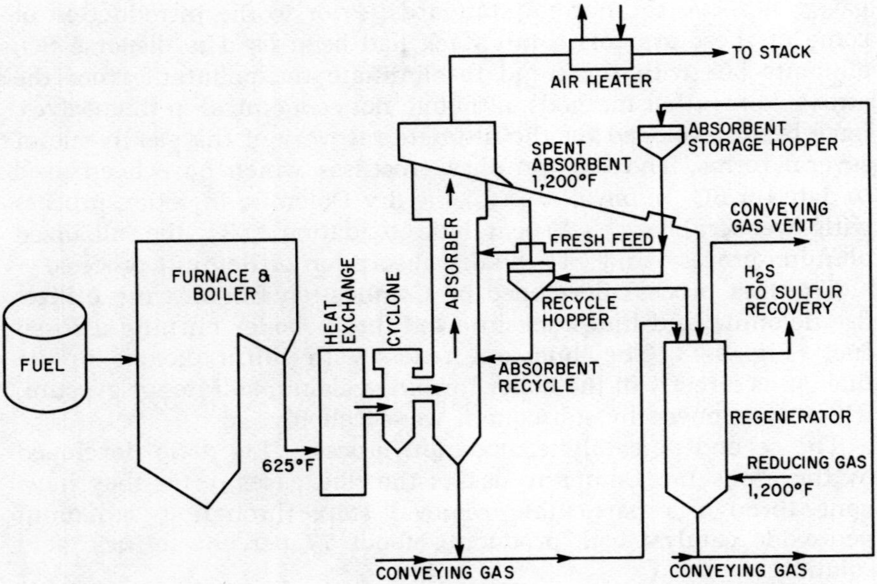

Fig. 9-6 Alkalized alumina process flow diagram. *(Courtesy Chemical Engineering Progress)*

The SO_2 is converted to the sulfate which is regenerated by heating to a temperature of about 700°C in the fluidized bed, using a reducing gas obtained by reforming natural gas or a suitable hydrocarbon feedstock. The product is H_2S gas which is processed in a Claus process plant to convert to elemental sulfur.

Finally, the Reinluft process (Fig. 9-7) absorbs SO_2 on a char formed by calcining coal at about 1100°F. The absorbent can be reactivated by treating the char with sulfuric acid and subsequent evaporation of the acid. The char can be absorbed by heating the gas containing 40–50 percent SO_2 which can be liquified or used to make sulfuric acid or sulfur.

While a number of other absorption processes have been developed for the treatment of a wide variety of flue gases, most are very specific in nature and are not applicable to the problems generally found in everyday waste gas processes. When a chemical reaction can be employed to successfully reduce pollution and it can be done at lower cost than other suitable methods it should be considered. The selection of gas absorption as a means of vapor or gas pollutant removal depends on so many factors that it would be impossible to justify a discussion of these economics here.

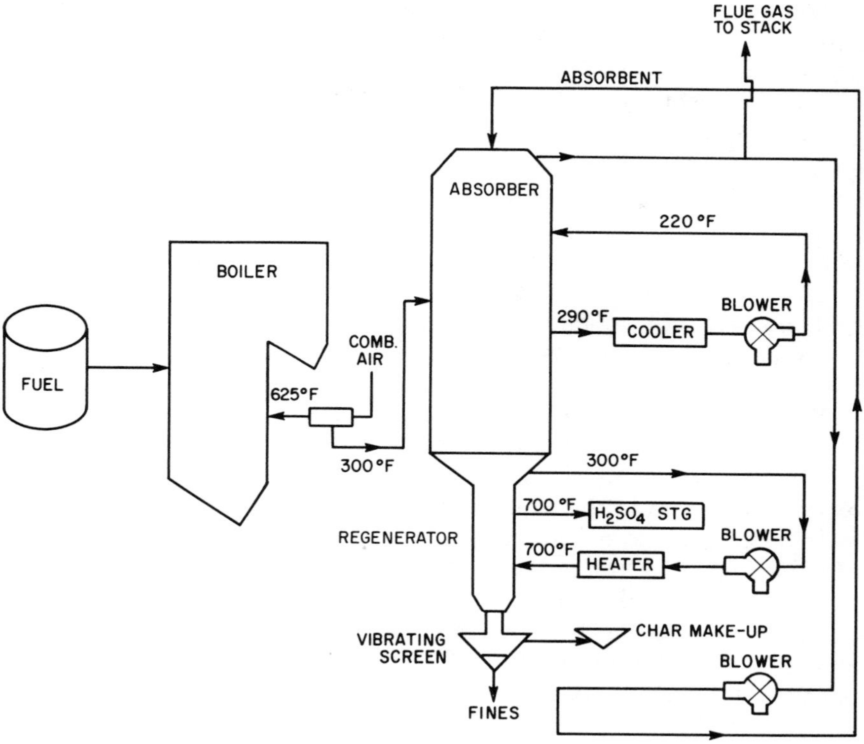

Fig. 9-7 Reinluft process flow diagram. *(Courtesy Chemical Engineering Progress)*

Permanganate and Chlorine Oxidation

Another process which should be discussed under absorption techniques is the use of potassium permanganate for the oxidation of various waste gases. This process actually involves a chemical reaction but it employs conventional gas-scrubbing or absorption equipment.

Potassium permanganate is a strong oxidizing agent. In a strong acid solution it gains 5 electrons, releasing 2½ oxygen atoms, and forms a water soluble manganous salt. The permanganate acid which is formed however is very corrosive and therefore requires expensive materials for the design of the scrubber. The preferred method of using permanganate is to dissolve it in a mildly alkaline solution where it will pick up only 3 electrons and generate 1½ oxygen atoms. It will produce an insoluble precipitate of manganese dioxide but the solution is noncorrosive and suitable for use in inexpensive scrubbing equipment.

The oxygen released from the dissolution of potassium permanganate is available for the oxidation of organic materials in the waste gas stream. It has found practical application in the removal of a wide variety of noxious odors from such sources as rendering plants, asphalt plants, and fish processing plants. The permanganate is added to the scrubbing liquid (normally water) on a continuous basis. Its strength is self-indicating. Potassium permanganate produces a bright purple color. As the oxygen is used the color fades indicating the need for more permanganate.

This process has not been widely used because of the operating cost; however it may be more economical than alternative methods for weak organic gaseous effluents that have heavy odor characteristics.

The alternatives to the treatment of such an effluent are odor counteraction and incineration, both of which can be very expensive for such wastes. Another disadvantage of the permanganate oxidation system is the precipitate of manganese dioxide. This can form a final disposal problem and probably requires the addition of some type of filter to the system.

Chlorine and chlorine dioxide can be similarly used for oxidation but take careful control to insure that the chlorine content of the exit gas is below minimums. Chlorine is generally the least expensive of the three oxidants.

CONDENSATION

An air pollution control method which is almost too obvious to mention is the process of condensation; yet it is a viable method of controlling atmospheric pollutants. Condensation is usually considered when the waste gas has substantial quantities of a condensable material such as air which is saturated with water or other vapor. It may also be considered as a method of pretreatment to other methods of removing solvent vapors. For example, an expensive solvent is vaporized in an oven with a discharge temperature of 200°F. The solvent is contained in a large quantity of air. By passing it through a surface condenser which is cooled by water at 80°F about half of the solvent can be economically recovered. The remainder must be recovered by activated carbon adsorption or by a refrigeration process, or it must be lost through subsequent incineration to meet air pollution regulations. It is still a good economical first step.

The condenser for such an operation should not be considered in the same way that a condenser is used for a steam boiler. The material that it must handle is not a pure vapor which will condense at a specific temperature, but a saturated gas stream consisting of mainly noncondensables. The solvent vapor will begin to condense along the saturation curve of that vapor in air (or other noncondensable gas). If the vapor is water then the curve would follow the psychrometric chart, and the same is true for any other vapor in a noncondensable carrier gas. The determining factors are (1) the economy of this method of recovery, and (2) the allowable limits of the pollutant in the exit air or gas stream.

The equipment that is available for this process is:

1. Tubular surface condenser (shell and tube)
2. Tubular air cooled condenser
3. Direct contact condenser

Shell and Tube Condensers

Little description is necessary here. The cooling medium is usually on the tube side with the vapor on the shell side. The cooling medium may be water, refrigerated water, or a refrigerant depending on the desired condensing temperature. Externally finned tubes are often used to reduce the surface area but in any case the overall heat transfer coefficient will be low (on the order of 10–15 Btu/hr/ft^2/°F).

Tubular Air Cooled Condensers

These are not usually practical for dilute vapor streams because the temperature difference is small and the area required large; however, they can be used more effectively where the stream contains large amounts of condensables. Generally such equipment is large in size and utilizes externally finned surface and large forced or induced draft fans. Overall coefficients are low, on the order of 2–7 Btu/hr/ft^2/°F.

Direct Contact Condensers

Spray condensers, jet condensers, and barometric condensers all utilize water or some other condensing fluid in direct contact with the vapor to be condensed. The temperature approach between the fluid and the vapor is very small, so the efficiency of the condenser is high but large volumes of the condensing fluid are necessary. If the vapor is soluble in the condensing fluid then the system is essentially an absorptive one. If the vapor is not soluble then it is a true condenser, where the temperature of the vapor must be below the dewpoint. Direct contact condensers are seldom used for the removal of organic solvent vapors because the condensate then contains an organic-water mixture which must be separated or treated before disposal. They are however the most effective method of removing heat from hot gas streams when the recovery of organics is not a consideration. For example, they can be used to handle waste gases with a high water content and low concentrations of SO_2. The removal of the water and some SO_2 as a weak acid will permit the final treatment for the residual SO_2 with more direct methods; or a scrubber-condenser utilizing caustic or a lime slurry will both condense the water and react with the SO_2.

ODOR COUNTERACTION

There are many industrial odors which are not necessarily toxic but which are definitely objectionable, sometimes to the point of being nauseating. Methods of detecting and measuring odors were described in Chapter 5.

While methods of odor masking or counteraction are not really considered air pollution control they should be mentioned as a possibility of improving a condition which otherwise may defy a simple solution. For example, a sewage treatment plant usually is not designed to smell like a lilac bush. Despite modern treatment methods it will always smell like a sewage treatment plant and when the wind is in the right (or wrong) direction, someone downwind will be the recipient of the odor. While this odor is objectionable it is not necessarily dangerous. The concentration of organic materials in the air is well below established minimums and except for the aesthetic aspect of the odor it does not affect the ecology of the neighborhood.

It would be almost impossible to solve the problem described above with presently available air pollution control methods. Primary and secondary treatment in municipal sewage treatment plants is usually accomplished in clarifiers that are open to the atmosphere so that some odor dissemination in the air is possible. Some municipalities have gone to odor counteraction to solve this problem.

It is a fact that strong odors tend to mask weaker ones. Therefore if we inject a chemical with a strong odor into a waste gas having a weak odor the olafactory nerve as the recipient of such odor would probably perceive only the stronger one. This is known as odor masking. In the same way, certain odors in appropriate relative concentrations will tend to destroy the odorous properties of both and the intensity of each is diminished. This is called odor counteraction or neutralization. Such methods are used in situations as the one described above for the sewage treatment plant. Similar situations occur with odors from bodies of water, landfill operations, ovens, storage tanks, and other places where fumes cannot be easily captured and ducted to an incinerator or some other form of treatment.

Since odors are gases, the material used as the odor counteractant must also be in gaseous or vapor form. This is done by dispersion. Chemical odor counteractants are atomized into the air through a nozzle designed to provide a fine spray. This is usually done at ground level with the counteractant chemical stored in a drum and pumped through the atomizing nozzle under pressure. It is not necessary however to work at ground level. It is obvious that such techniques are applicable at the top of a stack also. The odor counteractant is usually a concentrated chemical so that only minute amounts are required to counteract odors in large amounts of air.

Every odor needs its own particular counteractant. The chemical company that supplies the material will have to test the odor to be counteracted before making a recommendation for the proper counteractant. The counteractant should never be a material that desensitizes the olafactory nerve because without the sense of smell the individual is unable to identify toxic emissions that might be released in the same area.

ADSORPTION

Adsorption is a molecular surface phenomenon where molecules of a fluid contact and adhere to the surface of a solid. Adsorption processes demand important consideration in air pollution control since organic gases can be selectively captured or removed from air streams by the use of specific adsorbents.

In order to be a good adsorbent a material must have a large surface-to-volume ratio since the phenomenon occurs on the surface of the material. It is the magnitude of the surface area rather than the specific nature of the surface that is primarily responsible for the ability of activated carbon or other material to absorb large quantities of organic molecules.

Adsorbents

Adsorbents which are used for air pollution control consist of different types of material—activated carbon, silica gel, alumina, and bauxite. For practical considerations activated carbon is the only one which is used excessively for the removal of organic vapors from waste gases. While silica gel is capable of removing such materials it can only do so when no water is present, because silica gel is also used as a dehydrator and tends to pick up water much faster than organic molecules. Alumina and bauxite are also used mainly for dehydration of air and various gases.

It is entirely conceivable that silica gel could be used to remove organic gas from a waste stream if the latter was relatively dry. Regeneration would require dry heat and this would do no more than re-create the air pollution problem. Activated carbon, on the other hand, is regenerated by using steam which is subsequently condensed and separated from the organic material.

Anyone who has worked in a plant where air is used for instrumentation is familiar with the dual bed silica gel or activated alumina drier which is used for removing water from compressed air. This is an adsorption phenomenon and the bed is regenerated with dry heated air or gas. While water vapor emanating from a process is often suspect and will cause complaints from the uninitiated, most pollution regulations allow steam or water vapor to be vented to the atmosphere. Therefore for the purposes of this chapter we are considering organic vapors which are beyond the permissible level as specified by local, state, and national air pollution requirements and we are therefore concerned with the use of activated carbon as an adsorbing medium. Activated carbon is used as an adsorbent because it has a remarkably large surface-area-to-volume ratio, about 5 million sq ft/lb, and an extremely small pore size, on the order of 10–20 Å. There are several theories of what happens when a waste gas containing organic molecules passes through a bed of activated carbon. One is that the unbalanced forces of the surface molecules of the carbon are satisfied when other molecules become attached to the surface, or adsorbed. The forces responsible are commonly called van der Waals' forces. A second type, chemisorption, is encountered with less frequency. In this case the material to be adsorbed is bound to the surface by strong chemical forces and the process is irreversible. The carbon selected for vapor phase adsorption must have certain qualities. It must be sufficiently hard and abrasion-resistant to withstand the breakage which will occur in a bed of material through which gas is flowing. It should also have a high capacity for adsorbing the gas or vapor which it is meant to recover and it should easily release the adsorbate during the regeneration cycle. The raw materials used for the manufacture of vapor phase carbons consist of coal, by-product acid sludge coke, and coconut shells and other nut shells. The sizes normally used for vapor phase adsorption in the U.S. Sieve Series are 4 x 10, 6 x 16, and 12 x 30, although other sizes are available.

Activated Carbon Systems

Manufacturers of activated carbon have developed adsorption data from a wide range of organic materials which can provide the plant engineer excellent guidance concerning the mesh size and type of carbon which should be used for organic vapor removal. Manufacturers of adsorption equipment may also be able to supply similar information and a cross-check of recommendations is always available through such methods. If the organic vapor or mixture has not been handled by the equipment manufacturer or the supplier of the activated carbon, then laboratory tests of the material should be run before purchasing and installing equipment.

Applications. In selecting activated charcoal adsorption as a means of air pollution control it should be remembered that activated charcoal is more selective toward large molecules and those having higher boiling points under any given set of operating conditions. Since regeneration of the charcoal is accomplished by heating, it obviously would be difficult to adsorb any organic material on the surface of carbon at a temperature above its boiling point.

In order to determine whether an activated carbon adsorption system is applicable we must know certain information about the waste gas stream. We must know the concentrations of the various contaminants in the stream; operating conditions such as temperature, relative humidity, and boiling point of the various contaminants; and whether or not they are miscible or soluble in water.

Activated carbon adsorption is generally used to remove organic materials from waste gas streams when the following conditions prevail: (1) when the stream contains very small amounts of any organic solvent which cannot be discarded to atmosphere; (2) when the stream contains relatively high concentrations of solvent which should be reclaimed; (3) when it contains one or more organic materials which cannot be removed by any other pollution control method.

The first situation is rather remote since incineration can always be used for solvent-air mixtures; however, the economic feasibility of incineration as compared with activated carbon adsorption is probably predicated on the amount of solvent or organic vapor present in the stream. The second reason is the usual one for selecting activated carbon adsorption. In this case we are trying to reclaim the solvent or organic material because of its value and it can be reclaimed by this method. In case 3, where activated carbon adsorption is the only acceptable method, we can either discard the carbon and collected solvent or regenerate and separate the various solvents by distillation.

In order for a stream to be suitable for activated carbon adsorption it must be free of all particulate matter. Particulate matter will have a tendency to block off the passages between the carbon particles, increasing pressure drop till the flow stops. The waste air-solvent mixture temperature must be below the boiling point of the contaminants which are to be recovered. Gas velocities through the bed must be in accordance with those recommended by the activated carbon supplier. A typical curve is shown in Fig. 9-8. It is also desirable to keep the pressure drop as low as possible since this will affect the economics in terms of the cost of the fan horsepower which must be used to push or pull the vapors through the bed. Pressure drops for various carbon mesh sizes are shown in Fig. 9-9. The deeper the bed the better the transfer of the adsorbate from the gas to the carbon particle. However, this must be balanced against the pressure drop through the bed. Generally speaking, thin beds with a small mesh size carbon have the same efficiency as thick beds with a larger mesh size.

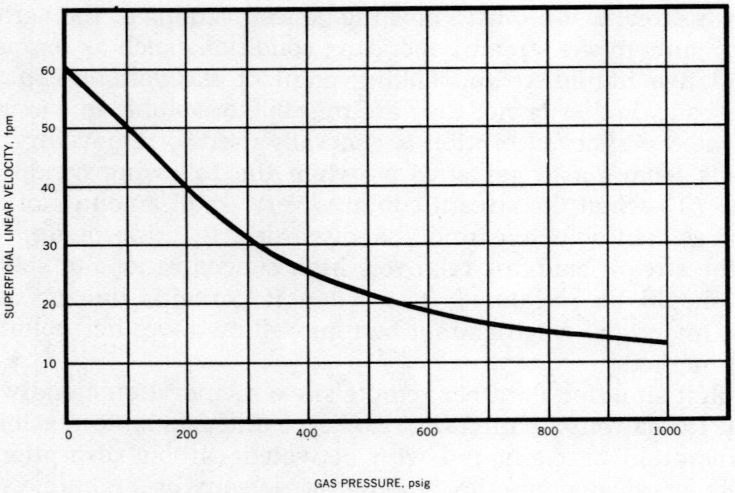

Fig. 9-8 Maximum recommended gas velocities for pressure systems. *(Courtesy Pittsburgh Activated Carbon Co.)*

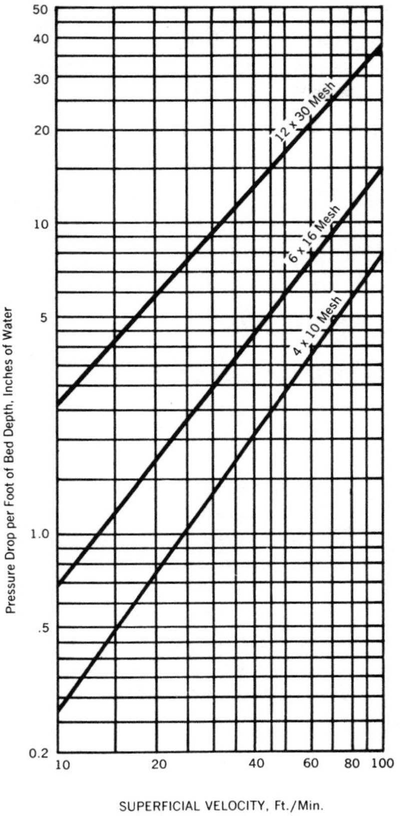

Fig. 9-9 Pressure drop as a function of superficial linear velocity for three different sizes of activated carbon. *(Courtesy Pittsburgh Activated Carbon Co.)*

System Types. There are two basic types of systems, regenerative and nonregenerative. For most industrial applications the regenerative type is utilized. The regenerative type is designed for the recovery of expensive solvents or vapors. The nonregenerative type is generally used where the vapor concentration is extremely small and the carbon and the adsorbed vapor can be discarded. The regenerative type of system must be designed to adsorb organic materials at the operating temperature and release them when the carbon temperature is raised above the boiling point of the solvent. Regenerative systems are obviously more expensive than nonregenerative types because they must include recovery or separation equipment.

Several applications for nonregenerative type systems are found in air conditioning systems, atomic power plants, atomic submarines, schools, and restaurants. These systems will remove small quantities of cigarette smoke, exhaust gases from internal combustion engines, and odors associated with the home, restaurant, etc. by filtering out organics and certain particulate media. Generally these are thin beds with low pressure drops on the order of one-quarter of 1 in. W.C. The carbon is packaged in replaceable cartridge form. The cartridge after being loaded with the pollutant is either discarded or sent back to the manufacturer for regeneration. Regeneration of this type of activated carbon is usually done at high temperatures in a furnace. The regenerative-type system which is mainly used in industrial applications is suitable for recovery of many organic vapors and solvents. Table 9-1 covers some of the applications in which regenerative systems are used.

TABLE 9-1 APPLICATIONS FOR REGENERATIVE SYSTEMS

Acetone	Isopropyl alcohol
Adhesive solvents	Ketones
Amyl acetate	Methyl alcohol
Benzene	Methyl chloroform
Benzol	Methylethylketone (MEK)
Brom-chlor methane (BCM)	Methylene chloride
Butyl acetate	Mineral spirits
Butyl alcohol	Mixed solvents
Carbon bisulfide	Monochlorobenzene
Carbon dioxide (controlled atmosphere)	Naphthas
Carbon tetrachloride	Paint manufacturing
Coating operations	Paint storage (vents)
Degreasing solvents	Pectin extraction
Diethyl ether	Perchlorethylene
Distilleries	Pharmaceutical encapsulation
Dry cleaning solvents	Plastic manufacturing
Drying ovens	Rayon fiber manufacturing
Ethyl acetate	Refrigerants (halocarbon)
Ethyl alcohol	Rotogravure printing
Ethylene dichloride	Smokeless powder extraction
Fabric coaters	Soya bean oil extraction
Film cleaning	Stoddard solvent
Fluorohydrocarbons	Tetrahydrofuran (THF)
Fuel oil	Toluene
Gasoline	Toluol
Halocarbons (some)	Trichlorethane
Heptane	Trichloroethylene
Hexane	Varnish storage (vents)
Hydrocarbons (aliphatic)	Xylene
Hydrocarbons (aromatic)	Xylol

Equipment

Equipment used in the regenerative type of system is generally described in Fig. 9-10. It may be a two-bed or a three-bed

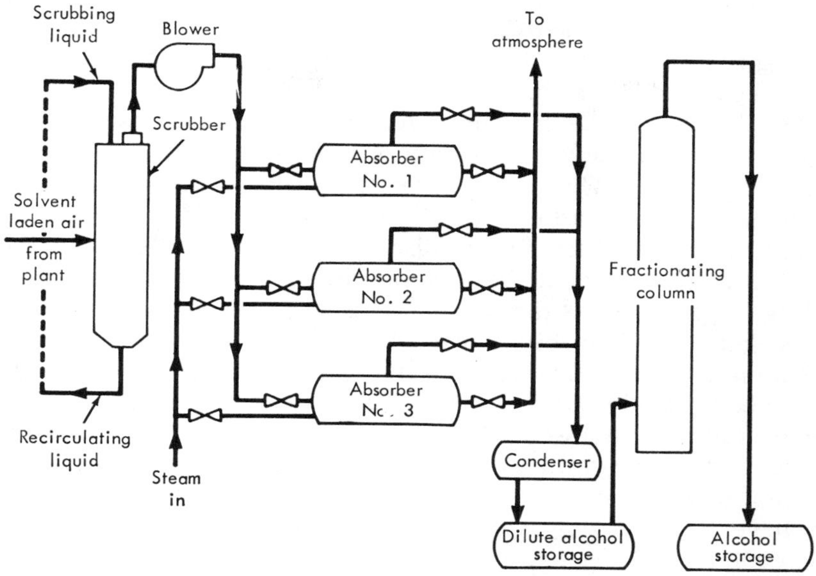

Fig. 9-10 Solvent recovery system. *(Courtesy Heating, Piping & Air Conditioning)*

version depending on the mode of operation, or it may be a single-bed system if the operation permits. The process of adsorption of vapor on the surface of the activated carbon bed is progressive. As the vapor-laden gas begins to pass through the bed it saturates those particles of carbon first contacted. As these become saturated with the organic vapor they no longer pick up the organic molecule. Then the particles deeper in the bed must begin to do their part in the adsorption process. Finally the bed becomes saturated with the organic molecules and there are no more unsaturated carbon particles in the waste gas stream. When this occurs there is a marked increase in the organic content of the effluent gas. This is known as the breakthrough point. It is the point at which it is time to regenerate the bed. The waste gas is then valved off and passed into another adsorber, if one is available—which would be the case in a two-bed or three-bed system. The adsorber is then saturated with superheated steam. The steam temperature must be above the boiling point of the solvent or organic vapor to be removed from the carbon. Saturated steam is preferred because it can be immediately condensed. It is passed through the bed until essentially all of the organic molecules are released from the activated carbon. In a standard adsorption system the maximum steam usage is at the beginning of the cycle. A well-designed plant will use 1–4 lb steam per pound of solvent recovered, depending on the size of the adsorber, since steam must be used not only to heat the carbon above the boiling point of the organic material to be recovered but also the entire adsorber.

The direction of desorption is the opposite of adsorption because this will prevent the accumulation of materials which might otherwise polymerize. The steam-solvent mixture is condensed in a surface condenser and is collected in the liquid phase. If the solvent and the water are immiscible they may be seperated by decantation. However if they are miscible then no such separation will be possible. After the regeneration step the carbon bed should be cooled and dried with solvent-free air. This is only necessary to obtain maximum recovery of water-miscible solvents. The surface of the bed should not be completely dried because it is good to leave some moisture on the surface of the carbon; but it is necessary to reduce the temperature of the bed because its absorptive capacity will not be as great if it is not cooled. In a normal cycle therefore, we have the following steps: (1) adsorption, (2) regeneration, (3) cooling, (4) drying.

The most common arrangement is the twin-bed adsorption system since the regeneration, cooling, and drying steps usually take no longer than the adsorption cycle. The one adsorber is ready to begin to adsorb the solvent by the time the other reaches breakthrough. In certain cases where very large adsorbers are used the cooling and drying cycle may take longer. Then a three-bed adsorption system is preferred, having one bed adsorbing, one regenerating, and one cooling and drying at the same time. A single-bed adsorber which is regenerative can only be used on a discontinuous basis. For example, if a process emits a waste gas containing organic vapors for 3–4 hr a day, a single carbon bed of sufficient thickness can be designed to hold the entire amount of solvent which will be discharged in this period of time. It can be regenerated, cooled, and dried during the shutdown period if this is long enough, and then put back on the absorption cycle when needed. The single-bed system should be applied where applicable since it is lower in first cost and in operating cost.

Packaged carbon absorption units are available to handle up to about 5,000 cfm of waste gas. Above this the unit is usually field-constructed and installed. Cycle time depends on the bed thickness and ranges from 15 to 20 min depending on the solvent concentration and the size of the bed.

Materials of Construction. Materials of construction of an activated carbon adsorption system will depend upon the organic vapors being removed. Generally carbon steel construction can be used for most applications. Stainless steel is used for esters and ketones because organic acids are formed by hydrolysis of these compounds. If chlorinated hydrocarbons or fluorocarbons are to be adsorbed highly corrosion-resistant nickel or rubber lining of the adsorber vessels may be required.

Safety Considerations

Activated carbon adsorption systems are designed to remove small quantities of organics from a waste gas. Occasionally concentrations of organics in air are high but regulations usually limit the concentration admissible to a treatment system to 25 percent of the L.E.L. Sometimes up to 50 percent of the L.E.L. is permissible but never higher. Explosion-proof electrical systems are usually required. This care is important to insure safe operation. There is a heat of adsorption of all solvents and this should be calculated to be sure that carbon bed temperatures are at or below the vaporization point of the solvent handled.

Economics

Most activated carbon adsorption systems involve a significant capital investment. Costs vary widely depending on the solvent or solvents to be recovered and the materials of construction. For carbon steel the cost averages about $3000 per 1,000 scfm of waste gas. Carbon costs about $0.42 lb and the carbon requirements are based on the solvent to be adsorbed and the quantity of solvent and air. Generally 10 lb carbon per pound of solvent is a reasonable capacity ratio.

COMBUSTION PROCESSES

The destruction of a waste gas or vapor by a combustion process is called incineration. Today, incineration is perhaps the most widely used process for the destruction of organic vapors emanating from manufacturing processes. The most common problem in handling a waste gas effluent is its total volume. Usually the contaminant level is quite low but the volume of the waste gases is high. Consequently we have a volume-oriented problem rather than a pollutant-oriented one. This is true in virtually every type of waste gas treatment apparatus whether it be incineration, absorption, adsorption, or other methods. For example, a single 100,000 lb/hr boiler, which is not large by today's standards, may give off 40,000–50,000 scfm of flue gas. If it is necessary for economic reasons to operate the boiler on either coal or residual fuel oil which contains sulfur, and if it is necessary to subsequently remove this sulfur, then the physical size and cost of the treatment apparatus is not dictated by the amount of sulfur present in the flue gas but rather the total volume of the gas that must be handled. This type of problem is present in most waste gas incineration systems; that is, the amount of pollutant is small but the amount of nontoxic carrier gas, usually air or inert gas, is high in comparison.

With incineration, as with other types of waste gas treatment equipment it is important that we know several things before we begin to design the system to handle the waste. These things are: (1) temperature, (2) volume, (3) chemical constituents, (4) dew point, and (5) permissible atmospheric tolerance levels for the pollutants in the gas. With the knowledge of these factors we can then select the method which is best for final disposal.

Assuming that we have selected incineration as the best disposal method for both process and economic reasons, then we must determine which type of incineration system is the most applicable. There are three basic types of waste gas incineration systems: (1) direct flame, (2) thermal, and (3) catalytic.

Direct Flame Incineration

Direct flame incineration is the method which is normally used when handling gaseous waste materials which are at or near their lower combustible limit when mixed with air. It is also used when the waste gas itself is a combustible mixture without the addition of air. In a well-designed commercial combustor or burner, gases having heating values as low as 90 Btu/ft^3 can be burned without auxiliary fuel. In some cases even lower heating value materials can be burned in this manner when they are preheated to 600 or 700°F. Blast furnace gas is a typical example of a low heating value fuel which sustains combustion. However, in the case of blast furnace gas we are talking about a fuel which is normally used in the steelmaking process rather than direct flame incineration. But if we were to have a waste gas with a similar calorific value and use it as a fuel in a burner, then the process would be incineration, especially if we were trying to destroy the contaminants or pollutants in the waste stream.

Hydrogen sulfide is a good example of waste gas where the heating value is high enough to sustain combustion when mixed with air. It is a highly toxic gas with a very disagreeable odor. While sulfur often can be recovered from H_2S waste gases there are many situations where the gas effluent has no value and must be incinerated. This is especially true of concentrations of H_2S below 100 percent. By incineration the toxicity and the odor of the H_2S can be eliminated. However, sulfur dioxide, which is a toxic gas, will be formed, requiring further treatment.

Hydrogen cyanide is perhaps a better example of a gas which is extremely toxic but may be burned in air in various quantities, producing an effluent which is acceptable to the atmosphere. Carbon monoxide is another deadly gas and a by-product of many partial combustion reactions which can also be burned by direct flame combustion. Direct flame systems are not only concerned with single gases but also with combinations of combustible gases and combinations of solvent vapors in air. In all such cases these wastes, if they are above the upper flammable limit, can be mixed with air and burned as a fuel in a conventional combustor or burner even though the process is considered incineration.

There are also a number of situations where the waste gas may contain constituents which are below the lower flammable limit. Then it may be necessary to add quantities of natural gas or other auxiliary fuel to sustain the combustion in the burner, but by passing the waste gas directly through the burner and utilizing the combustibles in the waste gas as fuel we can obtain complete destruction of the waste in a very short period of time. Direct flame systems usually operate at high temperatures, above 2000°F. Good mixing is achieved with the oxygen in the air if the burner is properly designed to produce such mixing and the resultant vapors should be carbon dioxide, nitrogen, and water vapor. In any direct flame incineration system the organic or solvent vapor serves as part of the fuel and it should contribute a significant portion of the total heat released to the system. If it does not, another type of incineration system should be chosen, either thermal or catalytic. As a general rule of thumb, direct flame combustion should be employed only where the amount of auxiliary fuel needed to sustain combustion is low and where the contaminant supplies at least 50 percent of the fuel value of the mixture.

The equipment for direct flame incineration is usually a conventional burner or combustor firing into some enclosure, or if practical, into the open. An open-fired combustor, especially if it is aimed in the vertical position, is generally called a flare. A flare is a special type of direct flame combustor which is usually found in petroleum refineries and petrochemical plants.

Flares. A flare is a highly specialized type of unsteady state combustor with an exposed flame burning into the atmosphere. It may be located hundreds of feet above grade, in which case it is known as a tower flare, or it may be utilized at ground level, in which case it is known as a ground flare. Flare combustion is used for combustible gases at energy concentrations within or above the flammable range because it is not designed to use auxiliary fuel. The primary use of flares is where great variations occur in the gas flow exceeding by many times the type of variation which would be found in industrial burner equipment. They can also be used where safety or emergency conditions dictate to protect plant and personnel.

Flares are seldom used for solvent vapors. They are used primarily for waste gas fuels which cannot be handled otherwise and materials which burn rapidly with a low percent of excess air yielding a high flame temperature. Flare burning leaves little ash or residue; however, it is susceptible to weather conditions to some extent, especially high winds which tend to disperse some of the gas stream before it can be burned.

There is a natural tendency for most gaseous hydrocarbons to generate smoke from combustion. The lower the hydrogen-to-carbon-weight ratio in the gas the greater the chance of producing black smoke. However gases with a hydrogen-to-carbon ratio of about 0.33 or above are easily burned without this problem. Such gases would be acetylene, propane, ethane, and methane.

There is a problem however, with waste gases having lower hydrogen-to-carbon ratios when operating standard flare equipment. Without special adaption they will probably create large quantities of smoke. The flaring of the low hydrogen-to-carbon-ratio hydrocarbons can be made smokeless by injecting steam into the flame close to the point of ignition. This promotes turbulence and provides better carbon-to-oxygen contact by inspirating excess air into the flare. It also reacts with the fuel to form oxygen and compounds that burn rapidly at lower temperatures and it retards polymerization of the fuel. For hydrogen-to-carbon ratios greater than 0.33 no steam is needed. The lower the ratio the greater the steam requirements.

Flares are basically open pipes which discharge combustible gas directly to the atmosphere, the end of the pipe containing a flame-holding device and a continuous pilot or pilots to ignite the waste gas. Air for combustion is supplied by the surrounding atmosphere (see Fig. 7-3).

Flares should be used with discrimination since they are not a cure-all problem for waste gas disposal. They are not suitable for other than industrial areas because they produce a bright light during night operation and in some cases a significant noise level.

Thermal Incineration

By far the greatest number of waste gas incineration problems involve a mixture of organic material and air in which the amount of organic is very small. This means that if this waste is injected directly through a burner along with auxiliary fuel such as natural gas, the amount of the latter required to achieve complete combustion is quite high. Most conventional industrial burners require temperatures of 2200°F or greater to sustain combustion whereas normal thermal incineration can be carried out at much lower temperatures, sometimes as low as 900°F, but generally between 1,000 and 1500°F. These weak mixtures of organic material and air will usually have very low heating values on the order of 1–20 Btu/ft^3. Common applications of this type may be found in drying ovens which drive off a solvent or plasticizer, a lithographing oven which drives off lithographing solvents, or a paint-baking oven which will drive off solvents normally associated with paint coatings.

In this type of process the waste gas is essentially air and contains enough oxygen to complete combustion of the organic contaminant. But in some cases where sufficient oxygen is not present it can be added by means of a fan or blower either by premixing with the fume or by injecting it into a secondary combustion chamber along with the fume. Such a situation might occur where an inert gas is being used as a carrier gas in an oven for safety purposes. In almost every case the waste gas-air mixture should be below 25 percent of the L.E.L. Most insurance regulations specify 25 percent of the L.E.L. as a safe mixture to convey to an incinerator or other device. The reason for this is that as the organic air mixture approaches the lower explosive limit it is possible to have flashbacks from the incinerator into the process equipment.

The Three T's. The three T's of combustion are paramount in thermal incineration systems. They are time, temperature, and turbulence, and the optimization of these three is extremely important to building the most economical thermal incinerator.

"Time" is the residence time in the incinerator or the time required for complete combustion of the waste material. "Temperature" is the operating temperature of the incinerator to which the waste gas must be raised to achieve complete combustion. "Turbulence" is the mixing within the incinerator of the waste gas with air which supplies oxygen for combustion.

The residence time in any incinerator must be extended if we operate at low temperatures with little turbulence. The temperature of any incinerator must be increased if the residence time is dropped and the turbulence is poor. The turbulence in an incinerator must be greater if we reduce either the residence time or the temperature. Therefore, the three are interdependent and an improvement in any one will allow us to reduce the other two. Rates of reaction are available for many chemical compounds when combined with air and they can be calculated, but empirical data for most materials derived from actual test results is the most reliable. Most equipment manufacturers of this type of waste gas equipment can provide such information. Since good turbulence or mixing is the easiest and cheapest of the three to supply in any combustion system it should always be considered first. Long residence time makes the incinerator larger and more expensive, and higher temperatures utilize more auxiliary fuel and result in higher operating cost.

Generally speaking, the residence time in most waste gas incineration systems varies between one-quarter of a second and one second with the average design being in the half-second range. The temperature range for thermal incineration of most organic and solvent vapors is between 1200 and 1500°F; however, lower temperatures, as low as 900°F, and higher temperatures, as high as 1800–2000°F, are required for certain materials.

Equipment. Equipment for thermal oxidation of gaseous waste varies quite widely depending upon the manufacturer, but several basic types of equipment are used today. The first is the line burner which is used with a fume containing sufficient oxygen for its own combustion. Line burners can be nothing more than a gas pipe with a number of holes which inject a fuel such as natural gas into the waste gas stream at the point of ignition. The waste gas therefore passes through the flame of the line burner and is heated to a temperature above the autoignition temperature of the organic constituents. More sophisticated arrangements of the line burner employ a series of baffle plates over the gas jets which promote better contact between the flames and the waste gas (see Fig. 9-11). The line burner is usually installed in the waste gas duct or an extension of the duct which is refractory lined or insulated beyond the point of ignition. Other systems utilize an external burner which can be either a natural, forced-draft, or aspirating type. In this system the burner is usually located at one end of a cylindrical or

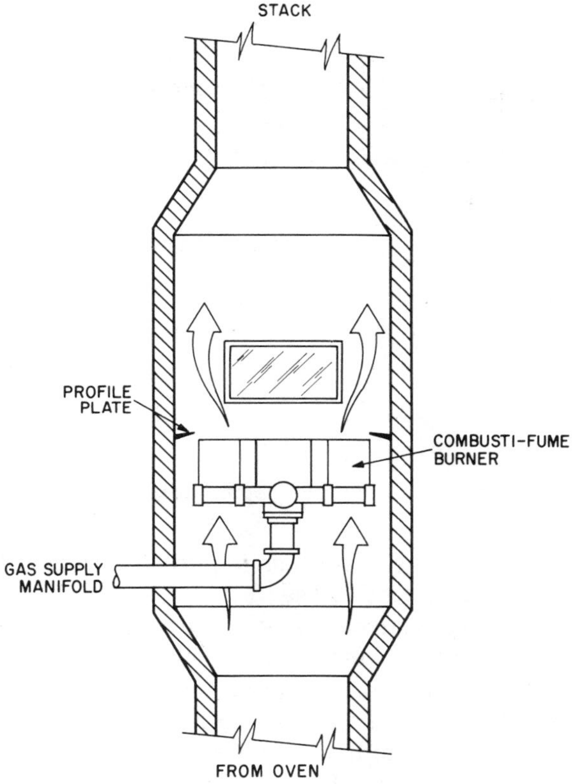

Fig. 9-11 Duct type fume burner. *(Courtesy Maxon Corp.)*

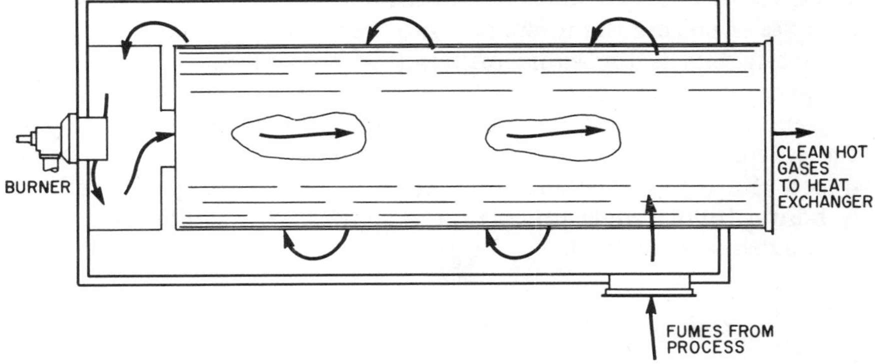

Fig. 9-12 Thermal incinerator.

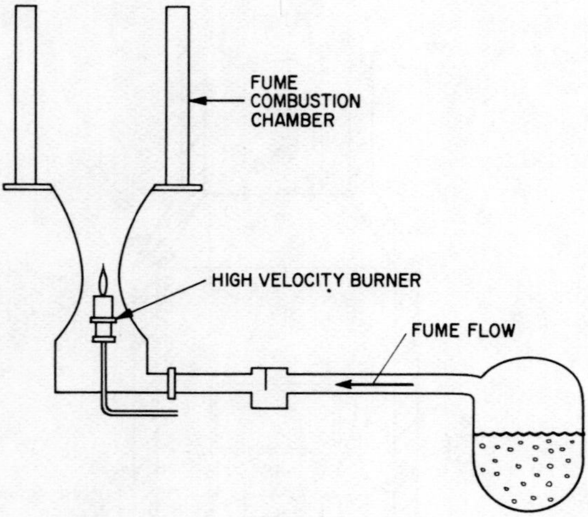

Fig. 9-13 Jet incinerator. *(Courtesy Thermal Research & Engineering Corp.)*

rectangular duct and the waste gas is passed into the duct through the burner flame or around it (see Fig. 9-12). Baffles or tangential entry of the waste gas give sufficient turbulence to mix the high-temperature products of combustion from a conventional burner with the waste gas so that a final temperature of the desired magnitude is achieved. Such units can be constructed of refractory lined carbon steel or stainless steel if the temperatures are not too high, and can be either horizontally or vertically disposed depending upon the most desirable physical arrangement of the system.

The third type of system could be called the jet burner. Here a high-velocity discharge burner is fired into the throat of a refractory-lined venturi section and the venturi action creates a slight suction capable of pulling the waste gas into the incinerator (see Fig. 9-13). Basically this is the same principle as the tunnel-type burner arrangement except that a waste gas fan is not required.

Generally it is more desirable to use a forced draft fan to push the waste gas into a thermal incinerator then it is to use an induced draft type because the latter will have to be capable of handling high temperatures unless some intermediate heat transfer device is employed.

Catalytic Incineration

Catalytic incineration is usually considered for gaseous wastes containing low concentrations of combustible materials and air. It is a system which is directly competitive with the thermal system although care should be taken to see that it is not used on high concentrations of organic materials.

Catalysts. A catalyst is defined as a material which promotes a chemical reaction without actually taking part in it and the catalyst does not change nor is it used up. While there are literally dozens of types of catalysts used in industry, those used for waste gas treatment or purification are relatively few. They are usually noble metals such as platinum or paladium dispersed on some form of catalyst support which can be easily distributed across the flow of waste gas. Since the temperature of the catalyst may go as high as 1500°F the material which supports the noble metal must be capable of withstanding these temperatures. The logical selection is generally alumina and the catalyst is dispersed on the surface of the alumina in some chemical compound of the metal.

Commercially available catalysts which are used for this purpose today come in four basic forms: the first is the wire screen, usually random in nature, which is placed directly in the path of the effluent gas. The second form of catalyst support consists of a number of alumina air foil shaped rods. The third is spherical or cylindrical pellets of alumina, and the fourth is an alumina honeycomb. In all cases the noble metal is deposited on this support which is located in the path of the waste gas. The catalyst support offers a tremendous amount of surface area on which the combustion reaction can take place. The catalyst support also has a very high percentage of free area, i.e., open area through the support. This is usually on the order of 90 percent and pressure drops are quite low, on the order of several tenths of an inch.

Operation. Most waste gases from industrial processes have low temperatures and therefore preheat burners are required to raise these to the reaction temperatures so that the catalyst can be effective. Because of the catalyst the reaction temperature in catalytic systems is lower than it is in thermal systems. Most catalytic reactions can be carried out at preheat temperatures of between 600 and 1000°F which, of course, results in a fuel saving when compared with thermal systems; however, it also involves a higher investment because of the initial cost of the catalyst. The preheat section of the catalytic incinerator looks very much like a thermal incinerator. A line burner or tunnel burner concept can be employed but usually the tunnel burner or single preheat burner is preferred. When the gases have been preheated to a temperature high enough to cause the reaction to occur on the surface of the catalyst the effluent gases should indicate complete combustion (see Fig. 9-14). The incineration reaction occurring on the surface

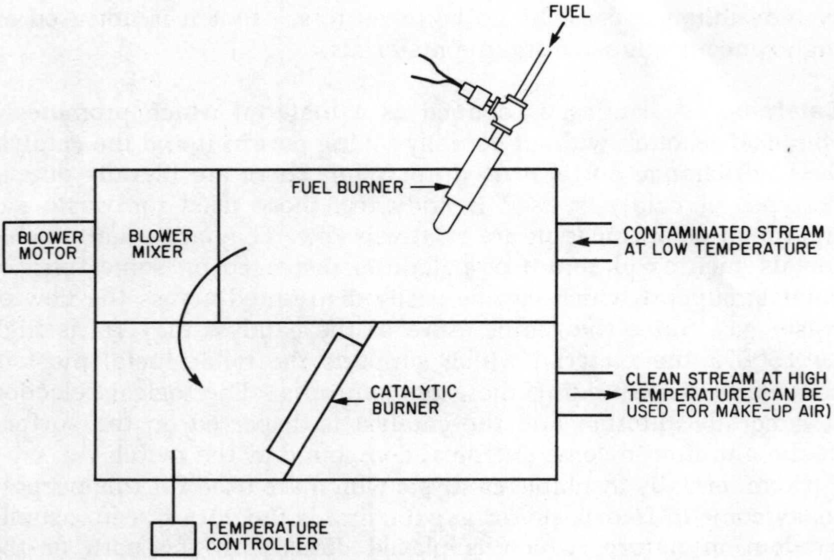

Fig. 9-14 Catalytic incinerator.

of the catalyst will cause the catalyst to glow, but unless the concentration of the waste gas is high there should be no visible flame. Combustion is completed on the surface of the catalyst and the effluent gases of carbon dioxide, nitrogen, and water vapor can be passed to the atmosphere. Most catalysts are suitable for maximum operating temperatures of 1500–1600°F. A high concentration of organic contaminant in the waste gas, even with minimal preheat, may release enough heat on the surface of the catalyst to cause burnout. Therefore catalytic systems should be carefully examined for their compatibility with the waste gas. They are most applicable to low concentrations of contaminants where the temperature rise across the catalyst will be on the order of 100–200°F. The temperature rise can be calculated based on the quantity of organic material in the waste gas multiplied by its calorific value.

Applications. Catalytic systems have been widely used for the oxidation of paint solvents, odors arising from chemical manufacturing, lithographing ovens, food preparation, wire-enameling ovens, and similar applications.

Catalyst Poisons. One difficulty with a catalyst is the fact that it can be contaminated or poisoned. While the noble metal is almost always recoverable, if the catalyst is poisoned the cost of replacing it in the system is a significant consideration. Materials such as phosphorous, silicon, and lead are known to shorten the life of catalysts. Iron and vanadium are also damaging to most catalysts used in the air pollution field. Arsenic is a catalyst poison. Halogen and sulfur compounds will tend to suppress the functionality of the catalyst and lower its life; however, they cannot be considered poisons.

In addition to poisons the catalyst may be blinded by dust or dirt. Obviously if there is particulate matter in the waste gas to be burned there is the possibility of it depositing on the surface of the catalyst. Wherever these deposits form, available surface area for combustion is decreased, which lowers the effectiveness of the catalyst and also its life. For example, in high-temperature paint-baking ovens, pigments exhausted with the waste gas could reduce the life of the catalyst. The average life of a catalyst under normal operating conditions is three to five years. It can be reactivated by returning the catalyst in exchange for a new catalyst charge.

Heat Recovery

Waste gas incineration systems are ideal for heat recovery and heat recovery systems can be applied to the direct flame, thermal, or catalytic types. In the direct flame type of incineration system the heat recovery would probably be a waste heat boiler since the temperatures generated by the direct flame incinerator or combustor are generally high. Boilers are sufficiently common and no detailed discussion of this type of system is required here. For the thermal incinerator or the catalytic incinerator however, a different approach is generally more economical because the incineration temperatures are in the range of 1200–1500°F and boilers are not efficient at these levels.

If incineration is selected as the disposal method it usually means that particulate matter, if present initially, has been removed by mechanical methods and corrosive or toxic gases have been eliminated because they cannot be discharged to the atmosphere. Therefore, after combustion we usually have inert gases containing carbon dioxide, nitrogen, and water vapor, and possibly some residual oxygen at temperatures from 800–1600°F.

The most immediate use for this heat is to preheat the waste gases prior to incineration, especially if they are free of corrosive and particulate matter. By using this heat we can preheat the incoming fumes to a temperature at least halfway between their exit temperature from the process equipment and the incineration temperature. For example, let us assume that we have fumes at 180°F coming from a process oven and that these fumes are well below the flammable limit. Experience indicates that the organic material in the fumes will be rapidly oxidized at 1300°F and this is the temperature which we select for incineration. The fumes from the oven pass through one side of the heat exchanger or recuperator where they are indirectly heated to 1050°F. The heated fumes then pass into the incinerator where they are heated the additional 250° to the incineration temperature of 1300°F by direct mixing with the products of combustion from the auxiliary fuel burner. The exhaust gases from the combustion reaction then pass over the other side of the heat exchanger supplying the heat to preheat the fumes and are finally vented to the stack at temperatures around 600°F instead of 1300°F which would not have been possible had we not used the recuperator (see Fig. 9-15).

The savings from the above are obvious. Fuel consumption is less than 25 percent of the fuel that would be used without the recuperator.

It must be kept in mind of course that the heat exchanger means a considerable capital investment and therefore heat recuperation

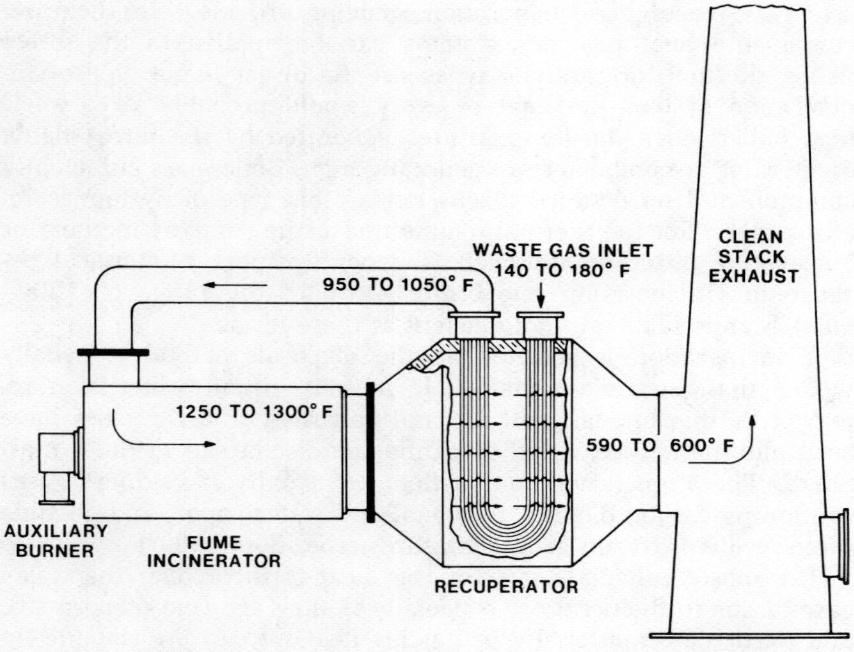

Fig. 9-15 Thermal incinerator with recuperator. *(Courtesy Thermal Research & Engineering Corp.)*

must be weighed against the cost of fuel along with all the other factors which are part of an economic evaluation. As a general rule, the addition of heat recuperation to any waste fume incineration system will usually more than double the original cost.

Heat exchangers in waste gas incineration systems can be used for other purposes. They can be used to preheat air for other processes, possibly even the oven from which the waste fumes might come. They can be used to preheat comfort air for a building as well.

Dual preheating roles are sometimes practical. As indicated in the previous example we have a stack temperature of 600°F. This has a significant amount of heat remaining in it and therefore can be employed to heat air for a low-temperature oven or drier or comfort air for a building after it has delivered its heat to the process. While it is usually not an efficient system, flue gas can be used for hot water generation or for the generation of low-pressure steam.

Equipment

The equipment for heat recuperation in a waste gas incineration system takes many forms. The heat exchanger can be a tubular type either in the form of straight tubes, U-tubes, or other configurations. It can be a plate-type exchanger which provides a large surface area per unit of volume or it can be a regenerative type having a series of metal plates which rotate through the hot gas stream to pick up heat and then discharge it while rotating through the low-temperature incoming fumes. Tubular and plate-type heat exchangers are generally preferred because there is no leakage from one side of the heat transfer system to the other. The regenerative type of system must always be pressurized in a direction that will prohibit the raw waste gas from escaping to the stack, since regenerative heat exchangers usually have a loss across the seal of at least 8 percent and often higher. Another type of regenerative heat exchanger which has been used in a few cases is the refractory checkerwall type. This is a cyclic system. Here a large refractory checkerwall or similar device absorbs heat from the hot gases exiting from the incinerator, and once the refractory is up to temperature, the stream is reversed by means of a switching valve to another mass of refractory in the form of a checkerwall, and the cold fumes are then run in reverse direction through the first or hot checkerwall and then into the incinerator. Such systems are usually very large and expensive and involve a number of control problems.

In any type of fume incineration system there are a number of opportunities to reclaim heat, and not always through the addition of a heat exchanger. The duct which carries the waste fumes may be used as a heat transfer surface by building a double duct and running the hot waste gases on the outside and the incoming fumes in the center duct. While this does not supply an excellent heat transfer surface in terms of efficiency, it is still by far the cheapest type which can be provided in most cases.

Heat transfer coefficients in either the tubular- or plate-type exchanger are usually quite low because it is gas-to-gas heat transfer. Heat transfer coefficients are on the order of 3–8 Btu/hr/ft^2/°F so the exchangers are, of necessity, quite large for the amount of heat transferred. Nevertheless they are economically justifiable in many cases.

Materials of Construction. Incineration systems and heat transfer recovery systems can be fabricated from a wide variety of materials. The selection of these materials depends upon corrosion, strength, and temperature. Therefore most incinerators are constructed of carbon steel material and are lined with appropriate alumina refractory to withstand the temperatures of the process. But where the temperature is low, stainless steel can be used without refractory lining. The heat exchangers must also be carefully designed for both operating pressure and temperature. The operating pressure is usually quite low, on the order of inches of water. Carbon steel is normally utilized for metal temperatures up to 600°F. Above this aluminized steel is satisfactory to about 1300°F and stainless steels for higher temperatures. Special materials are used only when the corrosive properties so dictate.

The refractories used in an incinerator are generally of the high alumina type but again depend on the operating temperature. Standard super-duty firebrick backed up by insulating brick or castables with similar alumina content are usually satisfactory. Some incineration systems utilize stainless steel backed up by an insulating material encased in a carbon steel enclosure.

Special Applications

There are several applications which deserve further discussion because they are not normally found in waste gas incineration systems but where fume incineration is nevertheless applicable. This is the destruction of chlorinated waste gases, sulfonated waste gases, and gases containing high percentages of nitrogen compounds.

Chlorinated Waste Gases. Chlorinated waste gases incinerated in a normal manner will usually produce a certain amount of hydrogen chloride gas as well as some free chlorine. The hydrogen chloride gas, as discussed under the absorption section in this chapter, can be absorbed in water, but the chlorine which is insoluble in water is more difficult to handle. Therefore it is desirable in the combustion reaction to change the chlorine into hydrogen chloride. This can be done in one of two ways: either natural gas or some other hydrocarbon fuel can be added to the combustion reaction with slightly less than theoretical quantities of air to provide enough hydrogen to react with all of the free chlorine present in the system, giving hydrogen chloride. Such a reaction for a typical waste gas might be as follows:

$$CHClCCl_2 + 3\tfrac{1}{2}\ O_2 + CH_4 = 3\ CO_2 + 3\ HCl + H_2O$$

The above reaction would occur if a trichloroethylene air vapor mixture were incinerated with excess natural gas. In order to destroy this organic vapor in air it is necessary to hydrolize the free chlorine, because there is not enough hydrogen in the original solvent molecule to complete its hydrolysis.

Another possibility to accomplish the same purpose would be the addition of steam to the system. However, this generally requires temperatures above 1800°F to achieve complete hydrolysis.

Sulfur Compounds. Sulfur compounds when burned in air will generate sulfur dioxide or sulfur trioxide depending upon the amount of excess air available and the temperature and this will exit with the products of combustion. Scrubbing is therefore dictated, either with caustic or lime to keep the sulfur dioxide or sulfur trioxide out of the effluent stream. This is discussed under the Absorption Section of this chapter.

Oxides of Nitrogen. One of the most difficult problems in any combustion system is not only the destruction of oxides of nitrogen but the creation of a minimum quantity, since they can be created in significant quantity even in a normal combustion reaction between fuel and air. Oxides of nitrogen or NO_x, as they are sometimes called, will form in increasing amounts as the temperature of the reaction increases and also as the amount of excess air in the reaction system increases (see Fig. 7-8). These two conditions are almost diametrically opposed to each other since the system having a high percentage of excess air will usually have a much lower temperature. In fume incineration systems however, the temperature is usually not high enough to create large volumes of NO_x. There is usually a high percentage of excess air, but because of the low temperature operation, the NO_x content of the effluent gas is usually minimal. Occasionally we are faced with the problem of handling nitrogen compounds, or more specifically oxides of nitrogen themselves, as effluent gases from some process. Then it is necessary to develop a more sophisticated system for their removal. For relatively high concentrations of NO_x the nitrogen oxide can be introduced with the combustion air into a conventional combustor or burner in stoichiometric quantities with fuel and air. The oxygen in the nitrogen oxide is used as an oxidant for the fuel. Natural gas is the best fuel to use in such a case since stoichiometric operation with liquid fuels may create smoke or other hydrocarbons.

For less concentrated emissions of NO_x catalytic reduction of the oxides of nitrogen with natural gas on the surface of a catalyst will sometimes provide the desired results; however, in this case also we must have a reducing atmosphere.

In contrast to conventional reducing gases such as natural gas, ammonia is able to reduce nitric oxides selectively in the presence of oxygen at a relatively high rate on the surface of a platinum catalyst. Again the problem requires the use of an excess amount of ammonia gas which can also be an atmospheric pollutant.

More recent work has utilized two-stage catalytic reduction for the oxides of nitrogen. This involves burning with a reducing atmosphere on the surface of a catalyst and then absorbing the heat from the first stage before adding excess air to complete the reaction of the products of combustion on the surface of the second stage of the catalyst. Most nitrogen oxides, however, are created from thermal combustion of conventional fuels or from nitric acid plants and therefore the problem in terms of handling the oxides of nitrogen in standard fume incineration equipment is minimal.

ECONOMICS

Waste disposal usually does not have an economic incentive associated with it. It is a problem which must be solved to satisfy local laws and maintain decent relations with the community. Therefore it cannot be considered in terms of a normal payout period as most process equipment is evaluated. It can however be made more palatable by close examination of the economics involved, not only in terms of initial capital investment but also in terms of operating cost. The selection of the most economical gaseous waste incineration method is not a simple problem and depends not only on the figures developed by the engineers evaluating the problem but upon many other considerations. The location of the plant, urban or rural, may dictate the degree of cleanup required to satisfy local ordinances. The present pollution abatement codes are not enough to consider when purchasing abatement equipment or constructing a system. The future possibilities of more restrictive legislation are equally important. For example, existing regulations may limit the emission level of a solvent vapor to a level that can easily be controlled by adding a small vapor condenser to the system for a cost of $3,000. If the process is experimental or the product is suspected to have a limited market life then this investment might be the best decision; however, if the process is expected to be in operation for a long time and there is indication that emission levels will be decreased for the particular solvent (and they will undoubtedly move in this direction) then other abatement methods must be considered. These other methods might be five or ten times the cost but will be cheaper in the long run. The cost of the solvent must be considered in evaluating recovery versus incineration. The possibility of utilizing another solvent of lower cost and selecting incineration as the best method exists, as does the possibility of using a more expensive solvent that may be recovered easily by water-scrubbing or carbon adsorption.

If the decision is to use incineration, there is seldom an economic choice between direct flame incineration and the other two alternatives, thermal or catalytic methods. Direct flame incineration is only applicable to high concentrations and the other methods to low.

If the decision must be made between thermal and catalytic methods then an economic evaluation can be made, but we must also make an evaluation between both methods with heat recovery and without. If heat recovery appears attractive then we must determine how much heat recovery is economical. This of course depends on the cost of fuel and electric power and also on the capital investment involved in the heat recovery device. The optimization of the heat recuperator is a difficult job unless we employ a computer. Most companies who manufacture incineration and recuperation systems today utilize a computer for the solution of such problems. To properly assess your problem the manufacturer must consider the following information, which you will need to provide:

Cost of fuel
Cost of electric power
Cost of other utilities: steam, cooling water, compressed air, or others that may be required

Cost of operating manpower plus fringes
Cost of space ($\$/ft^2$) if this is a factor
Amortization period
Interest rate
Taxes and insurance cost
Maintenance as a percent of capital investment
Overhead and administration as a percent of capital investment

With the above information the supplier can design a system which should be optimum for your plant in terms of the lowest annual cost.

The reason for such an elaborate optimization is obvious. It may be more prudent to buy a system having capital cost twice that of a competitive system because the cost of operation will overcome the capital advantage well within the amortization period, giving a much lower annual cost. This is especially true with incineration utilizing heat recovery.

A catalytic incineration system of any size will normally cost more than a thermal system to achieve the same end. This is due to the cost of the catalyst. If the catalytic system can reduce the fuel requirements by 20 percent compared with the thermal system, it is obvious that the former can be somewhat more expensive than the latter and still be equal on an annual cost basis depending on operating hours per year, fuel cost, and amortization period plus the other factors outlined above. Similarly a recuperator which will recover twice the heat of a smaller unit may be economical even if its capital cost is twice as great. There can be no shortcut or generalization to such economics. Each case must be individually evaluated considering all of the factors. For this reason the cost-comparison information presented in this chapter should be used with care and not as decision-making data.

REFERENCES

"Air and Water Pollution Control," HPAC Conference Proceedings, Van Nostrand Reinhold Co., New York, 1968.

Allenspach, M., "Cost Analysis of Fume Incinerators," Annual Meeting, Air Pollution Control Assoc., June 1969.

Argenbright, L. P., and Preble, B., "SO_2 from Smelters: Three Processes Form an Overview of Recovery Costs," *Environ. Sci. Tech.*, 4:7 (July 1970).

Bartok, W., Crawford, A. R., and Skopp, A., "Control of NO_x Emissions from Stationary Sources," *Chem. Eng. Prog.*, 67:2 (Feb. 1971).

Blanchard, T. A., and Morse, W. F., "Fume Incineration Research," *A. G. A. Monthly*, Oct. 1969.

Borgwalt, R. H., and Kittleman, T. A., "The Dry Limestone Process for Sulfur Dioxide Control: A Field Study of the Role of Overburning," Annual Meeting Air Pollution Control Assoc., June, 1969.
Brewer, G. L., "Fume Incineration," *Chem. Eng.*, Oct. 14, 1968.
Carlton-Jones, D., and Schneider, H. B., "Tall Chimneys," *Chemical Engineering,* Oct. 14, 1968.
Carson, J. E., and Moses, H., "The Validity of Several Plume Rise Formulas," *J.A.P.C.A.*, **19**:11 (Nov. 1969).
Cortelyou, C. G., "Commercial Processes for SO_2 Removal," *Chem. Eng. Prog.*, **65**:9 (Sept. 1969).
Danielson, J. A., *Air Pollution Engineering Manual,* USDHEW, Govt. Printing Office, Washington, D.C., 1967.
Edwards, F. J., "Catalytic vs. Thermal Air Pollution Control," *Contamination Control,* Sept. 1969.
Ellwood, P., "Versatility Is the Word for SO_2—Removal Process," *Chem. Eng.,* June 16, 1969.
Emanuel, A. G., "Potassium Permanganate Offers New Solutions to Air Pollution Control," *Air Eng.*, Sept. 1965.
Enneking, J. C., "Adsorption Control of Air Pollution," *Plant Eng.*, Dec. 25, 1969.
Falkenberry, H. L., and Slack, A. V., "SO_2 Removal by Limestone Injection," *Chem. Eng. Prog.*, **65**:12 (Dec. 1969).
Fawcett, R. L., "Air Pollution Potential of Phthalic Anhydride Manufacture," *J.A.P.C.A.*, **20**:7 (July 1970).
Ferguson, F. A., Semrau, K. T., and Monti, D. R., "SO_2 from Smelters: By-products Markets a Powerful Lure," *Environ. Sci. Tech.*, **4**:7 (July 1970).
Field, A. A., "Tall Chimneys," *Heating, Piping and Air Conditioning,* April, 1970.
Flower, F. B., "Incineration, A Method of Industrial Waste Disposal," presented at Industrial Waste Incineration Symposium, Parkersburg, W. Va., March 26, 1969.
Goar, B. G., "Today's Sulfur Recovery Processes," *Hydrocarbon Processing,* **47**:9 (Sept. 1968).
Hardison, L. C., "Gaseous Waste Disposal," Ohio Industry Gas Symp. May 18, 1967.
Hardison, L. C., "Techniques for Controlling the Oxides of Nitrogen," *J.A.P.C.A.*, **20**:6 June 1970).
Katell, S., and Plants, K. D., "Here's What SO_2 Removal Costs," *Hydrocarbon Processing,* **46**:7 (July 1967).
Land, P. E., Linna, E. W., and Earley, W. T., "Controlling Sulfur Dioxide Emissions from Coal Burning by the Use of Additives," Annual Meeting Air Pollution Control Assoc., June 1969.
Lee, D. R., "Activated Charcoal in Air Pollution Control" *Heating, Piping and Air Conditioning*, April 1970.
Manny, E. H., and Skopp, A., "Potential Control of Nitrogen Oxide Emissions from Stationary Sources," Annual Meeting, Air Pollution Control Assoc., June 1969.

Mattia, M. M., "Process for Solvent Pollution Control," *Chem. Eng. Prog.*, **66**:12 (Dec. 1970).

Newell, J. E., "Making Sulphur from Flue Gas," *Chem. Eng. Prog.*, **65**:8 (Aug. 1969).

"Odors from Industries Need Controls," *Environ. Sci. Tech.*, **3**:7 (July 1969).

Olden, K., R. D., and Margolin, E. D., "The Molten Carbonate Process for Sulfur Oxide Emissions," *Chem. Eng. Prog.*, **65**:11 (Nov. 1969).

Panofsky, H. A., "Air Pollution Meteorology," *Minerals Processing*, Sept. 1969.

Robertson, J. H., and Woodruff, P. H., "Incineration—The State of the Art," Water Pollution Control Assoc., Sept. 28, 1966.

Ross, R. D., "Industrial Waste Disposal," Van Nostrand Reinhold Co., New York, 1968.

Sappok, R. J., and Walker, P. L., "Removal of SO_2 from Flue Gases Using Carbon at Elevated Temperatures," *J.A.P.C.A.*, **19**:11 (Nov. 1969).

Smith, M. E., "Reduction of Ambient Air Concentrations of Pollutants by Dispersion from High Stacks," National Conference on Air Pollution, Dec. 13, 1966.

Sporn, P., and Frankenberg, T. T., "Pioneering Experience with High Stacks on the O.V.E.C. and A.E.P. Systems," International Clean Air Congress, London, Oct. 1966.

Stone, G. N., and Clarke, A. J., "British Experience with Tall Stacks for Air Pollution Control on Large Fossil-Fueled Power Plants," American Power Conference, Illinois Institute of Technology, April 27, 1967.

Thomas, F. W., Carpenter, S. B., Colbaugh, W. C., "Plume Rise Estimates for Electric Generating Stations," *J.A.P.C.A.*, **20**:3 (March 1970).

Turk, A., "Industrial Odor Control and Its Problems," *Chem. Eng.*, Nov. 3, 1969.

Werner, K., "Catalytic Oxidation of Industrial Waste Gases," *Chem. Eng.*, Nov. 4, 1968.

Westchester, J., "Prevention of Air Pollution by Fumes from Baking Finishes," *Metal Finishing*, Oct. 1966.

Wherle, A. A., "Attack Air Pollution with Fume Incineration," *Plant Eng.*, Nov. 12, 1970.

Yocum, J. E., and Duffee, R. A., "Controlling Industrial Odors," *Chem. Eng.*, June 15, 1970.

10 Air Pollution Considerations in Solid and Liquid Waste Disposal

INTRODUCTION

The disposal of solid and liquid wastes from industrial processes is of major concern to responsible engineers and plant managers. No longer can the transfer-to an off-the-site disposal contractor be assumed to relieve the industrial plant of any further responsibility. Any method that is utilized must serve the interests of the general public as well as the operating plant. Disposal methods must be designed for ultimate long-range conditions and not be of temporary nature.

SOLID WASTE DISPOSAL

The state of the art for solid waste disposal is not very advanced. There are more unresolved problems than solutions. Possible methods of solid waste disposal are given in Table 10-1. If land is available and the solid wastes do not contribute to ground water contamination, landfill is usually the most economical solution. Air voids which contribute to the start of fires can be avoided by using shredders before the wastes are filled. Compaction by spreading out the waste and with heavy equipment further improves the quality of disposal. An earthen cover of sufficient depth to prevent penetration by odors, rodents, and flies is often necessary. Uncovered fill is subject to fires from lightning, spontaneous

TABLE 10-1 SOLID WASTE DISPOSAL METHODS

Method	*Comments*
Dumping	Unsightly; possible fire, odor, and ground water contamination; cheap.
Landfill	Possible fire and ground water contamination. Improved if waste is shredded. Moderate in cost.
Incineration	Effective only if properly operated and equipped with air pollution corrective devices where needed. Expensive.
Pyrolysis	Similar to incineration but more expensive.
Reclaiming	Largely undeveloped and dependent on market for products. Economics unproven.
Composting	Bulky with odor problems. Moderate cost.

combustion, and small boys with matches. Where sufficient ground is unavailable for landfill, incineration is the most developed disposal method. Volume reduction is significant, so that the residue which is unburnable can be disposed of by dumping without creating pollution problems. Most solid waste incinerators require stack gas cleanup devices unless the waste is almost totally lacking in ash or paper. Controlled feeding and improved automatic combustion controls are a necessity. Figure 10-1 shows a unit suitable for general trash which is equipped with a venturi scrubber to clean up the stack gases. It is highly automated and because it is factory-fabricated offers economic advantages.

Pyrolysis is largely undeveloped for commercial disposal but there is evidence that partial pyrolysis or two-stage combustion units will significantly contribute to the solid waste disposal prob-

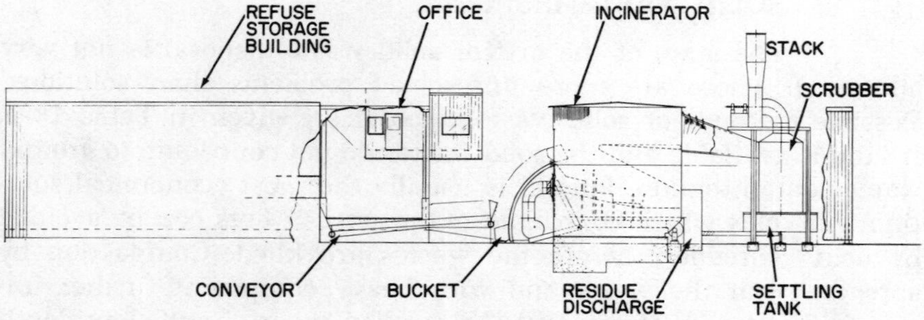

Fig. 10-1 Packaged solid waste incinerator with scrubber. *(Courtesy of Combustion Engineering Co.)*

Fig. 10-2 Two-stage solid waste incinerator. *(Courtesy Smokatrol, Inc.)*

lem while at the same time meeting air pollution requirements. Small units for cellulose type wastes, such as shown in Fig. 10-2, are readily available. By burning with a deficiency of air and with little turbulence, they pyrolyze the waste to gases which are then burned in a second chamber with additional air. Some units do not require scrubbers to meet air pollution emission codes and this factor more than offsets the increased furnace volume required to accommodate the two-stage combustion, which proceeds at a low burning rate per unit of furnace space.

A larger unit is shown in Fig. 10-3. While this unit may require scrubbers to prevent air pollution, it does offer the potential of accommodating batch feeding without the disastrous smoking effects that accompany conventional incinerators when they are overfed.

A potential hazard of all two-stage units is the danger of open

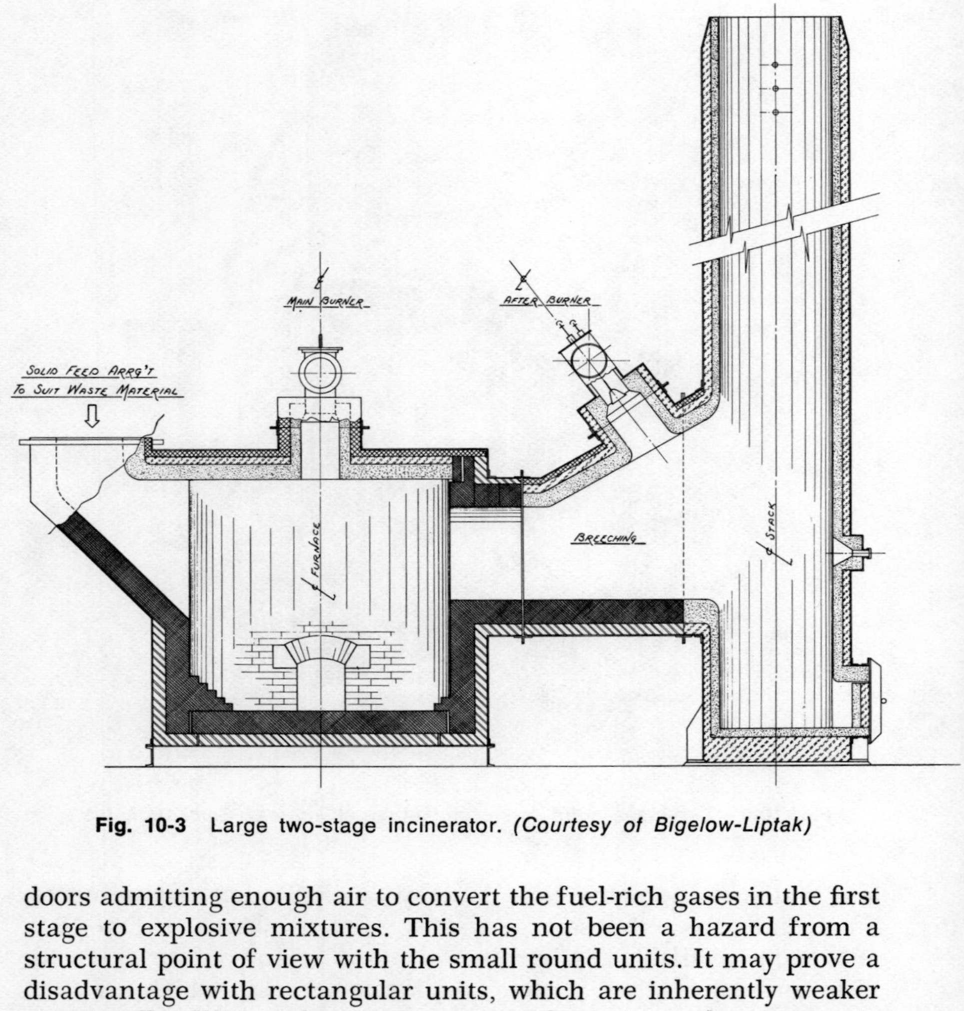

Fig. 10-3 Large two-stage incinerator. *(Courtesy of Bigelow-Liptak)*

doors admitting enough air to convert the fuel-rich gases in the first stage to explosive mixtures. This has not been a hazard from a structural point of view with the small round units. It may prove a disadvantage with rectangular units, which are inherently weaker structurally. With either type, personnel must avoid exposure to spurts of flame from open doors.

Reclaiming of solid wastes is an old method of disposal that the high cost of labor has largely caused to disappear. It remains to be seen if mechanization can restore its attractiveness. One of the principal difficulties is separation of the wastes. There are some notable examples of progress such as the automobile body grinders and compactors, but, by and large, neither reclaiming nor composting is developed enough yet for industrial use except in very specialized cases.

Where the available methods of landfill, incineration, and pyrolysis are used, air polution considerations involve more than just evaluating the discharges of chimneys and stacks. General odor control and housekeeping are required. From an esthetic point of view there is little to recommend an odor from a landfill or a waste storage pile any more than from the stack of an incinerator doing a poor job. Similarly, the unburned sheet of paper blowing down the street from a refuse dump is really no less air pollution than a half-charred piece of paper floating down from an incinerator stack. The control of stack discharges is amply covered in preceding chapters but the control of odors and housekeeping requires ingenuity and vigilance determined in connection with each individual situation.

LIQUID WASTE DISPOSAL

Fortunately liquid waste disposal is technically far ahead of solid waste disposal. Some possible methods for liquid waste disposal are shown in Table 10-2.

This Table is very comprehensive and the reader should refer to Reference 1 for more details. All of the methods given are tech-

TABLE 10-2 LIQUID WASTE DISPOSAL METHODS

Method	*Comments*
Filtration	Limited in application. May require flocculating and clarifiers.
Concentration	Evaporators, decanters, centrifuges, etc.
Solvent extraction	Liquid-liquid systems.
Biological oxidation	Anaerobic is slow and odorous; aerobic requires good control and creates a solid waste disposal problem (sludge).
Adsorbtion	Limited to small dilute systems. Large systems not yet commercially feasible.
Ion exchange	Restricted to special applications.
Dumping	Possible fire, odor or ground water contamination.
Storage	Restricted to small volumes and subject to container leakage.
Deep well	Dependent on proper geological factors
Incineration	Most feasible if combustion self-supporting
Reclaiming	Specialized but often feasible

nically feasible, with considerable overlapping as to economic factors. For example, an aqueous waste containing an organic oil might be (1) concentrated with the water layer then treated in an aerobic biological oxidation pond, (2) sent directly to a pond, (3) incinerated, or (4) discharged to a deep well. Any of many methods will work and might be the most satisfactory in a particular case. Each case must be evaluated on its own merits, being careful to consider all factors. It must also be recognized that the parameters are not fixed and may change with time. This is especially true with those methods that rely on natural phenomena. Careful evaluation of ecological effects may, in fact, demonstrate benefits from some liquid waste disposals.

The primary effects on air pollution depend on the minimizing of evaporation of organics, elimination of odors, a high degree of oxidation (where used), and removal of polluting gases (if existing in sufficient quantity). Too often incineration has been used where other methods would have greater overall utility. A fluorinated hydrocarbon might well be treated in some other fashion than by incineration, since the inevitable formation of hydrogen fluoride is a serious problem.

Aqueous wastes too strong for economic pond disposal and too weak to support their own combustion pose a special problem if dumping, deep-well, or chemical treatments are not possible. To completely oxidize organics, it is necessary that the liquid pass through a good combustion flame and the auxiliary fuel required may be very high.[6] Spraying the waste into hot combustion gases after a flame may only result in pyrolysis and vaporization without incineration, creating air pollution.

To reduce the high cost of auxiliary fuels, catalysis should be

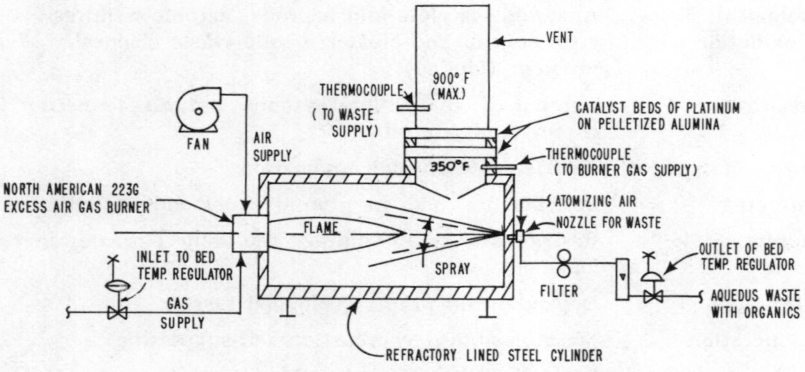

Fig. 10-4 Vapor catalysis. *(Courtesy E. I. Du Pont De Nemours & Co.)*

investigated for use. Figure 10-4 shows a successful system where a catalyst bed receives a vaporized organic-bearing aqueous waste and requires only about 20 percent of the auxiliary fuel needed for a direct flame incinerator.

Reclamation is an area that is often overlooked. Solvent recovery stills may be found in many industrial plants and as a service is widely available on a commercial basis. More emphasis must be placed on recycling of liquid wastes back into chemical processes themselves wherever possible. The inevitable tars must come out of many chemical processes but they should be minimized. Figure 10-5 shows a large reclaiming-type disposal unit where chlorinated hydrocarbons are first incinerated and then converted to usable HCl. Similar systems are common in the sulfuric acid industry where the spent acid from customers' plants is incinerated prior to reconversion back to H_2SO_4. Each industry must examine itself closely and

Fig. 10-5 Large HCl reclaiming incinerator. *(Courtesy E. I. Du Pont De Nemours & Co.)*

determine what it can do. Much reclaiming is being done but more is possible if liquids are to be economically disposed of without creating air pollution.

Of necessity, this chapter is brief. It is only an outline of the many possible solutions to the disposal of liquid and solid wastes. The list of references should be used for specific details on the many different solutions presented.

REFERENCES

General

1. Ross, R. D., *Industrial Waste Disposal,* Van Nostrand Reinhold Co., New York, 1968.
2. Perry, *Chemical Engineers' Handbook,* McGraw-Hill, Book Co., New York, 1963.

Liquid Waste

3. Eckenfelder, W. W., Jr., "Designing Biological Oxidation Systems for Industrial Wastes," *Wastes Eng.* **32** (1961), p. 238.
4. Gloyna, E. F., and Hermann, E. R., "Some Design Considerations for Oxidation Ponds," *J. Sanit. Eng. Div.,* ASCE, **82**:SA4, Paper 1047 (1956).
5. Howe, D. O., Miller, A. R., and Etzel, J. E., "Anaerobic Lagooning —A New Approach to Treatment of Industrial Wastes," *Purdue Univ. Eng. Bull.,* Ext. Ser., **115** (1964).
6. Monroe, E. S., Jr., "Burning Waste Water," *Chem. Eng.,* **75**:20 (1968).
7. Parsons, W. A., "Chemical Treatment of Sewage and Industrial Wastes," *Bull. No. 215, Natl. Lime Assoc.,* Washington, D.C., 1965.
8. Weston, R. F., and Stack, U T., Jr., "Fundamentals of Biological Treatment," *Tappi,* **46**:5 (1963), p. 166A.
9. "Waste Water Containing Oil," *Manual on Disposal of Refinery Wastes,"* Vol. 1, Ed. 6, American Petroleum Institute, New York, 1959.
10. Warner, D. L., "Deep Well Injection of Liquid Wastes," *E.H.S. Pub. No. 999, WP-21,* USDHEW, (1963).

Solid Wastes

11. Demarco, J. Keller, D. J., Leckman, J., and Newton, J. L., *Incinerator Guidelines—1969,* USDHEW, PHS, Washington, D.C., 1969.
12. Engdahl, R. B., *Solid Waste Processing—A State of The Art Report on Unit Operations and Processes,* USDHEW, PHS, Washington, D.C., 1969.
13. *Solid Waste Information Retrival System—Accession Bulletin,* USDHEW, PHS, Bureau of Solid Waste Management, Washington, D.C., (issued monthly).

Index

Index